Springer Actuarial

Editors-in-Chief

Hansjoerg Albrecher, University of Lausanne, Lausanne, Switzerland
Michael Sherris, UNSW, Sydney, NSW, Australia

Series Editors

Daniel Bauer, University of Wisconsin-Madison, Madison, WI, USA
Stéphane Loisel, ISFA, Université Lyon 1, Lyon, France
Alexander J. McNeil, University of York, York, UK
Antoon Pelsser, Maastricht University, Maastricht, The Netherlands
Ermanno Pitacco, Università di Trieste, Trieste, Italy
Gordon Willmot, University of Waterloo, Waterloo, ON, Canada
Hailiang Yang, The University of Hong Kong, Hong Kong, Hong Kong

This is a series on actuarial topics in a broad and interdisciplinary sense, aimed at students, academics and practitioners in the fields of insurance and finance.

Springer Actuarial informs timely on theoretical and practical aspects of topics like risk management, internal models, solvency, asset-liability management, market-consistent valuation, the actuarial control cycle, insurance and financial mathematics, and other related interdisciplinary areas.

The series aims to serve as a primary scientific reference for education, research, development and model validation.

The type of material considered for publication includes lecture notes, monographs and textbooks. All submissions will be peer-reviewed.

More information about this subseries at http://www.springer.com/series/15682

Michel Denuit • Donatien Hainaut • Julien Trufin

Effective Statistical Learning Methods for Actuaries III

Neural Networks and Extensions

 Springer

Michel Denuit
Université Catholique Louvain
Louvain-la-Neuve, Belgium

Donatien Hainaut
Université Catholique de Louvain
Louvain-la-Neuve, France

Julien Trufin
Université Libre de Bruxelles
Brussels, Belgium

ISSN 2523-3262 ISSN 2523-3270 (electronic)
Springer Actuarial
ISSN 2523-3289 ISSN 2523-3297 (electronic)
Springer Actuarial Lecture Notes
ISBN 978-3-030-25826-9 ISBN 978-3-030-25827-6 (eBook)
https://doi.org/10.1007/978-3-030-25827-6

Mathematics Subject Classification (2010): C1, C46, C22, C38, C45, C61, 62P05, 62-XX, 68-XX, 62M45

© Springer Nature Switzerland AG 2019
This work is subject to copyright. All rights are reserved by the Publisher, whether the whole or part of the material is concerned, specifically the rights of translation, reprinting, reuse of illustrations, recitation, broadcasting, reproduction on microfilms or in any other physical way, and transmission or information storage and retrieval, electronic adaptation, computer software, or by similar or dissimilar methodology now known or hereafter developed.
The use of general descriptive names, registered names, trademarks, service marks, etc. in this publication does not imply, even in the absence of a specific statement, that such names are exempt from the relevant protective laws and regulations and therefore free for general use.
The publisher, the authors, and the editors are safe to assume that the advice and information in this book are believed to be true and accurate at the date of publication. Neither the publisher nor the authors or the editors give a warranty, express or implied, with respect to the material contained herein or for any errors or omissions that may have been made. The publisher remains neutral with regard to jurisdictional claims in published maps and institutional affiliations.

This Springer imprint is published by the registered company Springer Nature Switzerland AG.
The registered company address is: Gewerbestrasse 11, 6330 Cham, Switzerland

Preface

Artificial intelligence, and machine learning in particular, does not have to be introduced anymore. Artificial intelligence shapes our lives, changes our habits, and opens new perspectives for the exploitation of new technologies related to data, automation, robotics, and many others. It is now widely accepted that artificial intelligence is both a threat and an opportunity and that it will have a major impact on our lives; the question is not whether it will happen but when and at what pace.

Among the revolutions induced by artificial intelligence, those related to the availability and use of data occupy a special place. Data are now ubiquitous: not only it is now possible to collect a huge amount of data and to store them, locally or on the cloud, but the increasing power of computers and networks of computers makes that processing of these data becomes possible, affordable, and efficient.

Processing data, and using them in decision processes, is now often called "data science". Data science is not new. Its roots lie mainly in machine learning, which is the science of building statistical models from data with few or no assumptions about the underlying process. This data-based view of modelling is not new either (as an example, principal component analysis has been invented in 1901) but has really gained interest in the 1980s, in parallel to the development of computers. Indeed data-based models become interesting when they can (1) process large numbers of data, (2) afford a large number of degrees of freedom, and (3) rely on efficient, nonlinear optimization to reach their objective. Artificial neural networks were developed in the 1980s with the idea that brain-like processing of data could be used to solve perception tasks much better, or more efficiently, than standard computers did. The idea and the early developments were brilliant, but the technology didn't follow fast enough: computational complexity (and its consequence on processing time) is really an issue, and computers of the end of the twentieth century were not powerful enough to let artificial neural networks solve more than simple problems.

Twenty years later, the technology has opened astonishing new perspectives. What is now called deep learning is nothing more than the natural evolution of the artificial neural networks known for 30 years but embedded in modern technology. Besides deep learning, many of the new advances in machine/statistical learning follow the same idea: a combination of sound statistical models with computing and storage power of the twenty-first century. It is this combination of advances that really makes the success of machine learning today.

Tens of books are published every year to cover this exciting and rapidly evolving discipline. Most of them, however, are textbooks that ambition to cover many, or all, domains of applications, with the drawback of often being too general, too academic, and not specific enough to an application domain. This book, *Effective Statistical Learning Methods for Actuaries*, is different. It covers the main advances in statistical learning, with an emphasis on the use of the models in actuarial science. This might seem as a simple adaptation of content with well-chosen examples in this domain. It is not: this book goes far beyond choosing specific examples. It really covers how the methods have to be adapted to the specific needs of actuaries. Because data used by actuaries are not data used by publicists or by big industries, methods differ. This is one of the current challenges in machine learning: going beyond standard methods and adapting them to specific needs. The specificity can be in the number of data, the balance in their distribution, their mathematical properties, and many others: shortly speaking, one dataset is not another one, and specificities must be taken into account.

This is certainly the major quality of this book: going beyond standard methods and showing how they can fit the specific needs from actuaries. It is the first of its kind and will certainly help students and practitioners to embrace this new technology. It is the third book in a series, the two first ones covering statistical models such as generalized linear and additive models and regression trees. This third book focuses on artificial neural networks and their extensions.

The book starts with the historical feed-forward neural networks. More than a page of history, feed-forward neural network models contain all the ingredients for understanding newer, possibly more complex models: types of data, learning principles, optimization of algorithms, overfitting, model selection, etc. This chapter also includes an in-depth discussion about the choice of loss function adapted to insurance portfolio modelling, the first example in the book of how to go beyond standard models when specific data must be analyzed.

The next chapter of the book introduces Bayesian models. Bayesian learning is an important concept that allows models to incorporate prior knowledge on the distribution of parameters. It is a small but fundamental step into making models less "blind", which is necessary when the complexity of the models grows, including the number of parameters and of degrees of freedom.

Deep learning, or deep neural networks, is certainly one of the hot topics today in machine learning. The book present deep learning as a natural extension of artificial neural networks and insists on the importance of regularization, which may take different forms. Deep networks are illustrated on claims frequency prediction, as an example of their huge possibilities.

Dimension reduction is an important branch of machine learning. It aims at reducing the number of variables or features in a problem, for better understanding, better predictions, and easier learning. Principal component analysis is a standard dimension reduction model, but as a linear model, it does not capture nonlinear relations between variables. In the context of mortality and longevity risk management, the next chapter of this book presents one of the many possibilities to perform nonlinear dimension reduction, the choice being justified by the proximity to neural networks and deep learning.

Chapter 5 covers self-organizing maps and k-means. These algorithms are unfortunately often neglected in the machine learning literature. In this book, the author presents these methods and their possibilities, including for handling categorical data. The effectiveness of self-organizing maps and k-means, together with their simplicity of use, justifies this chapter that merits attention, for example, in the context of insurance data analysis as illustrated.

The bias-variance dilemma is well known in statistics. It applies to machine learning algorithms as well and justifies a focus on ensemble methods that help in this context. This topic is split into two chapters, concentrating on bagging and on boosting, respectively. Once again, the methods are not only illustrated on but also adapted to specific problems in the context of actuarial sciences such as claim frequency and cost prediction.

Finally, the study of time series concludes the book. Time series are also specific types of data encountered in many domains of applications. After an introduction to traditional time series analysis methods, the book concentrates on nonlinear methods based on neural networks, including the efficient recurrent approaches and la long short-term memory model.

All the methods described in this book may be qualified as "state of the art." They also form a solid foundation for further developments and breakthroughs that are expected at short term. Indeed, expanding statistical learning, and artificial intelligence in general, to new types of data is probably the next moonshot thinking technology. For example, one can think to estimate losses based on an image of a damage, to exploit real-time data collected from cars, to search semi automatically for legal documentation, and many other disruptive technologies based on data that will help, but never replace, experts having to deal with complex situations.

Machine learning has been a strong area of research at UCLouvain in the last 30 years. Traditionally applied, inter alia, to industrial, image, and biomedical data, and to social networks, this book opens the way for new advances in statistical learning for actuaries. It is the first book of its kind, and I am convinced that generations of students, researchers, and practitioners will benefit from this high-level but accessible book.

Dean of the Polytechnic Faculty, UCLouvain,   Michel Verleysen
Ottignies-Louvain-la-Neuve, Belgium

Acknowledgments

Donatien Hainaut thanks for its support the Chair "Data Analytics and Models for Insurance" of BNP Paribas Cardif, hosted by ISFA (Université Claude Bernard, Lyon) and managed by the "Fondation Du Risque".

Contents

Chapter 1
Feed-Forward Neural Networks

This chapter introduces the general features of artificial neural networks. After a presentation of the mathematical neural cell, we focus on feed-forward networks. First, we discuss the preprocessing of data and next we present a survey of the different methods for calibrating such networks. Finally, we apply the theory to an insurance data set and compare the predictive power of neural networks and generalized linear models.

1.1 From Biology to Mathematics

Nature has always been an inspiration for scientists. Even the smallest animal has a more advanced capacity to treat data than most of personal computers. One approach for analyzing large data sets consists then to duplicate the adaptive behaviour and the learning capacity of living beings. In particular, the scientific community grants a particular attention to the neural system. This system is a complex network of cells, called neurons, working in parallel and capable to reorganize themselves during the learning phase.

As illustrated in Fig. 1.1, a neuron is composed of three parts: dendrites, a cell body and an axon. Incoming signals are received by dendrites trough a biochemical process that weights the information according to its importance. The cell body sums up incoming signal. If a threshold is reached, the cell reacts and transmits an output message down the axon. This output signal is next passed to the neighbouring neurons.

The study of biological neurons paved the way for neural networks. An artificial neuron is very similar to the biological one. Numerical information is weighted before sending to a cell body. This cell body is represented by an activation function. The numerical output is proportional to the value returned by this activation function. In an artificial neural network, the output signal serves as input information

© Springer Nature Switzerland AG 2019
M. Denuit et al., *Effective Statistical Learning Methods for Actuaries III*,
Springer Actuarial, https://doi.org/10.1007/978-3-030-25827-6_1

Fig. 1.1 Illustration of a
neuronal cell

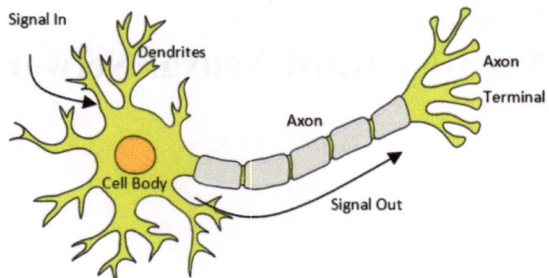

for several other neurons. The architecture of a neural net depends on its purpose. In a perceptron, also called feed-forward network, neurons are ordered in successive interconnected layers. Perceptrons are either used for classification or regression of data. In self-organizing maps (SOM), neurons are nodes of the grid. SOM were initially developed for dimensionality reduction. Recent developments of the technology offer new opportunities based on neural models for life and non-life insurers. The next three chapters aim to shed some light on new applications of artificial neural networks in actuarial sciences.

Perceptrons were developed by Rosenblatt (1958) and are networks with supervised learning. It means that we have to a priori identify the most relevant variables and to know the desired outputs for combinations of these variables. For example, forecasting the frequency of car accidents with a perceptron requires an a priori segmentation of some explanatory variables. The next three chapters focus on this type of neural models. This chapter presents generalities about neural cells and networks with an actuarial point of view. The theory is illustrated by an analysis of a portfolio of motorcycle insurances. A comparison of perceptrons to generalized linear models confirms their efficiency for predicting the frequency and average claims in function of insured profiles. The second chapter proposes a Bayesian neural networks. This technique allows finding confidence intervals for parameters of the neural network. The third chapter is about deep neural networks.

1.2   Neurons and Activation Functions

In neural models, the flow of information is carried by n real-valued vectors $\boldsymbol{x}_i = \left(x_{i1}, \ldots, x_{ip}\right)^\top$ of dimension p, for $i = 1, \ldots, n$. In an insurance context, the vector $\boldsymbol{x}_i$ contains the p variables representative of the ith policyholder. For the moment, we only consider $\mathbb{R}$-valued variables. The case of categorical variables will be discussed later. In the dendrites, input variables x_{ij} are weighted by $\omega_j \in \mathbb{R}$ and summed up, eventually with an offset, ω_0. The $(p + 1)$-vector of weights is denoted $\boldsymbol{\omega} = \left(\omega_0, \ldots, \omega_p\right)^\top$ In the cell body, this weighted sum passes through an activation function, noted $\phi(.)$. This function is continuous, strictly increasing and

Fig. 1.2 Illustration of an artificial neuron

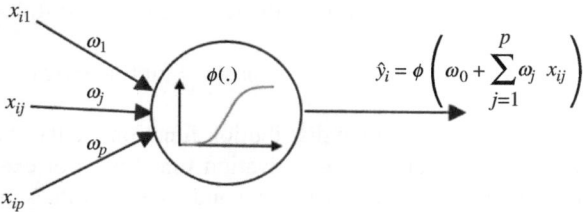

usually lower and upper bounded. The input information is squashed by $\phi(.)$ and the output signal $\widehat{y}_i$ is equal to:

$$\widehat{y}_i = \phi \left(\omega_0 + \sum_{j=1}^{p} \omega_j x_{ij} \right)$$

$$= \phi \left(\boldsymbol{\omega}^\top \begin{pmatrix} 1 \\ \boldsymbol{x}_i \end{pmatrix} \right)$$

for $i = 1, \ldots, n$. The complete process of information treatment is illustrated in Fig. 1.2. If we analyze insurance data, weights $\boldsymbol{\omega}$ are estimated in order to obtain estimate $\widehat{y}_i$ of e.g. the frequency of claims or the expected cost of claims caused by the ith policyholder.

If the transfer function is linear, the neural cell linearly regresses the covariates on the output signal. In order to capture more complex relations between input variables and responses, we rather choose non-linear functions for $\phi(.)$. The most common activation function is the sigmoid or logistic function:

$$\phi(z) = \frac{1}{1 + e^{-x}} .$$

This function is strictly increasing and is defined from $\mathbb{R}$ to $[0, 1]$. The appeal of this function comes from its threshold behavior which characterizes many types of responses of a system to changes in fundamental variables. An other activation function for the neurons is the hyperbolic tangent function, $tanh(.)$:

$$\phi(z) = \frac{e^x - e^{-x}}{e^x + e^{-x}} .$$

This function is defined on $\mathbb{R}$ and its response is in the interval $[-1, 1]$. In deep learning, the rectifier is a common activation function defined as the positive part of its argument:

$$\phi(z) = \max(z, 0) .$$

A smooth approximation to the rectifier is the "softplus function":

$$\phi(z) = \ln(1 + \exp z).$$

Probability cumulative distribution functions (cdf) of any random variable defined on $\mathbb{R}$ are also eligible as activation functions. For example, the standard Gaussian distribution is strictly increasing and takes its values in $[0, 1]$:

$$\phi(z) = \frac{1}{\sqrt{2\pi}} \int_{-\infty}^{z} e^{-\frac{1}{2}u^2} du$$

Furthermore, a standard Gaussian distribution does not have as wide dispersion as the sigmoid function. Given that the probability density function of a Gaussian random variable has thin tails, the responses for any input signal outside the interval $[-2, 2]$ is close to zero or one. The Gaussian cumulative distribution function belongs to the family of radial basis function. A radial basis function (RBF) is a real-valued function whose response depends only on the distance from the origin. If we note $x_i^\omega = \omega^\top \begin{pmatrix} 1 \\ x_i \end{pmatrix}$, the output of the activation function

$$\phi\left(\omega^\top x_i'\right) = \phi(\|x_i^\omega\|_2)$$

is a function of the Euclidean distance of x_i^ω from the origin. In radial basis network, the input of the Gaussian pdf is the distance from some other point c, specific to the neuron: $\|x_i^\omega - c\|_2$. In numerical illustrations, we mainly work with sigmoid or hyperbolic tangent activation functions.

1.3 Feed-Forward Networks

A neural network is a set of interconnected neurons and there is an infinity of possible layouts. In this section, we focus on feed-forward structures. As illustrated in Fig. 1.3, the information flows in one direction, forward, from input to output nodes. The intermediate layers of neurons between the information and output neurons are called hidden layers. In this configuration, there is no circular relation between neurons. Neural networks with circular connections are called recurrent. Feed-forward nets with only one hidden layer are called "shallow" networks. Whereas feed-forward net with more than one layers are called "deep neural network". Both types of networks belong to the family of multi-layer perceptrons.[1] We will use indifferently the terminology multi-layer perceptron or feed-forward network.

---

[1] A perceptron is a single artificial neuron using the Heaviside step function as the activation function. It was developed by Rosenblatt (1958) for image recognition.

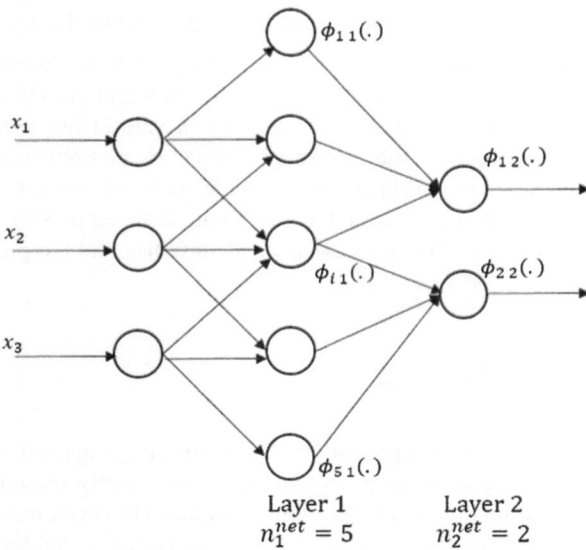

Fig. 1.3 Example of a feed-forward neural network with one hidden layer

We now introduce notations that will be used throughout chapters on neural network. We recall that the data set contains information about n policyholders and is coded in n real-valued vectors $x = (x_1, \ldots, x_p)^\top$ of dimension p (we temporarily drop the indices i of x_i for ease of reading). The architecture of a feed-forward network is defined by two features: the number of layers and the number of nodes within each layer. The number of neuronal layers (input and output neurons included) is denoted by n^{net}. Whereas the number of neurons in the ith layer is noted n_j^{net}. As illustrated in Fig. 1.3, the activation function of each neuron, $\phi_{i,j}(.)$ is then identified by two indices: one for the position in a layer and one for the layer. The output of the ith neurons in the input layer, denoted by $\widehat{y}_{i,1}$, is computed as follows:

$$\widehat{y}_{i,1} = \phi_{i,1}\left(\omega_{i0}^1 + \sum_{k=1}^{p} \omega_{i,k}^1 x_k\right) \quad i = 1, \ldots, n_1^{net}$$

where $\omega_{i,k}^1$ are the weights for the kth input signal, received by the ith neuron in the input layer. Whereas the output of the ith neurons in hidden and output layers j, denoted by $\widehat{y}_{i,j}$, is equal to:

$$\widehat{y}_{i,j} = \phi_{i,j}\left(\omega_{i0}^j + \sum_{k=1}^{n_{j-1}^{net}} \omega_{i,k}^j \widehat{y}_{k,j-1}\right) \quad i = 1, \ldots, n_j^{net}, \ j = 1, \ldots, n^{net}$$

where $\omega_{i,k}^j$ are the weights for the kth input signal received by the neuron (i, j). The vector of weights of the neuron (i, j) is denoted $\boldsymbol{\omega}_i^j$. Perceptron may be used for two purposes: classification and regression. Regression and classification are both related to prediction, where a regression predicts a value from a continuous set, whereas classification predicts the belonging to the class. In both cases, the weights are calibrated in order to minimize the distance between output signals of the network and real outputs. Calibration techniques are detailed in Sect. 1.6 but before we need to address the treatment of categorical variables and the preprocessing of variables.

1.4   Qualitative Variables

By construction, a perceptron represents a continuous mapping in all input variables. It requires that all inputs correspond to numeric, continuously valued variables and represents a continuous function in all input variables. However, insurance data set contains not only quantitative variables but also categorical variables. Treating this qualitative information requires adapting the architecture of the perceptron.

Categorical variables may be broken up into two types: ordinal and non-ordinal. Ordinal variable values can be ordered. An example of such variable is e.g. the month of the date of policy purchase. If ordinal variables are continuous in nature, we replace them by numeric variables and treat them as if they were continuously valued.

A non-ordinal categorical variable will be a label of a category without ordering property. An example of a non-ordinal categorical feature is the gender of the policyholder or the brand of the insured vehicle. In this case, categorical variables should be distinguished from the continuous independent variables. The simplest way to present categorical data to neural networks is using dummy variables. One binary variable is created for each modality of the categorical variable.

For example, imagine that among p explanatory covariates, the last one is a qualitative variable corresponding to the environment of the policyholder. If this variable possesses the three following modalities: 'Urban', 'Suburban' and 'Countryside', we create two dummy input variables as follows

$$x_{i,p} = \begin{cases} 1 & if \text{ Urban} \\ 0 & else\,. \end{cases}$$

$$x_{i,p+1} = \begin{cases} 1 & if \text{ Suburban} \\ 0 & else\,. \end{cases}$$

There is no need to define a third binary variable because if the policyholder does not live in a urban or suburban environment, he inevitably resides in the countryside. This permits categorical data to be input to a neural network (or any mathematical model) and effectively localizes the information about each categorical value. However, it does expand the number of input variables. If the number of variables increases dramatically, we can eventually reduce the dimensions of categorical variables with a self-organizing map. This technique of reduction in explained in Chap. 5.

1.5 Data Preprocessing

Activation functions introduced in Sect. 1.2 are defined on $\mathbb{R}$. Furthermore, sigmoid, hyperbolic tangent and Gaussian cumulative distribution functions quickly converge toward 0 (or -1) and 1 outside a relatively small interval centered around zero. Without scaling of initial data, the input signal may be far away from this interval. The response of neurons is then located far in the tails of activation functions, where their derivative with respect to the input signal is nearly null. Given that estimation algorithms are based on these derivatives, the calibration algorithm may never converge. Scaling data is then a necessary step. Without scaling, a great deal of information is likely to be lost, since neurons will transmit binary values for many values of the data set.

In practice, the most common scaling are linear. Imagine that we have n observations. If we wish to re-scale the jth variable x_{ij} for $i = 1, \ldots, n$ on the interval $[0, 1]$ we apply the following transformation:

$$x_{ij}^* = \frac{x_{i,j} - \min_{i=1,\ldots,n}(x_{i,j})}{\max_{i=1,\ldots,n}(x_{i,j}) - \min_{i=1,\ldots,n}(x_{i,j})} \quad i = 1, \ldots, n. \tag{1.1}$$

If we prefer to re-scale input variables on $[-1, 1]$, we instead use the following conversion:

$$x_{ij}^* = 2\frac{x_{i,j} - \min_{i=1,\ldots,n}(x_{i,j})}{\max_{i=1,\ldots,n}(x_{i,j}) - \min_{i=1,\ldots,n}(x_{i,j})} - 1 \quad i = 1, \ldots, n.$$

A last method for preprocessing data consists to center and norm the input signal. This is done by subtracting the mean and next dividing by the standard deviations of explanatory variables. If the sample mean and standard deviation of the jth variable are respectively denoted by $\bar{x}_j$ and s_j, the standardized version of data is:

$$x_{ij}^* = \frac{x_{ij} - \bar{x}_j}{s_j} \quad i = 1, \ldots, n.$$

In this case $x_{ij}^* \in \mathbb{R}$ but have an empirical distribution with zero mean and unit variance. This preprocessing of data is fundamental to ensure the success of the estimation procedure.

1.6   Fitting of Supervised Neural Networks

Perceptrons are used for classification and regression. In both cases, the weights are found by minimizing the distance between output signals of the network and real observed outputs. In this section, we present several estimation methods.

As previously, the information about the insurance portfolio is contained in n vectors $(x_i)_{i=1,...,n}$ of dimension p. We observe a vector of responses $y = (y_i)_{i=1,...,n}$. The neural network is described with notations introduced in Sect. 1.3. The criterion used to estimate the network is a loss function noted, $\mathcal{L} : \mathbb{R}^2 \to \mathbb{R}$, continuous and that admits a first order derivative. The vector of weights ω_i^j of neurons (i, j) for $i = 1, .., n_j^{net}$ and $j = 1, ..., n^{net}$ is denoted Ω. Ω is fitted in order to minimize the sum of losses:

$$\Omega = \arg\min_{\Omega} \frac{1}{n} \sum_{i=1}^{n} \mathcal{L}(y_i, \widehat{y}_{i,n^{net}}),\tag{1.2}$$

$\mathcal{L}$ may be the opposite of the log-likelihood, a deviance, a quadratic function or any other well-behaved penalty function. To lighten further notation, we denote the vector of estimators by $\widehat{y} = (\widehat{y}_i)_{i=1,...n} = (\widehat{y}_{i,n^{net}})_{i=1,...n}$ and the total of losses by $\mathcal{R}(\Omega) = \frac{1}{n} \sum_{i=1}^{n} \mathcal{L}(y_i, \widehat{y}_i)$.

1.6.1   Gradient Descent

The method to which we naturally think to solve problem (1.2) is the gradient descent. Firstly, we choose an initial vector of weights Ω_0. Given that data are re-scaled either in $[0, 1]$ or in $[-1, 1]$, weights of Ω_0 must be chosen on the same domains. If we use a second order Taylor expansion, the function $\mathcal{R}$ may be approached by the following sum

$$\mathcal{R}(\Omega) = \mathcal{R}(\Omega_0) + \nabla\mathcal{R}(\Omega_0)^\top (\Omega - \Omega_0)\tag{1.3}$$

$$+ \frac{1}{2} (\Omega - \Omega_0)^\top H(\mathcal{R}(\Omega_0)) (\Omega - \Omega_0) + \mathcal{O}(3),$$

where $\nabla\mathcal{R}(\Omega_0)$ and $H(\mathcal{R}(\Omega_0))$ are respectively the gradient vector and the Hessian matrix, evaluated at point Ω_0:

$$\nabla\mathcal{R}(\Omega_0) = \left(\frac{\partial}{\partial\omega_{ik}^j} \mathcal{R}(\Omega) \bigg|_{\Omega_0} \right)_{i=1,...,n_j^{net}, \, k=1,...,n_{j-1}^{net}, \, j=1,...,n^{net}},$$

$$H(\mathcal{R}(\Omega_0)) = \left(\frac{\partial^2}{\partial\omega_{ik}^j \partial\omega_{ut}^s} \mathcal{R}(\Omega) \bigg|_{\Omega_0} \right)_{i,u=1,...,n_j^{net}, \, k,t=1,...,n_{j-1}^{net}, \, j,s=1,...,n^{net}}.$$

Algorithm 1.1 Gradient descent procedure

Initialization:

Randomly attribute weights to each neurons: Ω_0.

Main procedure:

For $t = 0$ to maximum epoch, T

1. Calculate the gradient $\nabla \mathcal{R}(\Omega_t)$
2. Calculate the Hessian matrix $H\left(\mathcal{R}(\Omega_0)\right)$
3. Invert the Hessian and update the vector of weights:

$$\Omega_{t+1} = \Omega_t - H\left(\mathcal{R}(\Omega_t)\right)^{-1} \nabla \mathcal{R}(\Omega_t).$$

End loop on epochs

The vector of parameter Ω^* minimizing $\mathcal{R}(\Omega)$ cancels the first order derivative of the expansion (1.3):

$$0 = \nabla \mathcal{R}(\Omega_0) + H\left(\mathcal{R}(\Omega_0)\right)\left(\Omega^* - \Omega_0\right),$$

or

$$\Omega^* = \Omega_0 - H\left(\mathcal{R}(\Omega_0)\right)^{-1} \nabla \mathcal{R}(\Omega_0).$$

The gradient descent Algorithm 1.1 consists then to iteratively update the set of neurons weights according to this last relation. The number of iterations T is chosen such that the variation of the total loss $\mathcal{R}$ after T steps falls below a tolerance criterion.

If the total loss function $\mathcal{R}(\Omega)$ is not convex (and this is generally the case), the gradient descent algorithm can converge to a local minimum instead of the global one. The choice of Ω_0 is then crucial. In practice, we recommend to calibrate several times the neural network with different initial vectors of weights.

1.6.2 Back-Propagation

Another issue of the gradient descent algorithm is the inversion of the Hessian matrix. If this matrix is badly conditioned, we cannot invert it numerically. On the other hand, even if the matrix is well conditioned the inversion of a matrix of high dimension is time consuming and may be inefficient in practice.

An alternative solution is offered by the back-propagation Algorithm 1.2. In this procedure, we adjust the vector of weights Ω_t by a small step in the opposite direction of the gradient. The size of the step, noted ρ_t, is either constant or a decreasing function $g(\rho_0, t)$, inversely proportional to the epoch. If the step size is constant, we should choose a value that is small enough to avoid oscillations around the minimum solution (e.g. $\rho \in [0.01, 0.1]$). An example of decreasing function is

---

Algorithm 1.2 Back propagation procedure

Initialization:

   Randomly attribute weights to each neurons: Ω_0.

   Select an initial step size, ρ_0.

Main procedure:

   For $t = 0$ to maximum epoch, T

   1. Calculate the gradient $\nabla\mathcal{R}(\Omega_t)$
   2. Update the step size

$$\rho_{t+1} = g(\rho_0, t)$$

   3. Modify the vector of weights:

$$\Omega_{t+1} = \Omega_t - \rho_{t+1}\nabla\mathcal{R}(\Omega_t). \qquad (1.4)$$

   End loop on epochs

---

the exponential decay function:

$$\rho_t = g(\rho_0, t) = \rho_0 e^{-\alpha t},$$

where α regulates the speed of decay. In high dimensions, computing the gradient can be very time consuming. Solutions to reduce this time are detailed in Chap. 3 on deep learning.

1.6.3 Resilient Back-Propagation

A variant of the back-propagation algorithm is the resilient back-propagation, noted RPROP. This algorithm is a learning heuristic created by Riedmiller and Braun (1993). The RPROP takes into account only the sign of the partial derivative and not its size. For each weight, if the partial derivative of the total error function changes of sign compared to the last iteration, the update value for that weight is multiplied by a factor η^-, where $\eta^- < 1$. If the last iteration produced the same sign, the update value is multiplied by a factor of η^+, where $\eta^+ > 1$.

More precisely, the size of the weight change is exclusively determined by a weight-specific update, noted $\Delta_{ijk}^{(t)}$. If $\omega_{ik}^{j\,(t)}$ is the weight for the kth input signal of the ith neuron in the jth layer at iteration t then the variation of the weight between iteration $t - 1$ and t is equal to:

$$\omega_{ik}^{j\,(t)} - \omega_{ik}^{j\,(t-1)} = \begin{cases} -\Delta_{ijk}^{(t)} & if \ \left.\frac{\partial}{\partial\omega_{ik}^j}\mathcal{R}(\Omega)\right|_{\Omega_{t-1}} > 0 \\[4mm] +\Delta_{ijk}^{(t)} & if \ \left.\frac{\partial}{\partial\omega_{ik}^j}\mathcal{R}(\Omega)\right|_{\Omega_{t-1}} < 0 \\[4mm] 0 & else \end{cases}$$

The second step of RPROP learning consists to evaluate the new update values $\Delta_{ijk}^{(t+1)}$. They depend upon the sign of partial derivatives computed during the two last iterations:

$$\Delta_{ijk}^{(t+1)} = \begin{cases} \eta^+ \Delta_{ijk}^{(t)} & if \left. \frac{\partial}{\partial \omega_{ik}^j} \mathcal{R}(\Omega) \right|_{\Omega_{t-1}} \times \left. \frac{\partial}{\partial \omega_{ik}^j} \mathcal{R}(\Omega) \right|_{\Omega_t} > 0 \\ \eta^- \Delta_{ijk}^{(t)} & if \left. \frac{\partial}{\partial \omega_{ik}^j} \mathcal{R}(\Omega) \right|_{\Omega_{t-1}} \times \left. \frac{\partial}{\partial \omega_{ik}^j} \mathcal{R}(\Omega) \right|_{\Omega_t} < 0 \\ \Delta_{ijk}^{(t)} & else \end{cases}$$

where $0 < \eta^- < 1 < \eta^+$. η^+ is empirically set to 1.2 and η^- to 0.5.

To summarize, every time the partial derivative changes its sign, which indicates that the last weight update was too big and the algorithm has jumped over a local minimum, the update $\Delta_{ijk}^{(t)}$ is reduced by a factor η^-. If the derivative does not change of sign, $\Delta_{ijk}^{(t)}$ is slightly increased in order to speed up the convergence. The pseudo-code of the resilient back-propagation is presented in Algorithm 1.3. In this code, Δ_0 is the initial update value whereas Δ_{min} and Δ_{max} are respectively the minimum and maximum update values. The back-propagation and resilient back-propagation algorithms are available in most of neural network software (e.g. R package 'neuralnet').

Algorithm 1.3 Resilient back propagation procedure

Initialization:

   Randomly attribute weights to each neurons: Ω_0.

   $\forall i, j, k$ we set $\Delta_{ijk}^{(t)} = \Delta_0$

Main procedure:

   For $t = 0$ to maximum epoch, T

   1. Calculate the gradient $\nabla \mathcal{R}(\Omega_t)$

   If $\frac{\partial}{\partial \omega_{ik}^j} \mathcal{R}(\Omega_t) \times \frac{\partial}{\partial \omega_{ik}^j} \mathcal{R}(\Omega_{t-1}) > 0$ **then**

   $$\Delta_{ijk}^{(t)} = \min\left(\eta^+ \Delta_{ijk}^{(t-1)}, \Delta_{max}\right)$$

   Else if $\frac{\partial}{\partial \omega_{ik}^j} \mathcal{R}(\Omega_t) \times \frac{\partial}{\partial \omega_{ik}^j} \mathcal{R}(\Omega_{t-1}) < 0$ **then**

   $$\Delta_{ijk}^{(t)} = \max\left(\eta^- \Delta_{ijk}^{(t-1)}, \Delta_{min}\right)$$

   Else if $\frac{\partial}{\partial \omega_{ik}^j} \mathcal{R}(\Omega_t) \times \frac{\partial}{\partial \omega_{ik}^j} \mathcal{R}(\Omega_{t-1}) = 0$ **then**

   $$\Delta_{ijk}^{(t)} = \Delta_{ijk}^{(t-1)}$$

   End if

   $$\omega_{ik}^{j\,(t)} = \omega_{ik}^{j\,(t-1)} - \text{sign}\left(\frac{\partial}{\partial \omega_{ik}^j} \mathcal{R}(\Omega_t)\right) \Delta_{ijk}^{(t)}$$

   End loop on epochs

1.6.4 Simulated Annealing

The back-propagation or resilient back-propagation algorithms present the same default as the gradient descent method: they can converge to a local minimum instead of a global one. An alternative or complementary method to estimate weights of the neural network is simulated annealing. This is a meta-heuristic algorithm to approach global minimum in a large search space. This algorithm presents similarities with annealing in metallurgy which is a technique of heat control of a material to increase the size of its crystals and improve its resistance. This approach is close to Monte-Carlo methods, studied and tested in Chap. 2.

Algorithm 1.4 Simulated annealing procedure

Initialization:
        Randomly attribute weights to each neurons: Ω_0.
        Calculate the global loss function $\epsilon_0 = \mathcal{R}(\Omega_0)$.
Main procedure:
        For $t = 1$ to maximum epoch, T

                1. Random draw of a candidate $\tilde{\Omega} \sim N(\Omega_{t-1}, \Sigma_\Omega)$ and calculate $\tilde{\epsilon} = \mathcal{R}(\tilde{\Omega})$.
                2. Draw U from a $[0, 1]$ uniform distribution
                3. Compute the acceptance threshold

$$A(t) = \exp\left(\frac{-\gamma\,(\tilde{\epsilon} - \epsilon_{t-1})}{T(t)}\right)$$

            where $T(t) = \frac{T}{1+\ln(t)}$ is the "cooling schedule".
            If $(\tilde{\epsilon} - \epsilon_{t-1}) < 0$ **then**
                Accept the candidate vector, $\Omega_t = \tilde{\Omega}, \epsilon_t = \tilde{\epsilon}$.
            Else if $U \leq A(t)$
                Accept the candidate vector, $\Omega_t = \tilde{\Omega}, \epsilon_t = \tilde{\epsilon}$.
                Else
                    Reject the candidate vector, $\Omega_t = \Omega_{t-1}, \epsilon_t = \epsilon_{t-1}$.
                End if
            End if
        End loop on epochs

At each iteration t, we draw a random set of parameters $\tilde{\Omega}$ and recalculate the total loss $\mathcal{R}(\tilde{\Omega})$. If $\mathcal{R}(\tilde{\Omega})$ is lower than the current total loss $\mathcal{R}(\Omega_{t-1})$, the local optimal vector of parameter is set to $\Omega_t = \tilde{\Omega}$. In the opposite case, the vector of weights $\tilde{\Omega}$ does not reduce the total loss. However, the algorithm may accepts it as starting point for the next iteration with a certain probability. This probability, called the acceptance rate, is inversely proportional to $\mathcal{R}(\tilde{\Omega}) - \mathcal{R}(\Omega_{t-1})$ and decreasing with time. We generate a random number and if this number is below the acceptance rate, we accept $\tilde{\Omega}$ as starting point for the next iteration. This ingenious procedure is inspired from Monte-Carlo Markov Chain algorithm (Metropolis–Hastings algorithm) and allows escaping from a local minimum solution.

1.7 Choice of the Loss Function

The information about the insurance portfolio is contained in n vectors $(x_i)_{i=1,\ldots,n}$ of dimension p. Here, n is the number of policies whereas p is the number of available descriptive variables for a contract. We also observe a vector $y = (y_i)_{i=1,\ldots,n}$ of key quantities for the insurer. A key quantity is the realization of a ratio of a random variable Z_i, on an exposure v_i. Table 1.1 presents several examples of key quantities of interest for insurance companies. Notice that we assume the independence between variables Z_i for $i = 1, \ldots, n$.

The exposure is either the duration of the contract or the number of claims. For example, if we study the claim frequency, y_i is the ratio of the number of claims caused by the ith policyholder on the duration of the contract: $y_i = \frac{N_i}{v_i}$. If we are interested by the average claim cost, y_i is the ratio of the total amount of claims caused by the ith insured on the number of claims: $y_i = \frac{C_i}{N_i}$.

In the remainder of this chapter, we assume that Y_i are distributed according to an exponential dispersed (ED) distribution.

Definition 1.1 A random variable Y_i has an exponential dispersed distribution, if its probability density function admits the following representation:

$$f_{Y_i}(y; \theta_i, \phi) = \exp \left\{ \frac{y\theta_i - a(\theta_i)}{\phi/v_i} \right\} c(y, \phi, v_i) \tag{1.5}$$

where

- θ_i is a parameter that depends on i, while the dispersion parameter ϕ is identical for all i.
- The $a(\theta_i)$ is called the cumulant function and is C^2 with an invertible second derivative.
- The function $c(.)$ is independent from θ_i.
- ϕ is a positive constant called dispersion parameter, not depending on i. Notice that we comply here with the standard notations for ED distributions and that the dispersion parameter ϕ should not be confused with the activation function.

As reported in Table 1.2, many common statistical distributions, like the Normal or Poisson law, belong to this family. Reformulating their distribution as in Eq. (1.5)

Table 1.1 Example of key quantities of interest

Exposure v_i	Z_i	Key quantities $Y_i = Z_i/v_i$
Duration	Number of claims	Claim frequency
Earned premium	Claim cost	Loss ratio
Number of claims	Claim cost	(Average) Claim severity
Duration	Default or not	Default frequency
Lended amount	Lost amount	Loss given default

Table 1.2 Features of exponential dispersed (ED) distributions for claim numbers and claim sizes

	$f_{Y_i}(y)$	θ	ϕ	$a(\theta)$	$\mathbb{E}(Y_i)$	$V(z)$
Normal, $N(\mu_i, \frac{\sigma^2}{\nu_i})$	$\frac{\sqrt{\nu_i}}{\sigma\sqrt{2\pi}} e^{-\frac{\nu_i}{2}\left(\frac{y-\mu_i}{\sigma}\right)^2}$	μ_i	σ^2	$\theta^2/2$	μ_i	1
Gamma, $G(\nu_i\alpha, \nu_i\beta_i)$	$\frac{(\nu_i\beta_i)^{\nu_i\alpha}}{\Gamma(\nu_i\alpha)} y^{\nu_i\alpha-1} e^{-\nu_i\beta_i y}$	$-\frac{\beta_i}{\alpha}$	$\frac{1}{\alpha}$	$-\ln(-\theta)$	$\frac{\alpha}{\beta_i}$	z^2
Poisson, $Po(\lambda_i\nu_i)$	$e^{-\nu_i\lambda_i} \frac{(\nu_i\lambda_i)^{\nu_i y}}{(\nu_i y)!}$	$\ln(\lambda_i)$	1	e^θ	λ_i	z
Binomial, $Bin(\nu_i, p_i)$	$\binom{\nu_i}{\nu_i y} p_i^{\nu_i y}(1-p_i)^{\nu_i(1-y)}$	$\ln\frac{p_i}{1-p_i}$	1	$\ln(1+e^\theta)$	p_i	$z(1-z)$

allows us to standardize future developments. In particular, it is possible to link the expectation and variance of Y_i to the cumulant function $a(.)$. To show this last point, the next result is needed:

Proposition 1.1 *The moment generating exponent, denoted $\psi(t)$ of an ED random variable Y_i is equal to*

$$\mathbb{E}(e^{tY_i}) = e^{\psi_i(t)} \qquad \psi_i(t) = \frac{a(\theta_i + t\frac{\phi}{\nu_i}) - a(\theta_i)}{\frac{\phi}{\nu_i}} \tag{1.6}$$

Proof The moment generating function (mgf) is defined as $\mathbb{E}(e^{tY_i})$. This expectation is finite at least for $t \in \mathbb{R}$, in a neighborhood of zero. For continuous ED random variable and if we momentaneously forget the index i, the mgf is equal to an integral on the support of Y: $\mathbb{E}(e^{tY}) = \int_{dom(Y)} e^{tY} f_Y(y, \theta, \phi) dy$. This integral is developed as follows

$$\mathbb{E}(e^{tY}) = \int_{dom(Y)} \exp\left(\frac{y(\theta + t\frac{\phi}{\nu}) - a(\theta)}{\frac{\phi}{\nu}}\right) c(y, \phi, \nu) dy$$

$$= \exp\left(\frac{y(\theta + t\frac{\phi}{\nu}) - a(\theta)}{\frac{\phi}{\nu}}\right) \times$$

$$\int_{dom(Y)} \exp\left(\frac{y(\theta + t\frac{\phi}{\nu}) - a(\theta + t\frac{\phi}{\nu})}{\frac{\phi}{\nu}} + c(y, \phi, \nu)\right) dy$$

If the parameter space is open, it follows that at least for t in a neighborhood of zero, $\theta + t\frac{\phi}{\nu}$ is in the parameter space. The integrand is a an exponential dispersed density and the integral is equal to one.   □

From the expression (1.6), we infer the first two moments of Y_i by evaluating the derivative of the cumulant generating function with respect to t for $t = 0$:

$$\mathbb{E}(Y_i) = \left. \frac{\partial \psi(t)}{\partial t} \right|_{t=0} = a'(\theta_i)$$

$$\mathbb{V}(Y_i) = \left. \frac{\partial^2 \psi_i(t)}{\partial t^2} \right|_{\theta=0} = a''(\theta_i) \frac{\phi}{v_i}$$

In general, it is more convenient to view the variance as a function of the mean of Y_i. We have just seen that $\mathbb{E}(Y_i) = a'(\theta_i)$, since the cumulant function $a'(.)$ is invertible. Therefore, $\theta_i = h(\mathbb{E}(Y_i))$ where $h(.) := a'^{-1}(.)$ is the inverse of the first order derivative of the cumulant function. If we inject this relation into the expression of $\mathbb{V}(Y_i)$ to get what we call the variance function:

$$V(.) := a'' \left(a'^{-1}(.) \right) = a''(h(.)) .$$

Therefore, the variance of Y_i may be rewritten as

$$\mathbb{V}(Y_i) = V(\mathbb{E}(Y_i)) \frac{\phi}{v_i} .$$

For most of analyses, we mainly consider four ED distributions: the Poisson law and binomial laws for claim counts, the Normal and Gamma laws for claim amounts. We briefly recall their main features in Table 1.2 that presents their pdf, the cumulant function, the expectation and the variance function. We also report the correspondence between the parameters θ and ϕ in Eq. (1.5) and parameters in their usual formulation.

We aim to calibrate the neural network weights Ω such that its response $\widehat{y}_i$ to the information x_i is an estimator of $\mathbb{E}(Y_i)$ for a given predetermined distribution. Remember that the loss function $\mathcal{L}(y_i, \widehat{y}_i)$ is minimized in order to fit the neural net as follows:

$$\Omega = \arg \min_{\Omega} \frac{1}{n} \sum_{i=1}^{n} \mathcal{L}(y_i, \widehat{y}_i) .$$

In engineering applications, the most common loss function is the quadratic error. The neural net is estimated in order to minimize the quadratic spread between the real observation and the network output signal: $\mathcal{L}(y_i, \widehat{y}_i) = (y_i - \widehat{y}_i)^2$. However choosing such a criterion indifferently for claim counts or claim amounts introduces an estimation bias. Because we do not take into account the features of the statistical distributions of y_i during the estimation procedure. The solution consists to use as loss function minus the log-likelihood or the deviance. Our preference goes to the deviance mainly because the deviance allows comparing the goodness of fit of several models.

Let $l(\widehat{y}_i)$ denote the log-likelihood of the ith insurance contract response as a function of the estimator $\widehat{y}_i$ of $\mathbb{E}(Y_i)$. If the number of observations n is at least equal to the number of non-redundant parameters p, we can get a perfect fit by setting all $\widehat{y}_i = y_i$. This case is called the saturated model. This model is trivial and of no practical interest but since it perfectly fits data, its log-likelihood is the best one that we can obtain. The scaled deviance D^* is defined as the likelihood ratio test of the model under consideration against the saturated model:

$$D^*(y_i, \widehat{y}_i) = 2\left(l(y_i) - l(\widehat{y}_i)\right) .$$

Let us recall that we denote by $h(.)$ the inverse function of $a' = \frac{da}{d\theta}$, so that $\theta_i = h(\mathbb{E}(Y_i))$. Then, according to the definition of exponential dispersed distributions, we get that

$$D^*(y_i, \widehat{y}_i) = \frac{2}{\phi} v_i \left(y_i h(y_i) - a(h(y_i)) - y_i h(\widehat{y}_i) + a(h(\widehat{y}_i))\right)$$

By multiplying this expression by ϕ, we get the unscaled deviance $D(y_i, \widehat{y}_i) = \phi D^*(y_i, \widehat{y}_i)$ that we will use as loss function $\mathcal{L}(y_i, \widehat{y}_i)$ to estimate the feed-forward neural network. Table 1.3 presents the deviance functions that we use for defining the loss function, when Y_i are Normal, Gamma, Poisson or Binomial random variables. The total of losses is defined as the average unscaled deviance $\mathcal{R}(\Omega) = \frac{1}{n} \sum_{i=1}^{n} D(y_i, \widehat{y}_i)$.

Till now, we have assumed that the estimator is the output of the last layer of neurons : $(\widehat{y}_i)_{i=1,\dots n} = (\widehat{y}_{i,n^{net}})_{i=1,\dots n}$. However the domain of Y_i depends on the chosen distribution and is either equal to $\mathbb{R}$ for the Gaussian, $\mathbb{R}^+$ for the Gamma and Poisson, or $[0, 1]$ for the binomial law. For these three last distributions, we can transform the output signal of the neural network with a function $g(.)$ to ensure that the estimator is well in the domain of definition of Y_i that is,

$$\widehat{y}_i = g\left(\widehat{y}_{i,n^{net}}\right) \quad i = 1, \dots n .$$

Table 1.3 Deviance statistics for Normal, Gamma, Poisson and Binomial laws

	$a'(\theta)$	$h(y)$	Unscaled deviance, $D(y_i, \hat{y}_i)$	
Normal	θ	y	$v_i(y_i - \widehat{y}_i)^2$	
Gamma	$-\frac{1}{\theta}$	$-\frac{1}{y}$	$2v_i\left(\frac{y_i}{\widehat{y}_i} - 1 - \ln\left(\frac{y_i}{\widehat{y}_i}\right)\right)$	$y_i > 0$
			0	$y_i = 0$
Poisson	e^θ	$\ln y$	$2v_i(y_i \ln y_i - y_i \ln \widehat{y}_i - y_i + \widehat{y}_i)$	$y_i > 0$
			$2v_i \widehat{y}_i$	$y_i = 0$
Binomial	$\frac{e^\theta}{1+e^\theta}$	$\ln\left(\frac{y}{1-y}\right)$	$2v_i\left(y_i \ln\left(\frac{y_i}{\widehat{y}_i}\right) + (1 - y_i)\ln\left(\frac{1-y_i}{1-\widehat{y}_i}\right)\right)$	$y_i \in (0, 1)$
			$-2v_i \ln(1 - \widehat{y}_i)$	$y_i = 0$
			$-2v_i \ln(\widehat{y}_i)$	$y_i = 1$

Table 1.4 Transformation of the output signal

	Transform, $g(.)$	$\frac{\partial}{\partial \widehat{y}_{i,n^{net}}} D(y_i, \widehat{y}_{i,n^{net}})$
Normal	None	$\widehat{y}_i := \widehat{y}_{i,n^{net}}$	$2v_i \left(\widehat{y}_{i,n^{net}} - y_i \right)$
Gamma	Exponential	$\widehat{y}_i := \exp \left(\widehat{y}_{i,n^{net}} \right)$	$2v_i \left(1 - y_i e^{-\widehat{y}_{i,n^{net}}} \right)$
Poisson	Exponential	$\widehat{y}_i := \exp \left(\widehat{y}_{i,n^{net}} \right)$	$2v_i \left(e^{\widehat{y}_{i,n^{net}}} - y_i \right)$
Binomial	Logistic	$\widehat{y}_i := \frac{1}{1+\exp(-\widehat{y}_{i,n^{net}})}$	$2v_i \left(\frac{1-y_i}{1+e^{-\widehat{y}_{i,n^{net}}}} - y_i \frac{e^{-\widehat{y}_{i,n^{net}}}}{1+e^{-\widehat{y}_{i,n^{net}}}} \right)$

Table 1.4 presents three standard possible transformations that we use in numerical applications. On the other hand, the gradient $\frac{\partial}{\partial \omega_{ik}^j} \mathcal{R}(\Omega)$ is computed in two steps. First, we calculate analytically the derivative of the deviance with respect to the output signal $\widehat{y}_{i,n^{net}}$, with expressions reported in Table 1.4. Next, we calculate numerically the derivative of the output signal with respect to each weights of the neural network:

$$\frac{\partial}{\partial \omega_{ik}^j} \mathcal{R}(\Omega) = \frac{1}{n} \sum_{i=1}^n \frac{\partial}{\partial \widehat{y}_{i,n^{net}}} D(y_i, \widehat{y}_{i,n^{net}}) \frac{\partial \widehat{y}_{i,n^{net}}}{\partial \omega_{ik}^j}$$

for $i = 1, \ldots, n_j^{net}$, $k = 1, \ldots, n_{j-1}^{net}$ and $j = 1, \ldots, n^{net}$.

Actuaries are not only interested by predictions. They also want to evaluate the exposure of the insurance company to any adverse deviation of the frequency and amount of claims. Under the assumption that all claims caused by policyholders are independent, the risk exposure is assessed by a percentile of the distribution of Y_i. For the Poisson and Binomial cases, the estimators $\widehat{y}_i$ of $\mathbb{E}(Y_i)$ are also estimates of the unique parameters λ_i and p_i. The calculation of percentiles is then based on these estimates and does not present any particular difficulties.

The Normal and Gamma distributions are defined by two parameters and $\widehat{y}_i$ is the estimator either of $\hat{\mu}_i$ or of $\frac{\hat{\alpha}}{\hat{\beta}_i}$. In order to estimate the second parameter, we have to remember that a classic measure of the goodness-of fit of a statistical model is the (unscaled) Pearson's chi-square χ^2:

$$\chi^2 = \sum_{i=1}^n \frac{(y_i - \widehat{y}_i)^2}{\mathbb{V}(Y_i)} = \frac{1}{\phi} \sum_{i=1}^n v_i \frac{(y_i - \widehat{y}_i)^2}{V(\widehat{y}_i)} \tag{1.7}$$

where the function $V(.)$ is a variance function of Table 1.2. If $m = \text{card}(\Omega)$ is the number of parameters, the statistical theory tells us that χ^2 is approximately a $\chi^2(n - m)$ random variable. Hence, $\mathbb{E}(\chi^2) = n - m$. An estimator $\hat{\phi}$ of ϕ is then provided by:

$$\hat{\phi} = \frac{\phi \chi^2}{n - m} = \frac{1}{n - m} \sum_{i=1}^n v_i \frac{(y_i - \widehat{y}_i)^2}{V(\widehat{y}_i)}.$$

Table 1.5 Parameter
estimates for Normal,
Gamma, Poisson and
Binomial distributions

		Parameters	Estimates
Normal		μ_i	$\widehat{y}_i$
		σ^2	$\widehat{\phi}$
Gamma		α	$\frac{1}{\phi}$
		β_i	$\frac{1}{\phi \widehat{y}_i}$
Poisson		λ_i	$\widehat{y}_i$
Binomial		p_i	$\widehat{y}_i$

Based on this last relation and on definitions of $\mathbb{E}(Y_i)$, we report in Table 1.5 the estimators of parameters defining the Normal, Gamma, Poisson and Binomial distributions. Once that parameters are estimated, percentiles of the distribution of Y_i are easily computed.

1.8   Model Testing and Selection

This brief section presents two criterions for selecting a neural model and an asymptotic test for checking the relevance of explanatory variables. Each individual observation follows a ED distribution and according to definition (1.5), the log-likelihood is a function of $\boldsymbol{\theta} = (\theta_i)_{i=1,\dots,n}^T$:

$$l(\boldsymbol{\theta}, \phi, \boldsymbol{y}) = \frac{1}{\phi} \sum_{i=1}^{n} v_i (y_i \theta_i - a(\theta_i)) + \sum_{i=1}^{n} \ln c(y_i, \phi, v_i) \qquad (1.8)$$

If we remember that $\theta_i = h(\mathbb{E}(Y_i))$ and that $\widehat{y}_i$ is an estimator of $\mathbb{E}(Y_i)$, then an estimate of the log-likelihood is equal to:

$$\widehat{l}(\widehat{\boldsymbol{y}}, \widehat{\phi}, \boldsymbol{y}) = \frac{1}{\widehat{\phi}} \sum_{i=1}^{n} v_i (y_i h(\widehat{y}_i) - a(h(\widehat{y}_i))) + \sum_{i=1}^{n} \ln c(y_i, \widehat{\phi}, v_i) \qquad (1.9)$$

This log-likelihood estimate is used to evaluate two other important criteria to judge the overall quality of the model. The first one is the Akaike information criterion (AIC). If $m = \text{card}\{\Omega\}$ is the number of weights of the feed-forward perceptron, then the AIC value of the model is defined by:

$$\text{AIC} = 2m - 2\widehat{l}(\widehat{\boldsymbol{y}}, \widehat{\phi}, \boldsymbol{y}).$$

Given a set of candidate networks, the preferred model is the one with the lowest AIC. The AIC rewards goodness of fit (assessed by the likelihood function), but it also penalizes models with a large number parameters. The second criterion is the

Bayesian information criterion (BIC):

$$\text{BIC} = \ln(n)m - 2\widehat{l}(\widehat{\mathbf{y}}, \widehat{\phi}, \mathbf{y}).$$

As the AIC, the model with the lowest BIC is preferred. As the AIC, the BIC penalizes networks with a large set of weights. However, the penalty term is larger in BIC than in AIC.

Using the asymptotic property of the log-likelihood, we can perform a hypothesis test in order to decide whether to include an explanatory variable in a neural network. This test is done with a log-likelihood ratio test (LR) of the two models against each other, with and without the particular explanatory factor.

Let us consider two models M_r and M_s with respectively r and s parameters and such that $M_s \subset M_r$. Let $\widehat{y}_i^r$ and $\widehat{y}_i^s$ be respectively the estimate of y_i in models M_r and M_s. If we remember that the unscaled deviance, $D^*(.,.)$ is the difference between log-likelihoods of saturated and fitted models, the LR statistic for testing M_s against M_r is equal to:

$$LR(y_i, \widehat{y}_i^s, \widehat{y}_i^r) = D^*(y_i, \widehat{y}_i^s) - D^*(y_i, \widehat{y}_i^r)$$
$$= 2\left(l(\widehat{y}_i^r) - l(\widehat{y}_i^s)\right).$$

Under the assumption that models M_r and M_s have the same dispersion parameter ϕ, the following sum

$$\chi_{s,r} = \sum_{i=1}^{n} LR(y_i, \widehat{y}_i^s, \widehat{y}_i^r)$$

is approximately χ^2 distributed with $r - s$ degrees of freedom (see Wilks 1938). Excepted for the Poisson and Binomial distributions (for which $\phi = 1$), we have to estimate the parameter of dispersion ϕ to calculate $\chi_{s;r}$. If this estimator is denoted $\widehat{\phi}$, the statistics becomes

$$\chi_{s,r} = \frac{1}{\widehat{\phi}} \sum_{i=1}^{n} \left(D(y_i, \widehat{y}_i^s) - D(y_i, \widehat{y}_i^r)\right)$$

where $D(.,.)$ is the unscaled deviance as presented in Table 1.3. As suggested by Ohlsson and Johansson (2010), $\widehat{\phi}$ should be computed in the larger model. Otherwise the estimate would include the variation that is explained by additional variables. This variation could not be considered as random in case of rejection of the model M_s.

1.9   Confidence Intervals

We can estimate the sensitivity of the log-likelihood to its parameters Ω with the score function. The score is the derivative of $\widehat{l}(\widehat{y}, \widehat{\phi}, y)$ with respect to weights $\omega_j \in \Omega$ (we momentously replace the notation ω_{ik}^j by ω_j to lighten developments). Deriving in chain the log-likelihood leads to the following expression for the score:

$$
\frac{\partial l}{\partial \omega_j} = \sum_{i=1}^{n} \frac{\partial l}{\partial \theta_i} \frac{\partial \theta_i}{\partial \omega_j} = \frac{1}{\phi} \sum_{i=1}^{n} v_i \left(y_i - a'(\theta_i) \right) \frac{\partial \theta_i}{\partial \omega_j} \tag{1.10}
$$

$$
= \frac{1}{\phi} \sum_i v_i \left(y_i - a'(\theta_i) \right) \frac{\partial \theta_i}{\partial \mathbb{E}(Y_i)} \frac{\partial \mathbb{E}(Y_i)}{\partial \omega_j} .
$$

Given that $\mathbb{E}(Y_i) = a'(\theta_i)$, we have $\partial \mathbb{E}(Y_i) / \partial \theta_i = a''(\theta_i) = V(\mathbb{E}(Y_i))$. The derivative of the inverse relation is $\partial \theta_i / \partial \mathbb{E}(Y_i) = 1 / V(\mathbb{E}(Y_i))$. As $\widehat{y}_i$ is an estimate of $\mathbb{E}(Y_i)$ and if we remember that $\widehat{y}_i = g\left(\widehat{y}_{i,n^{net}} \right)$ (where $g(.)$ is defined in Table 1.4) then an estimator of the score is provided by:

$$
\frac{\widehat{\partial l}}{\partial \omega_j} = \frac{1}{\widehat{\phi}} \sum_i v_i \frac{\left(y_i - a'(\widehat{\theta}_i) \right)}{V(\widehat{y}_i)} \frac{\partial \widehat{y}_i}{\partial \omega_j}
$$

$$
= \frac{1}{\widehat{\phi}} \sum_i v_i \frac{(y_i - \widehat{y}_i)}{V(\widehat{y}_i)} \frac{\partial g\left(\widehat{y}_{i,n^{net}} \right)}{\partial \widehat{y}_{i,n^{net}}} \frac{\partial \widehat{y}_{i,n^{net}}}{\partial \omega_j} .
$$

The derivative $\frac{\partial \widehat{y}_{i,n^{net}}}{\partial \omega_j}$ can easily be computed numerically.

For Poisson and Binomial distributions, maximizing the log-likelihood is trivially equivalent to minimizing the unscaled deviance. We can then assume that fitted weights are close to maximum likelihood estimator (MLE), $\widehat{\Omega}$ of real weights Ω. Under this assumption, we can rely on the theory of asymptotic behaviour of maximum likelihood estimators to build confidence intervals for weights. For networks with a high number of nodes, this assumption may be not satisfied. The calibration algorithm can indeed converges to a local minimum of the deviance loss function. In this case, the bias between real and estimated parameters is important and we cannot rely anymore on the MLE asymptotic theory to build confidence intervals. However, for small networks, we can expect the convergence of weights toward MLE. In this case, we use the Fisher information matrix, denoted by $\mathcal{I}(\Omega)$ in order to approach confidence intervals. This matrix is defined as minus the expectation of the Hessian of the log-likelihood function:

$$
\mathcal{I}(\Omega) = \left(-\mathbb{E} \left[\frac{\partial^2}{\partial \omega_j \partial \omega_k} l(Y) \right] \right)_{\omega_j, \omega_k \in \Omega}
$$

where Y is a vector of random variable distributed as $(Y_i)_{i=1,\ldots,n}$. If we consider a Normal approximation, real weights Ω are asymptotically distributed as $N\left(\widehat{\Omega},\, \mathcal{I}(\widehat{\Omega})^{-1}\right)$, where $\widehat{\Omega}$ is the set of weights fitted with the back-propagation algorithm. In view of Eq. (1.8), the second order derivative of $l(.)$ is:

$$\frac{\partial^2 l}{\partial \omega_j \partial \omega_k} = \frac{1}{\phi} \sum_i v_i \left(\left(-\frac{\partial}{\partial \omega_k} \left(a'(\theta_i) \right) \right) \frac{\partial \theta_i}{\partial \omega_j} + \left(Y_i - a'(\theta_i) \right) \frac{\partial \theta_i}{\partial \omega_j \omega_k} \right).$$

given that $\mathbb{E}(Y_i) = a'(\theta_i)$, the Fisher information may be rewritten as follows:

$$\mathcal{I}(\Omega) = \frac{1}{\phi} \sum_i v_i \left(\frac{\partial}{\partial \omega_k} \left(a'(\theta_i) \right) \frac{\partial \theta_i}{\partial \omega_j} \right)$$

$$= \frac{1}{\phi} \sum_i v_i \left(\frac{\partial \mathbb{E}(Y_i)}{\partial \omega_k} \frac{\partial \theta_i}{\partial \mathbb{E}(Y_i)} \frac{\partial \mathbb{E}(Y_i)}{\partial \omega_j} \right).$$

Since $\partial \mathbb{E}(Y_i)/\partial \theta_i = a''(\theta_i) = V(\mathbb{E}(Y_i))$ and $\widehat{y}_i$ estimates $\mathbb{E}(Y_i)$, we approximate the Fisher information matrix by

$$\hat{I} = \frac{1}{\phi} \sum_i \frac{v_i}{V(\widehat{y}_i)} \left(\frac{\partial \widehat{y}_i}{\partial \omega_k} \frac{\partial \widehat{y}_i}{\partial \omega_j} \right). \tag{1.11}$$

If we respectively denote by α and $\Phi^{-1}(\alpha)$ the level of confidence and the α-percentile of a standard Normal random variable, then the interval of confidence for Ω is approached by

$$\Omega \in \left[\widehat{\Omega} - \Phi^{-1}(\alpha) \sqrt{\mathrm{diag}\left(\hat{I}^{-1} \right)}, \; \widehat{\Omega} + \Phi^{-1}(\alpha) \sqrt{\mathrm{diag}\left(\hat{I}^{-1} \right)} \right]. \tag{1.12}$$

A less accurate alternative but numerically more efficient consist to calculate numerically the observed Fisher information:

$$\mathcal{I}^*(\Omega) = \left(-\frac{\partial^2}{\partial \omega_j^2} l(\boldsymbol{y}) \right)_{\omega_j \in \Omega} \tag{1.13}$$

where $\boldsymbol{y} = (y_i)_{i=1,\ldots,n}$ is this time the vector of realization of $(Y_i)_{i=1,\ldots,n}$. In a notable article, Efron and Hinkley (1978) argued that the observed information should be used in preference to the expected information when employing Normal approximations for the distribution of maximum-likelihood estimates. Replacing $\hat{I}^{-1}$ by $\mathcal{I}^{*-1}$ in Eq. (1.12) is an acceptable solution to approximate confidence intervals.

Notice that results mentioned in this section are valid under the assumption that fitted weights are close to their maximum likelihood estimator (MLE). For large networks, the back-propagation algorithm may converge toward a local minimum and in this case the accuracy of confidence intervals found with the Fisher's information is not warranted. Furthermore, the Fisher's information matrix may not be invertible for numerical reasons. In this case, we use the Tikhonov regularization. This techniques consists to add a constant diagonal matrix to $\mathcal{I}^*(\Omega)$ as follows:

$$\tilde{\mathcal{I}}^*(\Omega) = \mathcal{I}^*(\Omega) + \kappa \; \mathrm{Id}(m),$$

where $\mathrm{Id}(m)$ is a $m \times m$ identity matrix and $\kappa \in \mathbb{R}^+$ is a small positive constant. However, there exists a Bayesian alternative method to calculate confidence intervals. This approach, based on Monte Carlo simulations is studied in Chap. 2.

1.10 Generalized Linear Models (GLM) in a Nutshell

In numerical illustrations, we compare neural networks to generalized linear models (GLM). GLM extend the ordinary linear regression. Firstly, GLM are compatible with all exponential dispersed (ED) distributions and is then not limited to the Normal law. Secondly, a non-linear relation is introduced between the mean of responses and covariates. The GLM theory was introduced by Nelder and Wedderburn (1972). In the nineties, GLM became a standard in the insurance industry. For a detailed presentation, we refer to the book of Ohlsson and Johansson (2010) or Denuit et al. (2019).

As in Sect. 1.7, the information about policyholders is contained in n vectors $(x_i)_{i=1,...,n}$ of dimension p. This vector contains both real numbers (e.g. age, power of a vehicle) and binary modalities (e.g. sex, region). The response is a vector $y = (y_i)_{i=1,...,n}$ of key quantities which is the realization of a ratio of a random variable Z_i on an exposure v_i. The responses are distributed according to an exponential dispersed law, as presented in Definition 1.1 of Sect. 1.7. To lighten further developments, the expectation of Y_i is denoted by $\mu_i := \mathbb{E}(Y_i)$ and $\widehat{y_i}$ is an estimator of μ_i.

Linear regression models are based on the assumption that the mean of Y_i is a linear function of explanatory variables:

$$\mathbb{E}(Y_i) := \mu_i = \boldsymbol{\beta}^\top x_i \quad i = 1, \dots, n$$

where $\boldsymbol{\beta}$ is a $\mathbb{R}^p$ vector of regression weights. GLMs allow the left hand side to be a monotone and then invertible function $g(.)$. This function is called the *link function*, since it relates the mean of the response Y_i to the linear structure through :

$$g(\mu_i) = \boldsymbol{\beta}^\top x_i \quad i = 1, \dots, n.$$

If the function $g(.)$ is the identity function, we retrieve the linear regression (but extended to ED distributions). This model is also called *additive*. Another popular link function is the *logarithm link*,

$$g(\mu_i) = \ln (\mu_i) = \boldsymbol{\beta}^\top \boldsymbol{x}_i \quad i = 1, \dots, n .$$

In this case, the mean of Y_i is a product of transformed explanatory variables

$$\mu_i = \prod_{j=1}^{p} \exp \left(\beta_j x_{i,j} \right) \quad i = 1, \dots, n .$$

For this reason, the log link GLM is also called *multiplicative* model. Notice that in this case, μ_i is restricted to $\mathbb{R}^+$. If Y_i is a Bernoulli variable, representing e.g. the default of a loan, μ_i is the probability of default and is in the interval $[0, 1]$. In this case, we use a *logit link* function:

$$g(\mu_i) = \ln \left(\frac{\mu_i}{1 - \mu_i} \right) = \boldsymbol{\beta}^\top \boldsymbol{x}_i \quad i = 1, \dots, n .$$

This link guarantees that the mean will stay between zero and one. The corresponding GLM is named the *logistic regression*. Table 1.6 lists the usual link functions and their inverse.

To summarize, a generalized linear model is specified by

- an exponential dispersed distribution for response variables $(y_i)_{i=1:n}$,
- a link function defining a relation between expected responses and covariates, $g(\mu_i) = \boldsymbol{\beta}^\top \boldsymbol{x}_i$ for $i = 1, \dots, n$.

The vector of weights $\boldsymbol{\beta}$ is estimated by log-likelihood maximization. The set of optimal weights is obtained by cancelling the derivative of the log-likelihood with respect to β_j for $j = 1, \dots, p$. Since individual observations are ED distributed and

Table 1.6 Some common link functions and their inverses

Link function $g(.)$	$g(\mu_i)$	$g^{-1}(\boldsymbol{\beta}^\top \boldsymbol{x}_i)$
Identity	μ_i	$\boldsymbol{\beta}^\top \boldsymbol{x}_i$
Inverse	μ_i^{-1}	$\left(\boldsymbol{\beta}^\top \boldsymbol{x}_i \right)^{-1}$
Log	$\ln (\mu_i)$	$\exp \left(\boldsymbol{\beta}^\top \boldsymbol{x}_i \right)$
Logit	$\ln \left(\frac{\mu_i}{1-\mu_i} \right)$	$\frac{\exp\left(\boldsymbol{\beta}^\top \boldsymbol{x}_i\right)}{1+\exp\left(\boldsymbol{\beta}^\top \boldsymbol{x}_i\right)}$
Probit	$\Phi^{-1}(\mu_i)$	$\Phi \left(\exp \left(\boldsymbol{\beta}^\top \boldsymbol{x}_i \right) \right)$

$\Phi(.)$ is the cdf of a standard Normal distribution

according to ED Definition 1.1, the log-likelihood as a function of θ_i is equal to:

$$l(\theta, \phi, y) = \frac{1}{\phi} \sum_{i=1}^{n} v_i \left(y_i \theta_i - a(\theta_i) \right) + \sum_{i=1}^{n} \ln c(y_i, \phi, v_i). \tag{1.14}$$

The parameter of overdispersion, ϕ, does not influence the maximization of l with respect to θ. Let us recall that $\theta_i = h\left(\mathbb{E}(Y_i)\right)$ where $h(.) := a^{'-1}(.)$ is the inverse of the first order derivative of the cumulant function. The weights $\boldsymbol{\beta}$ being related to θ_i by the following equality

$$\theta_i = h\left(g^{-1}\left(\boldsymbol{\beta}^\top \boldsymbol{x}_i \right) \right).$$

If we denote $\eta_i := \boldsymbol{\beta}^\top \boldsymbol{x}_i$ then the derivative of the log-likelihood $l(\theta, \phi, y)$ with respect to β_j is by successive derivation:

$$\frac{\partial l}{\partial \beta_j} = \sum_{i=1}^{n} \frac{\partial l}{\partial \theta_i} \frac{\partial \theta_i}{\partial \beta_j}$$

$$= \frac{1}{\phi} \sum_{i=1}^{n} \left(v_i y_i - v_i a^{'}(\theta_i) \right) \frac{\partial \theta_i}{\partial \beta_j}$$

$$= \frac{1}{\phi} \sum_{i=1}^{n} \left(v_i y_i - v_i a^{'}(\theta_i) \right) \frac{\partial \theta_i}{\partial \mu_i} \frac{\partial \mu_i}{\partial \eta_i} \frac{\partial \eta_i}{\partial \beta_j}.$$

By the relation $\mu_i = a^{'}(\theta_i)$, we have $\partial \mu_i / \partial \theta_i = a^{''}(\theta_i) = V(\mu_i)$ where $V(.)$ is the variance function (see Table 1.2 for a list of common variance function). Since the derivative of the inverse relation is the inverse of the derivative, we infer that $\partial \theta_i / \partial \mu_i = 1 / V(\mu_i)$. Moreover, we have that

$$\frac{\partial \mu_i}{\partial \eta_i} = \left(\frac{\partial \eta_i}{\partial \mu_i} \right)^{-1} = \frac{1}{g^{'}(\mu_i)} \quad \text{and} \quad \frac{\partial \eta_i}{\partial \beta_j} = x_{i,j}.$$

The derivative of the log-likelihood with respect to β_j is called the **score function** and is defined as

$$\frac{\partial l}{\partial \beta_j} = \frac{1}{\phi} \sum_{i=1}^{n} v_i \frac{y_i - \mu_i}{V(\mu_i) g^{'}(\mu_i)} x_{i,j} \quad j = 1, \ldots, p.$$

When we set these equations to zero, we get the maximum log-likelihood (ML) system:

$$\sum_{i=1}^{n} v_i \frac{y_i - \mu_i}{V(\mu_i) g^{'}(\mu_i)} x_{i,j} = 0 \quad j = 1, \ldots, p \tag{1.15}$$

where $\mu_i = g^{-1}\left(\boldsymbol{\beta}^\top x_i\right)$. Equation (1.15) admit a simple matrix representation. Let us define a $n \times n$ diagonal matrix $\mathbf{W} := diag\{\tilde{v}_i\,;\ i = 1,\dots n\}$ where

$$v_i = \frac{v_i}{V(\mu_i)g'(\mu_i)} \quad i = 1,\dots,n$$

and a $p \times n$ matrix $X := (x_1,\dots,x_n)$. Then the ML system can be rewritten as follows

$$\mathbf{XW}y = \mathbf{X}^\top \mathbf{W}\boldsymbol{\mu} \text{ where } \boldsymbol{\mu} = g^{-1}\left(\boldsymbol{\beta}^\top \mathbf{X}\right). \tag{1.16}$$

Except in a few cases, the ML system of equations must be solved numerically with e.g. a Newton Raphson algorithm. The estimate of $\boldsymbol{\beta}$ is a vector denoted by $\widehat{\boldsymbol{\beta}}$ and the estimator of y_i is $\widehat{y}_i = g^{-1}(\widehat{\boldsymbol{\beta}}x_i)$. In order to estimate the parameter of overdispersion, we use the same method as the one developed in Sect. 1.7. Since the statistics

$$\chi^2 = \frac{1}{\phi} \sum_{i=1}^{n} v_i \frac{(y_i - \widehat{y}_i)^2}{V(\widehat{y}_i)} \tag{1.17}$$

is approximately a $\chi^2(n-m)$ random variable where m is the number of parameters, therefore $\mathbb{E}(\chi^2) = n - m$. An estimator $\widehat{\phi}$ of ϕ is then provided by:

$$\widehat{\phi} = \frac{\phi\chi^2}{n-m} = \frac{1}{n-m} \sum_{i=1}^{n} v_i \frac{(y_i - \widehat{y}_i)^2}{V(\widehat{y}_i)}.$$

The correspondence between $(\widehat{y}, \phi)$ and parameters defining the Normal, Gamma, Poisson and Binomial distributions is reported in Table 1.5. The calculation of standard deviations and confidence intervals for estimates is performed in a similar way to the procedure presented in Sect. 1.9.

1.11 Illustration with Shallow Neural Networks

To illustrate this chapter, we fit several single layer feed-forward networks to an insurance database. This type of structure is also called shallow neural networks. We will see in a next section that this category of network can in theory approximate any non-linear functions. The performances of deep neural networks (in other words, with more than one hidden layers) are explored in the next chapter. The data comes from the Swedish insurance company *Wasa*, before its fusion with *Länsförsäkringar Alliance* in 1999. The database is available on the companion website of the book of Ohlsson and Johansson (2010) and contains information about motorcycles insurances over the period 1994–1998. Each policy is described by quantitative and

Table 1.7 Rating factors of motorcycle insurances

Rating factors	Class	Class description
Gender	M	Male
	K	Female
Geographic area	1	Central and semi-central parts of Sweden's three largest cities
	2	Suburbs plus middle-sized cities
	3	Lesser towns, except those in 5 or 7
	4	Small towns and countryside
	5	Northern towns
	6	Northern countryside
	7	Gotland (Sweden's largest island)
Vehicle class	1	EV ratio –5
	2	EV ratio 6–8
	3	EV ratio 9–12
	4	EV ratio 13–15
	5	EV ratio 16–19
	6	EV ratio 20–24
	7	EV ratio 25–

Source: Ohlsson and Johansson (2010)

categorical variables. The quantitative variables are the insured's age and the age of his vehicle. The categorical variables are: the policyholder's gender, the geographic zone and the category of the vehicle. The category of the vehicle is based on the ratio power in KW $\times$ 100/vehicle weight in kg $+$ 75, rounded to the nearest integer. The database also reports the number of claims, the total claim costs and the duration of the contract for each policies. Table 1.7 summarizes the information provided by categorical variables. The database counts $n = 62{,}436$ policies (after removing contracts with a null duration). The total exposure is equal to 65,217 years and 693 claims are reported.

In a first attempt, we work with a limited number of covariates and shallow networks for pedagogical purposes. We consider two quantitative variables, the owner's age and age of the vehicle, and one categorical variable, the gender. We scale the ages of owners and vehicles in order to convert their domain on the pavement $[0, 1] \times [0, 1]$. The vector of covariates $x = (x_1, x_2, x_3)$ for the ith policyholder contains the following information

$$x_1 = \frac{age(i) - \min(age)}{\max(age) - \min(age)},$$

$$x_2 = \frac{vehicle\,age(i) - \min(vehicle\,age)}{\max(vehicle\,age) - \min(vehicle\,age)},$$

$$x_3 = \begin{cases} 1 & if\ woman \\ 0 & otherwise \end{cases}.$$

Firstly, we regress the frequency of claims on these covariates. The output y is then the number of claims caused by the ith policyholder divided by the exposure v_i, which is here is the duration of the contract. We work with a single hidden layer of neurons with logistic activation functions. The activation functions of input and output layers are linear. We consider configurations with 2 to 5 neurons in the hidden layer. The link function between estimates of claims frequency and the output signal is exponential. For a network with 4 neurons in the hidden layer, the relation between covariates and estimated frequency is summarized by the following equations:

$$\widehat{y}_{i,1} = \phi \left(\omega_{i0}^1 + \sum_{k=1}^{3} \omega_{ik}^1 x_k \right) \quad i = 1, 2, 3, 4$$

$$\widehat{y} = \exp \left(\omega_{10}^2 + \omega_{11}^2 \widehat{y}_{1,1} + \omega_{12}^2 \widehat{y}_{2,1} + \omega_{13}^2 \widehat{y}_{3,1} + \omega_{14}^2 \widehat{y}_{4,1} \right).$$

The loss function chosen to estimate neural networks is the Poisson deviance. At the best of our knowledge, most of neural net softwares do not use the deviance statistics as loss function and do not manage exposures. The results that we present in the rest of this chapter are obtained with a customized version of the "neuralnet" R package in order to take into account these exposures. The networks are calibrated with the resilient back-propagation algorithm. This algorithm is initialized with random Normal weights centered around zero and of unit variance. The algorithm stops when the absolute variation of deviance between two successive iterations is less or equal than 0.10.

In machine learning, the choice of a neural network architecture is done by splitting the data into a training and a validation set. The training data set is used for calibrating a series of neural networks and the best model minimizes the losses computed with the validation data set. This simple technique is an efficient way to avoid what we call overfitting. A model that overfits a data set, explains very well the observations used for training the network, but fails to fit additional data or predicts future observations reliably. In practice, an overfitted model contains more parameters than can be justified by the data. In actuarial sciences this approach is difficult to implement mainly because claims are rare events by nature. For example, the data set from *Wasa* contains 62,436 contracts but only 693 claims are reported. Therefore, splitting the database for validation forces us to arbitrarily allocate sparse and valuable information about claims between the training and validation sets. For this reason, we prefer to calibrate the neural networks to the whole data set of contracts. As in classical statistical inference, the selection of a model is motivated by the deviance, the Akaike and Bayesian information criterions (AIC and BIC). We can eventually use the training-validation set approach for confirming the choice of a model, but we prefer to forecast frequencies with a network fitted to the whole database. Notice that we discuss alternative strategies of validation in Sect. 1.12.

Table 1.8 Deviances, Akaike and Bayesian information criteria for GLM and different configurations of the neural network fitted to the whole dataset

Model	# of hidden neurons	# of weights	Deviance	AIC	BIC
NN(2)	2	9	5993.48	7364.05	7463.51
NN(3)	3	13	5978.06	7358.63	7503.30
NN(4)	4	17	5947.21	7337.78	7527.66
NN(5)	5	21	5939.97	7340.53	7575.62
GLM		4	6037.40	7423.97	7460.13

Table 1.9 Deviances, Akaike and Bayesian information criteria for GLM and different configurations of the neural network

Model	# of hidden neurons	Training set			Validation set deviance
		Deviance	AIC	BIC	
NN(2)	2	4088.84	5025.73	5163.35	1907.26
NN(3)	3	4077.50	5024.39	5121.26	1912.26
NN(4)	4	4062.24	5019.13	5201.52	1893.99
NN(5)	5	4052.06	5018.95	5244.76	1894.23
GLM		4146.17	5069.06	5103.80	1922.28

The database is split in a training and validation set

We test 4 configurations of networks with a single hidden layer containing 2 to 5 neurons. We use a logarithm link function between the claims frequency and the output of the network, to guarantee the positivity of estimates. Table 1.8 reports the deviances, AIC and BIC of these networks. We compare these statistics to a generalized linear model estimated on the same data set. Even with a simple configuration, the deviance obtained with a neural network is lower than the one of the generalized linear model. According to the AIC and BIC, the best model is the neural network with four hidden neurons. To confirm this intuition, we may split data into a training set (70% of observations) and a validation set. The training and validation sets count respectively 43,705 and 18,731 policies. The number of claims observed in these samples are respectively 472 and 221 claims. Statistics of goodness of fit on these data sets are reported in Table 1.9. Here, networks with 4 and 5 neurons in the hidden layer display similar AIC on the training set and the same deviance on the validation set.

The principle of parsimony leads us to select the network with 4 neurons for further developments. Figure 1.4 shows this structure. Weights ω_i^j of each neurons (i, j), estimated with the whole data set, are reported above the arrows in this graph. This model counts 17 parameters, 4 four times more than a GLM. We also present in Table 1.10 these weights and their asymptotic standard deviations computed with the Fisher and observed Fisher information matrix. We draw two conclusions from this table. Firstly, the standard deviations may be high compared to parameter estimates. Such high standard deviations can be explained by the limited number of claims observed in the training set and by numerical instability (we use the Tikhonov

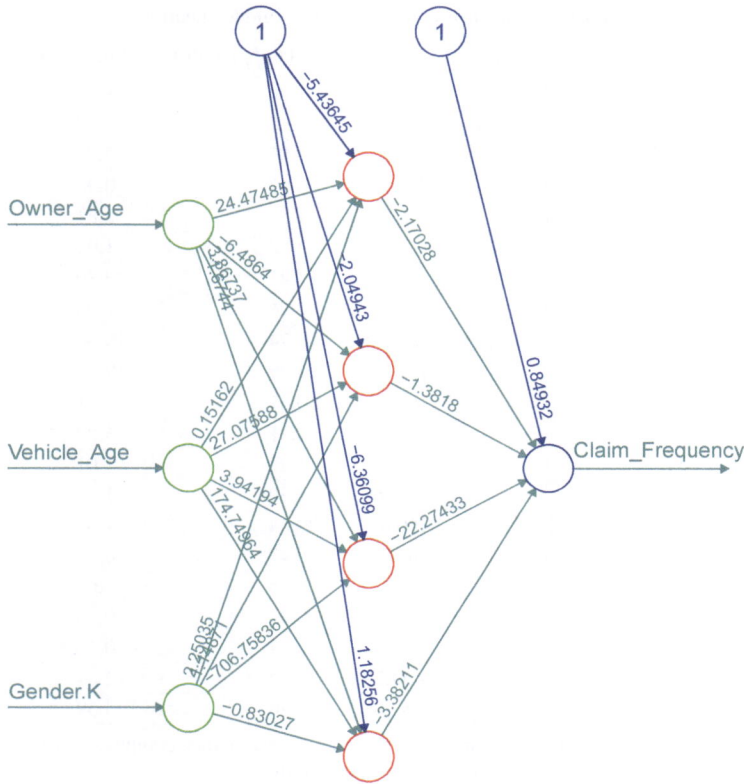

Fig. 1.4 Example of a feed-forward neural network, with 4 hidden neurons

regularization technique to obtain invertible information matrix). Secondly, standard deviations computed with the observed Fisher information are most of the time lesser than their equivalent evaluated with the Fisher information matrix.

The two first subplots of Fig. 1.5 compare expected claims frequencies computed with the NN(4) network, for male and female drivers of a 4 years old vehicle. The frequency of claims caused by women is clearly lesser than men's claim frequency and decreases sharply between 20 and 30 years old. The claims frequency for male drivers decreases with age from 20 to 45 years old. The curve of frequencies for men changes of curvature around 30 year old whereas the curve for women is strictly convex. In comparison, estimates computed with the GLM decrease exponentially. The two last subplots of Fig. 1.5 show claims frequencies for 32 years old drivers in function of the vehicle age. We observe that the frequency falls sharply after 1 year. Such a reduction is explained by the fact that a driver of a new motorcycle needs a learning period to fully master his vehicle. The GLM fails to identify this excess of claims frequencies for new motorcycles and overestimates the claims frequency for female drivers compared to the NN(4) model.

Table 1.10 Weight estimates for the neural network with 4 hidden neurons

ith neuron	jth layer	kth input	$\omega_{i,k}^j$	Std $(\omega_{i,k}^j)$ with $\hat{I}$	Std $(\omega_{i,k}^j)$ with I^*
1	1	1	−5.12	1.26	1.12
1	1	2	21.72	5.17	4.67
1	1	3	0.13	3.09	2.89
1	1	4	2.99	0.87	0.95
2	1	1	−4	1.5	2.81
2	1	2	−3.72	2.13	1.93
2	1	3	37.53	13.41	22.39
2	1	4	−1.06	0.96	1.04
3	1	1	−55.81	17.74	10.54
3	1	2	71.17	24.29	12.92
3	1	3	16.01	10.21	24.81
3	1	4	2.12	4.31	5.82
4	1	1	1.83	1.28	1.01
4	1	2	−2.16	1.29	1.08
4	1	3	160.4	29.64	22.68
4	1	4	−0.62	0.47	0.41
1	2	1	2.43	5.76	3.65
1	2	2	−2.17	0.23	0.28
1	2	3	−1.45	0.34	0.46
1	2	4	−54.32	31.62	12.04
1	2	5	−5.18	5.76	3.59

The two last columns reports the standard deviations of estimators computed with the Fisher information matrix $\hat{I}$ and the observed Fisher information matrix I^*

The first results are clearly encouraging: neural networks seem able to better discriminate the claims frequency of policyholders than a GLM. In a second series of tests, we estimate neural networks with the whole data set and for all available explanatory variables: owner's age, age of the vehicle, gender, geographic area and vehicle class. We consider three configurations with a single hidden layer (from 3 to 5 neurons) and two networks with two hidden layers (counting 2 or 3 neurons). Networks are calibrated with the resilient back-propagation algorithm and initial weights are drawn from a standard Normal distribution. We use a log link function. Since we consider all covariates, the number of weights increases considerably (e.g. 69 weights for the network with 4 intermediate neurons) and the performance of learning depends upon the starting point of the back-propagation procedure. For this reason, we run for each configuration the calibration algorithm with a dozen of different initial weights and select the best models in term of deviances.

Table 1.11 reports the deviance and AIC of tested networks and of a generalized linear model fitted to the same data set. The lowest AIC is achieved by a perceptron with 4 hidden neurons whereas the NN(3,3) network has the lowest deviance. Whatever the configuration, neural networks lead to a lower deviance than a GLM. We will see in Chap. 3 on deep learning how to estimate reliable networks with

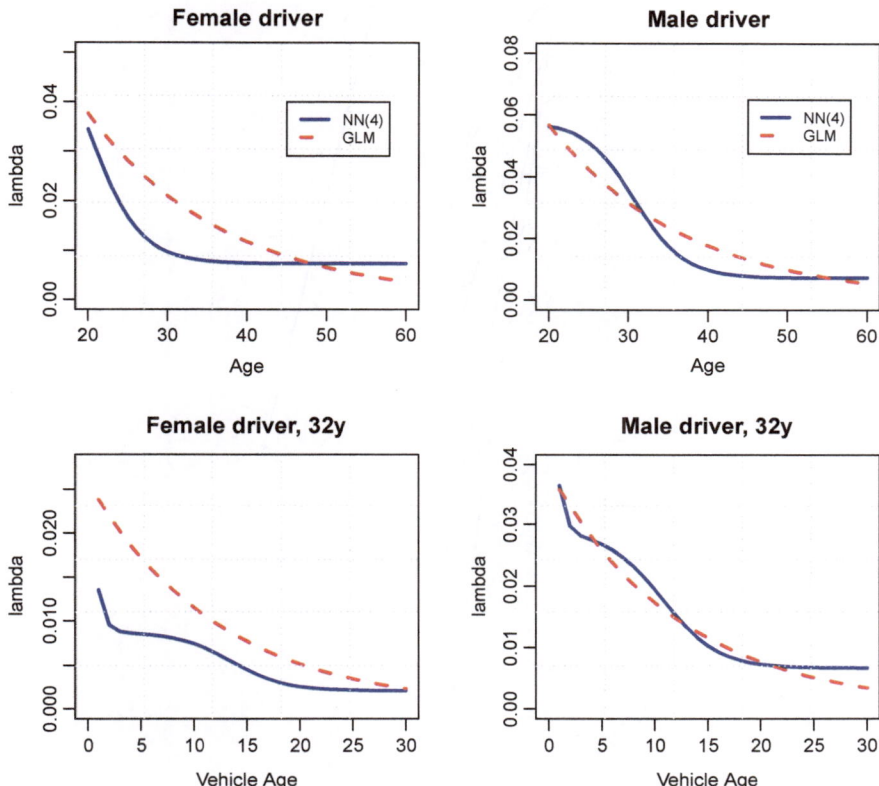

Fig. 1.5 Forecast frequency of claims for a male and female driver of a 4 years old vehicle

Table 1.11 Deviance, AIC and BIC of models fitted to the full data set

Model	# of hidden neurons	# of weights	Deviance	AIC	BIC
NN(3)	3	52	5664.36	7116.93	7587.11
NN(4)	4	69	5564.86	7051.43	7675.32
NN(5)	5	86	5546.91	7067.48	7845.08
NN(2,2)	2 × 2	41	5684.57	7115.14	7485.86
NN(3,3)	3 × 3	57	5499.72	7080.29	8129.15
GLM		16	5781.66	7162.23	7306.90

NN(3), NN(4) and NN(5) counts one single hidden layer with 3 to 5 neurons. NN(2,2) and NN(3,3) are perceptrons with 2 hidden layers containing respectively 2 and 3 neurons

several hidden layers and many neurons. But at this stage, we choose to work with the NN(4) model based on the principle of parsimony. This network has 69 weights and its architecture is shown in Fig. 1.6.

Figures 1.7, 1.8 and 1.9 compare expected claims frequencies for different profiles of drivers with the NN(4) networks. Whatever their characteristics, the age

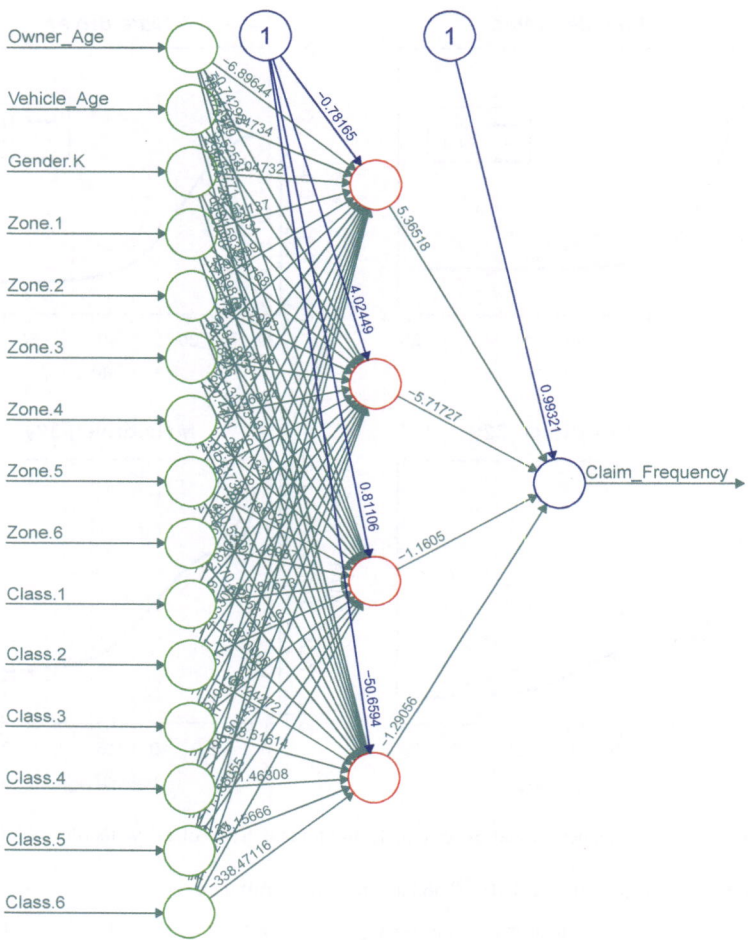

Fig. 1.6 Feed-forward neural network, with 4 hidden neurons, using all explanatory variables

is a key factor of the claims frequency. Before 30 years old, female drivers of class 3 motorcycle and living in the countryside (geographic area number 4) report on average less accidents than male drivers. Men driving a class 3 vehicle in a urban environment (geographic area number 1) have a higher claim frequency than those living in the countryside. The same observation holds for female drivers. The curves of frequencies obtained with a GLM present less convexity than these computed with the neural network. The spread between frequencies computed with the GLM and neural network reaches several percent for young drivers.

The left graph of Fig. 1.8 reveals that women driving a class 3 motorcycle in one of the largest Swedish cities report three times more accidents than women driving in the countryside. For this category of drivers, the GLM and neural networks both

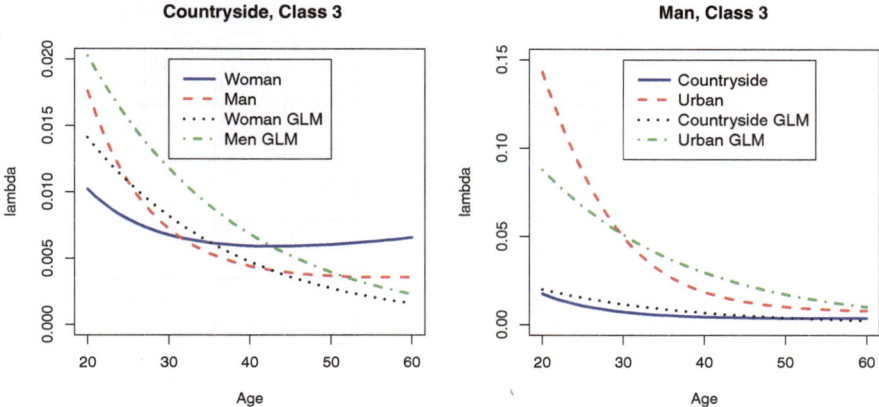

Fig. 1.7 Expected frequency of claims for drivers of a 4 years old vehicle, computed with the NN(4) model and GLM

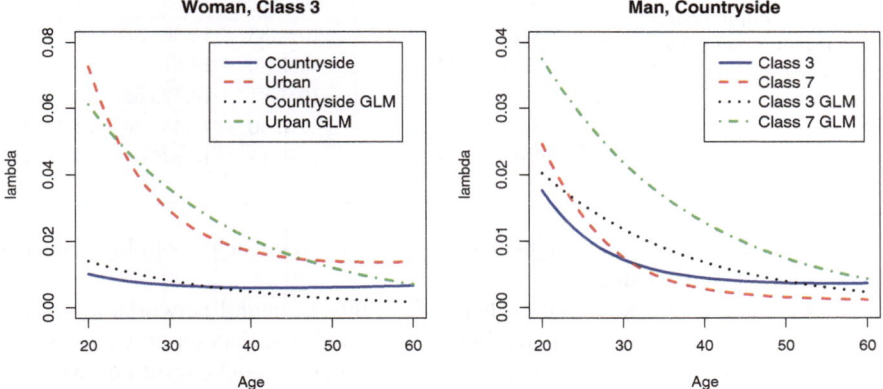

Fig. 1.8 Expected frequency of claims for drivers of a 4 years old vehicle, computed with the NN(4) model and GLM

yield very similar estimates. The right graph of the same Figure emphasizes the importance of the vehicle power. Before 30 years old, drivers of a class 7 motorcycle in the countryside are more riskier profiles than class 3 drivers. The GLM forecast significantly higher claims frequency than the neural network.

From the left graph of Fig. 1.9, we conclude that drivers in the Northern countryside (geographic area number 6) and younger than 40 years are more regularly involved in an accident than other drivers in the Swedish countryside. This is explained by the weather conditions that are less favorable in Northern regions than in the rest of the country. Notice that the GLM fails to discriminate these two categories of policyholders. The last graph emphasizes that compared to men, female drivers of class 7 motorcycles in the countryside causes on average

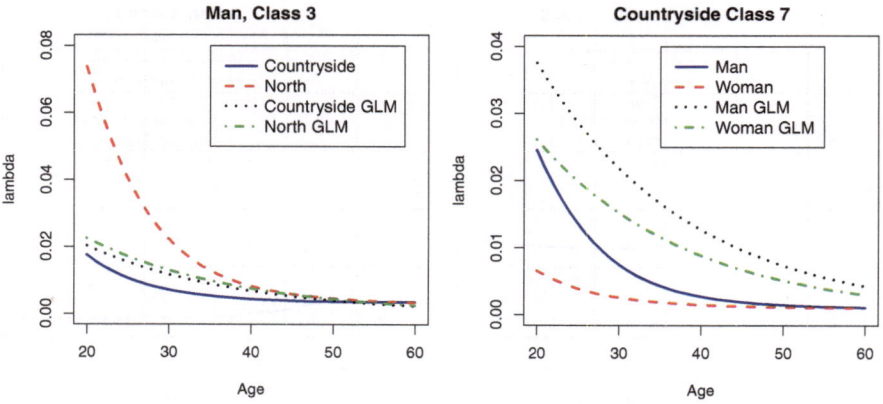

Fig. 1.9 Expected frequency of claims for drivers of a 4 years old vehicle, computed with the NN(4) model and GLM

Table 1.12 Deviances, AIC and BIC of neural models fitted to claims, with 2 to 5 hidden neurons and GLM

Network	# of weights	Deviance	AIC	BIC
NN(2)	29	1091.88	14,502.88	14,642.42
NN(3)	42	1068.20	14,512.83	14,719.89
NN(4)	55	1031.81	14,523.65	14,798.23
NN(5)	68	920.30	14,454.76	14,796.86
GLM	14	1243.15	14,679.63	14,742.65

less claims. The GLM yields higher expected claims frequencies both for male and female drivers in this category.

To conclude this section, we evaluate the ability of neural networks to regress the expected size of claims on policyholder's features. The data set only counts 693 policies that have submitted a positive claim. In order to avoid overfitting, we have selected the three explanatory variables that are the most pertinent for explaining claims costs: the age of the vehicle, the power class of the motorcycle and the geographic area of the driver. We assume that claims have a Gamma distribution. To guarantee the positivity of estimates, we use a log link function between the average claim and the output of the network.

Table 1.12 presents the deviances and AIC of shallow neural models with 2 to 5 neurons in the hidden layer. Whatever the configuration, we obtain a lower deviance with a neural network than a GLM model. Since the lowest AIC is achieved by the network with only two neurons, we select this configuration for further analysis. The network counts 29 weights and is shown in Fig. 1.10.

Figures 1.11 and 1.12 compare average claim amounts in NOK, for different categories of insured and vehicles. From the left plot of Fig. 1.11, we conclude that claims costs are higher in the countryside than in cities, at least for a motorcycle older than 3 years. Whereas average claims costs in the countryside decrease regularly with vehicle age, we observe a clear change of trend for claims in a urban

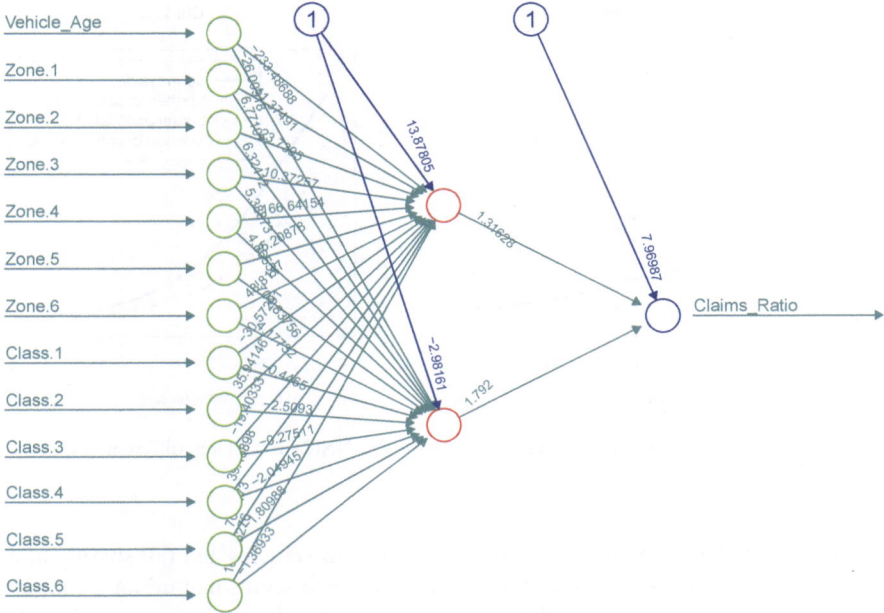

Fig. 1.10 Architecture of the feed-forward neural network for claims prediction

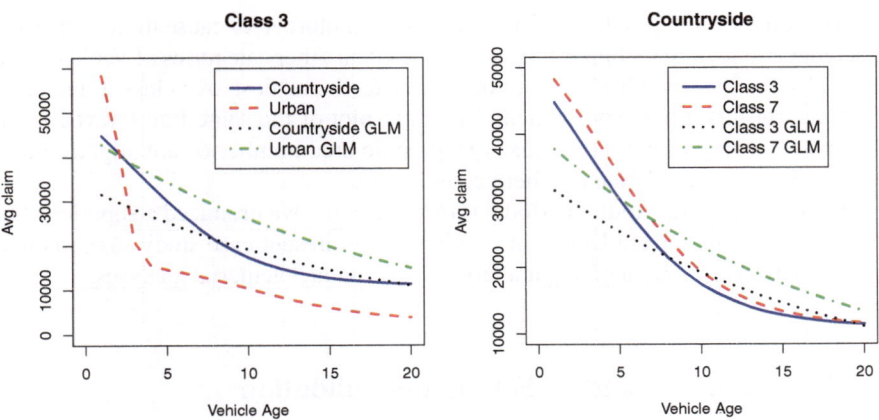

Fig. 1.11 Estimates of the average claim size with the NN(2) model for different profiles of policyholders

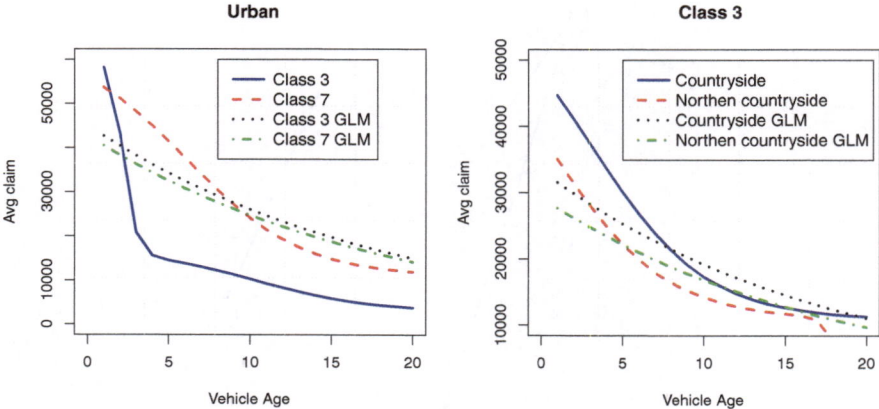

Fig. 1.12 Estimates of the average claim size with the NN(2) model for different profiles of policyholders

environment: claim amounts are very high for recent vehicles and fall sharply after 3 years. The GLM does not detect this trend. Finally, it seems that urban claims are on average less expensive than claims in the countryside. From the right graph of the same figure, we deduce that claims involving a class 7 vehicle are more expensive than these of class 3, in the countryside. Again, the GLM model forecasts higher claims costs than the neural network.

The left plot of Fig. 1.12 confirms that class 7 motorcycles cause more expensive accident in one of the 3 biggest Swedish cities than other categories of vehicles. We also observe that the GLM fails to discriminate claims costs for class 3 and class 7 drivers in this urban environment. The right plot of the same figure reveals that claims occurring in Northern areas (geographic area number 6) are slightly more expensive than these in the Southern countryside.

Even if a General Additive Model (GAM) would have produced a superior fit to GLM, as we know from Denuit et al. (2019), the present case study suggests that neural networks offer an efficient alternative to conduct actuarial analyses.

1.12 Overfitting and K-Fold Cross Validation

The classical method for validating different configurations of neural networks and to control overfitting consists to split the data set $\mathcal{D}$ into one training sample $\mathcal{D}_B = \{(\boldsymbol{x}_i, y_i) : i \in \mathcal{B}\}$, and one test sample $\mathcal{D}_B^c\{(\boldsymbol{x}_i, y_i) : i \in \mathcal{B}^c\}$ where $\mathcal{B}$ is a subset of $\{1, \ldots, n\}$ and $\mathcal{B}^c$ is its complementary set. As done in the previous section, usually 30% of the initial data set serves as testing set. Weights of the neural net are estimated with the training set. The most common criterion to detect overfitting is the average mean square error of prediction (MSEP) computed on the test sample:

$$MSEP = \frac{1}{|\mathcal{B}^c|} \sum_{i \in \mathcal{B}^c} v_i \, (\widehat{y}_i - y_i)^2 \,. \tag{1.18}$$

However, we have seen that for claims frequencies, the MSE is quite small given that claims are rare events. Another measure of the goodness of fit is the mean deviance, computed on the out-of-sample test:

$$MD = \frac{1}{|\mathcal{B}^c|} \sum_{i \in \mathcal{B}^c} D(y_i, \widehat{y}_i) \,,$$

where $D(.)$ is the unscaled deviance as defined in Sect. 1.7 . More generally, we may replace the deviance by any other error function, $E\,(y_i, \widehat{y}_i)$.

As underlined in the numerical illustration, when we work with insurance data sets, the implementation of this validation approach is sometimes difficult due to the lack of observations. As for regression trees studied in Trufin et al. (2019), we can perform a K-fold cross-validation. We choose a fixed integer $K \geq 2$ (often 10) and partition $\{1, \ldots, n\}$ randomly into K disjoint subsets $\mathcal{B}_1, \ldots, \mathcal{B}_K$ of approximately the same size. This provides for every $k = 1, \ldots, K$, a subset of data

$$\mathcal{D}^{(-\mathcal{B}_k)} = \{(x_i, y_i) \,:\, i \notin \mathcal{B}_k\} \subset \mathcal{D}$$

on which we train a neural network. If the goodness of fit is evaluated by the error function $E\,(y_i, \widehat{y}_i)$, the K-fold cross-validation error, denoted by CE_K, is given by

$$CE_K = \frac{1}{n} \sum_{k=1}^{K} \sum_{i \in \mathcal{B}_k} E\left(y_i, \widehat{y}_i^{(k)}\right)$$

where $\widehat{y}_i^{(k)}$ is the estimated response with the kth neural network. A comparison of cross validation errors yield by different architectures of neural networks is the best way to determine which structure is optimal. However, this cross validation is often time consuming and the number of validation sets is limited to a dozen.

In Sect. 1.11, we have fitted feed-forward neural networks with 3 and 4 intermediate neurons to data from the insurance company *Wasa* in order to predict the claims frequency of motorcycle drivers. Tables 1.13 and 1.14 presents the results of the cross validation procedure, with 10 subsets of validation. The average deviances on $\mathcal{D}^{(-\mathcal{B}_k)}$ and $\mathcal{B}_k$ for the net with three neurons, noted NN(3), are respectively equal to 5035.70 and 608.77. Whereas the same statistics for the net with four neurons, NN(4), are 5032.53 and 567.72. These results confirm that this last configuration has a better explanatory power than the NN(3) network, without overfitting. However, the average AIC of the NN(3) network is slightly lower than the one of the NN(4) network.

Table 1.13 Results of the cross validation procedure applied to the neural net with 3 intermediate neurons fitted to claims frequencies

3 intermediate neurons, NN(3)

K	Deviance	Log. lik.	AIC	Deviance, CE_K
1	5015.73	−3109.54	6323.07	602.38
2	5023.98	−3120.05	6344.09	630.37
3	5051.54	−3130.13	6364.27	591.42
4	4985.51	−3095.12	6294.24	651.56
5	5024.35	−3113.85	6331.69	631.91
6	5091.14	−3159.63	6423.26	571.59
7	5057.13	−3131.54	6367.08	579.09
8	5026.63	−3109.99	6323.97	637.71
9	5032.59	−3121.97	6347.93	626.07
10	5048.43	−3134.96	6373.93	565.7
Average	5035.70	−3122.67	6349.35	608.77

Table 1.14 Results of the cross validation procedure applied to the neural net with 4 intermediate neurons fitted to claims frequencies

4 intermediate neurons, NN(4)

K	Deviance	Log. lik.	AIC	Deviance, CE_K
1	5037.92	−3120.63	6379.26	608.36
2	5063.46	−3141.4	6420.8	529.97
3	5072.3	−3149.52	6437.03	521.19
4	5040.04	−3128.39	6394.77	575.7
5	5013.26	−3106.38	6350.76	616.06
6	5038.04	−3133.08	6404.15	551.33
7	5031.61	−3130.48	6398.95	560.07
8	5003.28	−3114.62	6367.24	577.95
9	4955.57	−3081.84	6301.68	626.73
10	5069.86	−3150.22	6438.43	509.87
Average	5032.53	−3125.65	6389.30	567.72

1.13   Why Does a Shallow Network Perform Well?

The mathematics in this section are more advanced and readers more interested in applications may consider skipping this section. Shallow networks are single layer neural networks and the theoretical background of this approach is based on the results of Cybenko (1989), Hornik et al. (1989) and Hornik (1991). We review the main elements of these articles. Let us denote by B^p the Borel tribe that is the smallest filtration containing all open sets of $\mathbb{R}^p$. The activation function of a neuron, also called a squashing function, is defined as follows:

Definition 1.2 A function $\phi(.)$: $\mathbb{R} \rightarrow [a, b]$ where $a, b \in \mathbb{R}$ is a squashing function if it is non-decreasing, $\lim_{x \to \infty} \phi(x) = b$ and $\lim_{x \to -\infty} \phi(x) = a$

As seen in Sect. 1.2, a squashing function is e.g. a sigmoid function ($a = 0$, $b = 1$) or the hyperbolic tangent function ($a = -1$, $b = 1$) . We introduce a first class of functions representative of a family of neural networks:

Definition 1.3 For any (Borel) measurable function $\phi(.) : \mathbb{R}^p \to \mathbb{R}$ and $p \in \mathbb{N}$, the class $\Sigma^p(\phi)$ is the class of functions $f : \mathbb{R}^p \to \mathbb{R}$ such that for $x = (x_1, \ldots, x_p)$

$$f(x) = \sum_{j=1}^{q} \beta_j \phi \left(\omega_j^\top x + \omega_{j,0} \right) \tag{1.19}$$

where $\omega_j \in \mathbb{R}^p$, $\beta_j \in \mathbb{R}$, $q \in \mathbb{N}$, and $\omega_{j,0} \in \mathbb{R}$.

If the function $\phi(.)$ is a squashing function, Eq. (1.19) describes the output of a single hidden layer feed-forward network, without squashing at the output layer. The weights β_j correspond to network weights from hidden to output layers.

The set of Borel measurable functions from $\mathbb{R}^p$ to $\mathbb{R}$ is denoted by M^p. By definition, the classes $\Sigma^p(\phi)$ belong to M^p for any Borel measurable function $\phi(.)$. Furthermore if $\phi(.)$ is continuous then $\Sigma^p(\phi)$ is in the set C^p of continuous function of dimension p. For the following developments, we have to remind the concept of *denseness*.

Definition 1.4 A subset S of a metric space X endowed with a measure ρ is ρ-*dense* in a subset T if for every $\epsilon > 0$ and $t \in T$, there exists $s \in S$ such that $\rho(s, t) \leq \epsilon$.

Furthermore, let K be a compact subset of $\mathbb{R}^p$. We define a ρ_K-measure as follows

$$\rho_K(f, g) := \sup_{x \in K} |f(x) - g(x)| .$$

The ρ_K-measure allows us to define the uniform denseness in the set of continuous functions of dimension d:

Definition 1.5 A subset $S \subset C^p$ is *uniformly dense on compacta* in C^p if for every compact subset $K \in \mathbb{R}^p$, S is ρ_K-dense in C^p.

As we are concerned by the ability of a neural network to approximate a non linear function of C^p, we introduce another metric with a probabilistic interpretation, in the following definition:

Definition 1.6 Given a probability measure $\nu(.)$ on $(\mathbb{R}^p, B^p)$, the metric ρ_ν from $M^p \times M^p$ to $\mathbb{R}^+$ is defined by

$$\rho_\nu(f, g) = \inf \{\epsilon > 0 : \nu (\{x : |f(x) - g(x)| > \epsilon\}) < \epsilon\}$$

Two functions are close in this metric if and only if there is only a small probability that they differ significantly. In the extreme case that f and g are ν-equivalent

$\rho_v(f, g)$ is equal to zero. We reproduce the theorem 2.4 of Hornik et al. (1989) and refer the interest reader to their article for a proof.

Theorem 1.1 *For every squashing function* $\phi(.)$, *every* p *and every probability measure* v *on* $(\mathbb{R}^p, \mathbb{B}^p)$, *the class* $\Sigma^p(\phi)$ *is uniformly dense on compacta in* C^p *and* ρ_v*-dense in* M^p.

In other words, single hidden layer in the class of $\Sigma^p(\phi)$ feedforward networks can approximate any measurable function arbitrarily well, regardless of the squashing function used and regardless of the dimension of the input space p. In this sense, $\Sigma^p(\phi)$ networks are universal approximators of any function in C^p.

The natural question that arises is why do we then use Multiple Layers Neural Networks (also called deep networks)? Even if shallow networks are universal approximators, the number of hidden neurons required to achieve a reasonable level of accuracy may be substantial. In pattern recognition, the functions to approximate are high-dimensional and non-linear. In this case, working with deep networks is justified by the fact that each layer composes its own level of non-linearity with the previous one. In shallow networks, non-linear relations are summed up and not composed.

1.14   Further Readings on Feed-Forward Networks

The motivation for neural networks dates back to McCulloch and Pitts (1943) and Widrow and Hoff (1960). Rosenblatt (1958) developed an electronic device that was simulating the perceptron, at Cornell Aeronautical Laboratory. The emergence of neural networks in the eighties is due to Parker (1985) and Rumelhart et al. (1986) who introduced the back-propagation algorithm. The simulated annealing is a pure stochastic search method, originally due to Metropolis et al. (1953) and inspired from the theory of statistical mechanics. In Chap. 2, we introduce and test the MCMC algorithm that is a more advanced random search procedure. In our applications, the activation function is a sigmoid. In radial basis networks, the linear combination of input signals is sent to several neurons with Gaussian activation functions. Contrary to perceptrons, radial basis networks have a single hidden layer. Differences between radial networks and perceptrons are studied by Haykin (1994). Since the eighties, neural networks are applied to patterns recognition. We refer the reader to Bishop (1995) or Ripley (1996) for more details on this topic. Neural networks are also used in finance for time-series forecasting or for bankruptcy prediction. We refer to McNelis (2005) for some of these applications. Hastie et al. (2009) study the combination of machine learning methods like penalization, bagging or boosting with neural networks. More recently, some aspects of neural networks applied to non-life insurance are presented in the manuscript of Wuthrich and Buser (2017).

References

Bishop C (1995) Neural networks for pattern recognition. Clarendon Press, Oxford

Cybenko G (1989) Approximation by superpositions of a sigmoidal function. Math Control Signals Syst 2:303–314

Denuit M, Hainaut D, Trufin J (2019) Effective statistical learning methods for actuaries: GLMs and extensions. Springer, Berlin

Efron B, Hinkley DV (1978) Assessing the accuracy of the maximum likelihood estimator: observed versus expected FisherInformation. Biometrika 65(3):457–487

Hastie T, Tibshirani R, Friedman J (2009) The Elements of statistical learning: data mining, inference, and prediction, 2nd edn. Springer, New York

Haykin S (1994) Neural networks: a comprehensive foundation. Prentice-Hall, Saddle River

Hornik K (1991) Approximation capabilities of multilayer feed-forward networks. Neural Netw 4:251–257

Hornik K, Stinchcombe M, White H (1989) Multi-layer feed-forward networks are universal approximators. Neural Netw 2:359–366.

McCulloch W, Pitts W (1943) A logical calculus of the ideas imminent in nervous activity. Bull Math Biophys 5:115–133

McNelis PD (2005) Neural networks in finance: gaining predictive edge in the market. Advanced finance series. Elsevier Academic Press, Cambridge

Metropolis N, Rosenbluth AW, Rosenbluth MN, Teller AH, Teller E (1953) Equation of state calculations by fast computing machines. J Chem Phys 21:1087–1092

Nelder JA, Wedderburn RWM (1972) Generalized linear models. J R Stat Soc Ser A 135(3):370–384

Ohlsson E, Johansson B (2010) Non-life insurance pricing with generalized linear models. Springer, Berlin

Parker D (1985) Learning logic, technical report TR-87. MIT Center for Research in Computational Economics and Management Science, Cambridge

Riedmiller M, Braun H (1993) A direct adaptive method for faster backpropagation learning: the RPROP algorithm. In: Proceedings of the IEEE international conference on neural networks. IEEE, Piscataway, pp 586–591

Ripley BD (1996) Pattern recognition and neural networks. Cambridge University Press, Cambridge

Rosenblatt F (1958) The perceptron: a probabilistic model for information storage and organization in the brain. Psychol Rev 65:386–408

Rumelhart D, Hinton G, Williams R (1986) Learning internal representations by error propagation. In: Parallel distributed processing: explorations in the microstructure of cognition. MIT Press, Cambridge, pp 318–362

Trufin J, Denuit M, Hainaut D (2019) Effective statistical learning methods for actuaries: tree-based methods. Springer, Berlin

Widrow B, Hoff M (1960) Adaptive switching circuits, IRE WESCON convention record, vol 4. pp 96–104

Wilks SS (1938) The Large-sample distribution of the likelihood ratio for testing composite hypotheses. Ann Math Stat 9:60–62

Wuthrich M, Buser C (2017) Data analytics for non-life insurance pricing. Swiss Finance Institute Research paper no.16-68. Available on SSRN https://ssrn.com/abstract=2870308

Chapter 2
Bayesian Neural Networks and GLM

The learning of large neural networks is an ill-posed problem and there is generally a continuum of possible set of admissible weights. In this case, we cannot rely anymore on asymptotic properties of maximum likelihood estimators to approximate confidence intervals. Applying the Bayesian learning paradigm to neural networks or to generalized linear models results in a powerful framework that can be used for estimating the density of predictors. Within this approach, the uncertainty about parameters is expressed and measured by probabilities. This formulation allows for a probabilistic treatment of our a priori knowledge about parameters based on Markov Chain Monte Carlo methods. In order to explain those methods that are based on simulations, we need to review the main features of Markov chains.

2.1 Markov Chain

A neural network (or a generalized linear model) is a non-linear function that squashes an input vector $x_i \in \mathbb{R}^p$ of information into an output signal $\hat{y}_i$ and which is defined by a vector of weights Ω. The dimension of Ω is noted m. In a Bayesian set-up, these weights are realizations ω of a multivariate random variable, with a density $\pi(\omega)$ and defined on a space of parameters $\mathcal{X} \subset \mathbb{R}^m$. The MCMC algorithm builds a discrete time Markov chain that converges in distribution toward $\pi(\omega)$.

Let us recall that a stochastic process is a sequence of random variables or vectors defined on some known state space. A Markov chain is a stochastic process in which future states are independent of past states given the present state where a stochastic process is a consecutive set of random (not deterministic) quantities defined on some known state space. Think of $\mathcal{X}$ as our parameters space. "Consecutive" implies a time component, indexed by t. Consider a draw of Ω_t to be a state at iteration t.

© Springer Nature Switzerland AG 2019
M. Denuit et al., *Effective Statistical Learning Methods for Actuaries III*,
Springer Actuarial, https://doi.org/10.1007/978-3-030-25827-6_2

Fig. 2.1 Example of a
Markov chain, defined on a
discrete set
$\Omega_t \in \mathcal{X} = \{\omega_L, \omega_I, \omega_H\}$

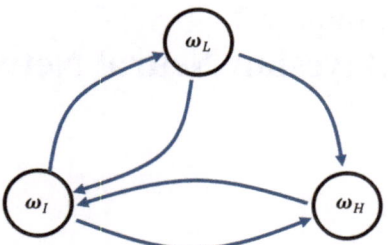

The next draw Ω_{t+1} is dependent only on the current draw Ω_t, and not on any past
draws.

In the rest of this chapter, Ω_t represents a random vector and we denote by ω a
realization of Ω_t. The transition between two consecutive steps of time is defined
by the transition kernel.

Definition 2.7 A transition kernel is a function K defined on $\mathcal{X} \times \mathcal{B}(\mathcal{X})$[1] such that
$\forall \omega \in \mathcal{X}$, $K(\omega, .)$ is a probability measure
$\forall A \in \mathcal{B}(\mathcal{X})$, $K(., A)$ is measurable.

When $\mathcal{X}$ is discrete the transition kernel is a matrix with elements

$$P_{\omega_t \omega_{t-1}} = P(\Omega_t = \omega_t | \Omega_{t-1} = \omega_{t-1}) \quad \omega_{t-1}, \omega_t \in \mathcal{X}.$$

In the continuous case, the kernel also denotes the conditional density such that
$P(\Omega_t \in A | \Omega_{t-1} = \omega) = \int_A K(\omega, d\omega')$. We are now able to define a Markov chain.

Definition 2.8 Given a transition kernel K, a sequence $\Omega_0, \Omega_1, \ldots \Omega_t$ of random
variables is a Markov chain denoted by $(\Omega_t)_{t \geq 0}$ if for any t the conditional
distribution of Ω_{t+1} given $\omega_t, \omega_{t-1}, \ldots, \omega_0$ is the same as the distribution of Ω_t
given ω_{t-1}:

$$P(\Omega_{t+1} \in A | \omega_0, \omega_1, \ldots, \omega_t) = P(\Omega_{t+1} \in A | \omega_t) = \int_A K(\omega_t, d\omega). \quad (2.1)$$

The chain is time-homogeneous if the distribution of $\Omega_{t_1}, \ldots, \Omega_{t_k}$ given Ω_{t_0} is the
same as the distribution of $\Omega_{t_1-t_0}, \Omega_{t_2-t_0}, \ldots, \Omega_{t_k-t_0}$ given Ω_0 for every k and
every $(k+1)$-uplet such that $t_0 \leq t_1 \leq \ldots \leq t_k$.

Example 2.1 Figure 2.1 shows an example of a Markov chain, defined on a discrete
set: $\Omega_t \in \mathcal{X} = \{\omega_L, \omega_I, \omega_H\}$ where $\omega_L, \omega_I, \omega_H \in \mathbb{R}^p$. The arrows represent the
possible transitions between two successive steps of Ω_t.

---

[1]$\mathcal{B}(\mathcal{X})$ is the sigma algebra defined on $\mathcal{X}$.

In this case, the transition kernel is given by the following matrix:

$$K(\omega_t, \omega_{t-1}) = P(\Omega_t = \omega_t | \Omega_{t-1} = \omega_t) = \begin{array}{c} \\ I \\ L \\ H \end{array} \begin{array}{ccc} I & L & H \\ 0 & p_{IL} & p_{IH} \\ p_{LI} & 0 & p_{LH} \\ 1 & 0 & 0 \end{array}$$

where p_{xy} are probabilities of transition between states.

Example 2.2 The autoregressive model AR(1) provides a simple illustration of Markov chains on a continuous state space. if

$$\Omega_t = \theta \Omega_{t-1} + \epsilon_t \quad \theta \in \mathbb{R} \tag{2.2}$$

where $\epsilon_t \sim N(0, \sigma I_m)$ is a multivariate white noise of variance $\sigma^2 I_m$ where I_m is the identity matrix of dimension m. By construction Ω_t is independent from $\Omega_{t-2}, \Omega_{t-3}$, conditionally on Ω_{t-1}. The kernel function is in this case given by $K(\omega_{t-1}, \omega_t) \sim N(\theta \omega_{t-1}, \sigma I_m)$.

According to Eq. (2.1), a Markov chain has a short memory: Ω_t only depends upon the last realization of the chain, $\Omega_{t-1} = \omega_{t-1}$. A direct consequence of this limited memory is that the expectation of any function $h(.)$ of $\Omega_{t+1}, \ldots, \Omega_{t+k}$ conditionally to the information up to time t is given by

$$\mathbb{E}\left(h(\Omega_{t+1}, \ldots, \Omega_{t+k}) | \omega_0, \ldots, \omega_t\right) = \mathbb{E}\left(h(\Omega_{t+1}, \ldots, \Omega_{t+k}) | \omega_t\right) ,$$

provided that the expectation exists.

The distribution of Ω_0, the initial state of the chain, plays an important role. In the discrete case, K is a transition matrix and given an initial distribution $\alpha_0 = \omega_0$ the marginal probability distribution of Ω_1 is obtained from the matrix multiplication:

$$\alpha_1 = \alpha_0^\top K$$

and for Ω_t by repeated multiplication $\Omega_t \sim \alpha_t = \alpha_0^\top K^t$. If the state space of the Markov chain is continuous, the initial distribution of Ω_0 is also denoted by α_0 . In later developments, we need the kernel for k transitions which is defined as

$$K^k(\omega, A) = \int_{\mathcal{X}} K^{k-1}(\omega', A) K(\omega, d\omega') .$$

The Chapman Kolmogorov equation is a consequence of properties of the kernel function.

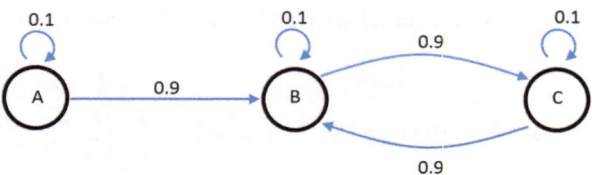

Fig. 2.2 Example of an irreducible Markov chain

Proposition 2.2 *Chapman Kolmogorov equation: for every* $(m, k) \in \mathbb{N}^2$, $\boldsymbol{\omega} \in \mathcal{X}$ *and* $A \in \mathcal{B}(\mathcal{X})$

$$K^{m+k}(\boldsymbol{\omega}, A) = \int_{\mathcal{X}} K^m(\boldsymbol{\omega}', A) K^k(\boldsymbol{\omega}, d\boldsymbol{\omega}').$$

In an informal sense, the Chapman Kolmogorov equation states that to get from $\boldsymbol{\omega}$ to A in $m + k$ steps, you pass through some $\boldsymbol{\omega}'$ on the k th step. In the discrete case, the integral is interpreted as a product of matrix.

The property of irreducibility is a first measure of the sensitivity of the Markov chain to the initial conditions α_0. In the discrete case, the chain is irreducible if all states communicate.

Definition 2.9 Given a measure φ on $\mathbb{R}^p$, the Markov chain Ω_t with transition kernel $K(\boldsymbol{\omega}_{t-1}, \boldsymbol{\omega}_t)$ is φ **irreducible** if for every $A \in \mathcal{B}(\mathcal{X})$ with $\varphi(A) > 0$ there exists t such that $K^t(\boldsymbol{\omega}, A) > 0$ for all $\boldsymbol{\omega} \in \mathcal{X}$. The chain is strongly φ irreducible if $t = 1$ for all measurable A.

Example 2.2 Continued: When $\Omega_t = \theta \Omega_{t-1} + \epsilon_t$ where ϵ_t is a white noise on $\mathbb{R}^m$, the chain is irreducible for the Lebesgue measure. But if ϵ_t is uniform on $[-1, 1]^m$ and $|\theta| > 1$, the chain is not irreducible anymore. Indeed,

$$\Omega_{t+1} - \Omega_t \geq (\theta - 1)\Omega_{t-1} - 1 \geq 0$$

for $\Omega_t \geq \frac{1}{\theta - 1}$. The chain is thus monotonically increasing and cannot visit previous values.

Example 2.3 Figure 2.2 presents an example of irreducible Markov chain, defined on a discrete set $\mathcal{X}$. The chain is irreducible because we cannot get to A from B or C regardless of the number of steps we consider.

If $\mathcal{X}$ is discrete, the period of a state $\boldsymbol{\omega}$ of chain is the minimal number of time steps before we could expect to see again the chain in this state[2]:

$$d(\boldsymbol{\omega}) = g.c.d. \left\{ u \geq 1 \,;\, K^u(\boldsymbol{\omega}, \boldsymbol{\omega}) > 0 \right\},$$

---

[2] *g.c.d.* : greatest common divisor.

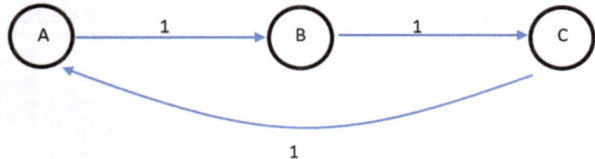

Fig. 2.3 Example of a periodic Markov chain. The period of the chain is 3

and a chain is aperiodic if it has period 1 for all states. Figure 2.3 shows an example of periodic Markov chain defined on a discrete state space.

The extension to continuous Markov chain is the following:

Definition 2.10 A φ-irreducible chain (Ω_t) has a life cycle of length d on A if d is the g.c.d. of

$$\{u \geq 1, \, such \, that \, K^u(\omega, A) \geq 0\}$$

$\forall \omega \in \mathcal{X}, \forall A \in \mathcal{B}(\mathcal{X})$.

From an algorithmic point of view, irreducibility ensures that every set A is visited by the Markov chain but this property is too weak to guarantee that Ω_t often enters in A . Let us denote by $\eta_A = \sum_{t=1}^{\infty} I_A(\Omega_t)$, the number of visits of Ω_t to A. The following definition characterizes the frequency of visits to A:

Definition 2.11 In a finite state space $\mathcal{X}$, a state $\omega \in \mathcal{X}$ is **transient** if the average number of visits to ω, denoted by $\mathbb{E}(\eta_\omega)$, is finite and **recurrent** if $\mathbb{E}(\eta_\omega) = \infty$.

An increased level of stability is attained if the marginal distribution of Ω_t is independent from t. This is required to ensure the existence of a probability distribution π such that $\Omega_{t+1} \sim \pi$ if $\Omega_t \sim \pi$. MCMC methods are based on this requirement.

Definition 2.12 A finite measure π is **invariant** for the transition kernel $K(.,.)$ and the chain Ω_t if

$$\pi(B) = \int_{\mathcal{X}} K(\omega, B)\pi(d\omega) \quad \forall B \in \mathcal{B}(\mathcal{X}).$$

When there exists an invariant probability measure for a φ irreducible chain, the chain is called "positive". The invariant distribution is also referred to as **stationary** if π is a probability measure.

Example 2.2 Continued: If $\Omega_t = \theta \Omega_{t-1} + \epsilon_t$, the kernel is $N(\theta \omega_{t-1}, \sigma I_m)$ and the stationary distribution $N(\mu, \tau)$ is stationary for the AR(1) chain only if

$$\mu = \theta \mu \quad \tau^2 = \tau^2 \theta^2 + \sigma^2$$

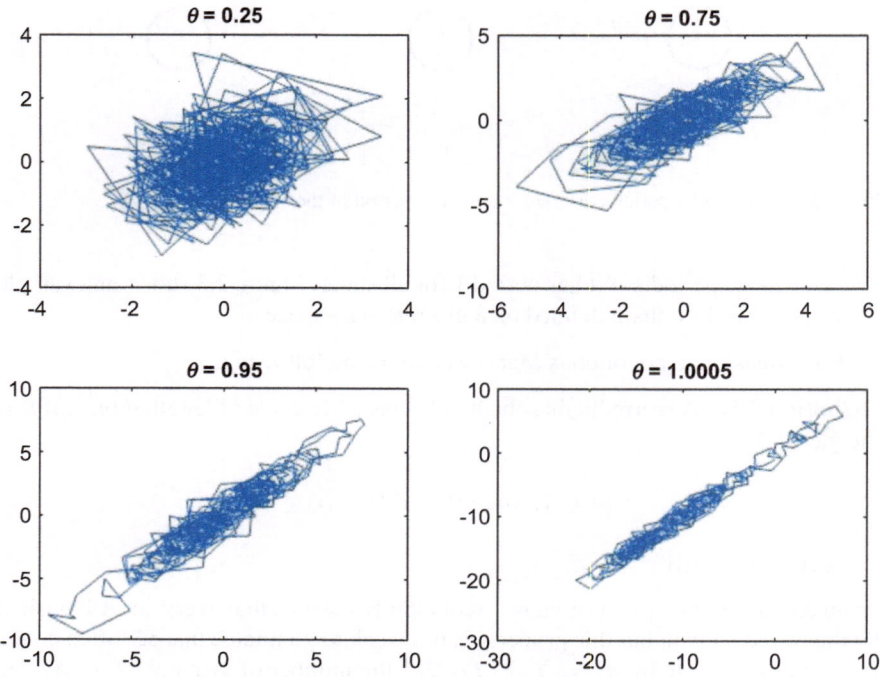

Fig. 2.4 Trajectories of four autoregressive AR(1) chains with $\sigma = 1$

This implies that $\mu = 0$ and

$$\tau^2 = \sigma^2/(1 - \theta^2)$$

which is possible only for $|\theta| < 1$. Figure 2.4 exhibits the sample path of pairs (Ω_{t-1}, Ω_t) of four univariate AR(1) chain, as defined by Eq. (2.2) and when $\sigma = 1$. We observe that all states of chains with $\theta = 0.25$, 0.75, 0.95 are recurrent. We clearly observe that the states of the chain becomes transient when θ increases. When $|\theta| >> 1$, the chain does not show any stability and diverges.

The next result formalizes the intuition that the existence of a invariant measure prevents the probability mass from escaping to infinity:

Proposition 2.3 *If the chain* $(\Omega_t)_{t\geq0}$ *is* **positive** *(invariant + φ irreducible) then it is* **recurrent**.

Proof Let us recall that η_ω is the number of visits by $(\Omega_t)_{t\geq0}$ to ω. If $(\Omega_t)_{t\geq0}$ is transient, there exists a covering of $\mathcal{X}$ by uniformly transient sets A_j with bounds

$$\mathbb{E}\left(\eta_{A_j}\right) \leq M_j \quad \forall j \in \mathbb{N}.$$

From the invariance of π, we know that

$$\pi(A_j) = \int K(\omega, A_j)\pi(d\omega) = \int K^t(\omega, A_j)\pi(d\omega)$$

then for every $k \in \mathbb{N}$

$$k\pi(A_j) = \sum_{t=0}^{k} \int K^t(\omega, A_j)\pi(d\omega) \le \int \mathbb{E}(\eta_{A_j})\pi(d\omega) \le M_j$$

since $\mathbb{E}(\eta_{A_j}) = \sum_{t=0}^{\infty} K^t(\omega, A_j)$. When $k \to \infty$, this shows that $\pi(A_j) = 0$, for every $j \in \mathbb{N}$. It is hence impossible to obtain an invariant probability. $\square$

Considering a Markov chain $(\Omega_t)_{t \ge 0}$, it is natural to establish the limit behaviour of Ω_t. The existence and uniqueness of an invariant distribution π makes that distribution a natural candidate for the limiting distribution. In this case, if Ω_t converges to this invariant distribution and the chain is **ergodic**. More precisely, we have:

Definition 2.13 A chain $(\Omega_t)_{t \ge 0}$ that is invariant, irreducible, aperiodic and positive recurrent is said ergodic. Then with probability 1, the sum $S_t(h) = \frac{1}{t} \sum_{k=1}^{t} h(\Omega_k)$, for any function $h(.)$ defined on $\mathcal{X}$ converges to the corresponding expectation:

$$S_t(h) \to \int_{\mathcal{X}} h(\omega)\,\pi(d\omega)$$

The proof of this convergence theorem may be found in Robert and Casella (2004). The stability inherent to stationary chains can be related to another property called the reversibility.

Definition 2.14 A stationary Markov chain $(\Omega_t)_{t \ge 0}$ is **reversible** if the distribution of Ω_{t+1}, conditionally on $\Omega_{t+2} = \omega$ is the same as the distribution of Ω_{t+1} conditionally on $\Omega_t = \omega$.

We will see later that this property is related to the existence of a stationary distribution.

Definition 2.15 A Markov chain with transition kernel K satisfies the **detailed balance condition** if there exist a function f such that

$$K(\omega', \omega)f(\omega') = K(\omega, \omega')f(\omega) \tag{2.3}$$

for every (ω, ω').

The balance condition is not necessary for f to be a stationary measure associated to $K(.,.)$ it provides however a sufficient conditions that is used in the Metropolis–Hastings algorithm. More generally,

Proposition 2.4 *Suppose that a Markov chain with kernel K satisfies the detailed balance condition (2.3) with a function π. Then*

1) The density π is the invariant density of the chain,
2) The chain is reversible.

Proof (1) comes from the following relation

$$
\int_{\mathcal{X}} K(\omega, B)\pi(\omega)d\omega = \int_{\mathcal{X}} \int_{B} K(\omega, \omega')\pi(\omega)d\omega' \, d\omega
$$
$$
= \int_{\mathcal{X}} \int_{B} K(\omega', \omega)\pi(\omega')d\omega' \, d\omega
$$
$$
= \int_{B} \pi(\omega')d\omega'
$$

as $\int K(x, y)dy = 1$. The proof of (2) follows from the existence of the kernel and invariant density. In this case, the detailed balance condition and reversibility are the same.   □

2.2 MCMC

A neural network or a generalized linear model is a non-linear function that converts a vector of covariates $x_i \in \mathbb{R}^p$ into an output vector $\widehat{y}_i$, which is an estimator of $\mathbb{E}(y_i)$. In a Bayesian framework, this function, that we denote by $f(x_i \mid \Omega)$, is parameterized by a random m−vector Ω of weights. Realizations of Ω and the probability density function of Ω are respectively denoted by ω and $\pi(\omega)$.

The MCMC algorithm builds a discrete time Markov $(\Omega_t)_{t \geq 0}$ chain that converges in distribution toward $\pi(\omega)$, the stationary distribution:

Definition 2.16 A Markov chain Monte Carlo (MCMC) method for the simulation of a statistical distribution $\pi(\omega)$ is any method producing an ergodic Markov chain $(\Omega_t)_{t \geq 0}$ whose stationary distribution is $\pi(\omega)$.

The main difficulty is to produce valid transition kernels associated to an arbitrary stationary distribution. However, the Metropolis–Hastings algorithm is an elegant solution to this issue. Once that the kernel is determined, the Markov chain is simulated and after a burn-in period, the empirical distribution of the sample converges to π, the target density.

We select an arbitrary transition probability that is denoted by $q(\omega_t | \omega_{t-1})$. In practice, we choose a distribution that is easy to simulate, and admits a closed form solution and eventually symmetric: $q(\omega_{t-1} | \omega_t) = q(\omega_t | \omega_{t-1})$. Next we apply the Algorithm 2.1.

Algorithm 2.1 Metropolis Hastings algorithm

Main procedure:

    For $t = 0$ to maximum epoch, T

        1. Simulate $\omega' \sim q\left(\omega'|\omega_t\right)$
        2. Take

$$\Omega_{t+1} = \begin{cases} \omega' & with\ probability\ \rho(\omega_t, \omega') \\ \omega_t & with\ probability\ 1 - \rho(\omega_t, \omega') \end{cases}$$

        where $\rho(\omega_t, \omega')$ is the acceptance probability

$$\rho(\omega_t, \omega') = \min\left\{ \frac{\pi(\omega')}{\pi(\omega_t)} \frac{q\left(\omega_t|\omega'\right)}{q\left(\omega'|\omega_t\right)}, 1 \right\} \tag{2.4}$$

    End loop on epochs

In the next proposition, we examine the Metropolis kernel and find that it satisfies the detailed balance condition.

Proposition 2.5 *Let* $(\Omega_t)_{t \geq 0}$ *be the chain computed by the Metropolis Hastings algorithm. For every conditional distribution whose support includes* $\mathcal{X}$, *the support of the target distribution,*

(a) The kernel of the chain satisfies the detailed balance condition with the function π.
(b) π *is a stationary distribution of the chain*

Proof The transition kernel of $(\Omega_t)_{t \geq 0}$ is the following:

$$K(\omega_t, \omega') = \rho(\omega_t, \omega')q(\omega'|\omega_t) + (1 - r(\omega_t))\delta_{\omega_t}(\omega')$$

where $r(\omega_t) = \int \rho(\omega_t, \omega')q(\omega'|\omega_t)d\omega'$ and $\delta_{\omega_t}(.)$ is the Dirac mass at point ω_t. By construction, it is straightforward to check that

$$\rho(\omega_t, \omega')q(\omega'|\omega_t)\pi(\omega_t) = \rho(\omega', \omega_t)q(\omega_t|\omega')\pi(\omega')$$

$$(1 - r(\omega_t))\,\delta_{\omega_t}(\omega')\pi(\omega_t) = \left(1 - r(\omega')\right)\delta_{\omega'}(\omega_t)\pi(\omega')$$

which together establish the detailed balance condition. Statement (b) follows from Proposition 2.4.   □

Since convergence usually occurs regardless of our starting point, we can usually pick any feasible (for example, picking starting draws that are in the parameter space) starting point. However, the time it takes for the chain to converge varies depending on the starting point. As a matter of practice, most people throw out a certain number of the first draws, known as the burn-in. This is to make our draws closer to the stationary distribution and less dependent on the starting point.

However, it is unclear how much we should burn-in since our draws are all slightly dependent and we do not know exactly when convergence occurs.

2.3   Bayesian Inference

As in the previous chapter, we consider a portfolio of n insurance policies in which each contract is described by a p vector of covariates, noted $x_i = (x_{i1}, \ldots, x_{ip})^\top$. The quantity of interest (e.g. the number or the amount of claims caused by the ith policyholder) is denoted y_i. The exposure is either the duration of the contract or the number of claims and is denoted by v_i. We still denote by $\widehat{y_i} = f(x_i \mid \Omega)$, the output signal of a neural network (or of a generalized linear model) parameterized by a random m−vector Ω of weights. Realizations of Ω and the probability density function of Ω are respectively denoted by ω and $\pi(\omega)$. Furthermore, we adopt the following conventions:

- The log-likelihood of $y = (y_i)_{i=1,\ldots,n}$ for a realization ω of Ω is noted $p(y \mid \omega)$.
- Conditionally to observations y, the m−vector of weights Ω is distributed according to $p(\omega \mid y)$, the posterior distribution.
- The prior distribution of Ω is noted $p(\omega)$.

In a Bayesian framework, we aim to estimate parameters Ω given the measurement signals y. Using the Bayes rule, we try to determine the posterior distribution of Ω which is equal to

$$\underbrace{p(\omega|y)}_{posterior} = \frac{p(y|\omega)p(\omega)}{\int_\mathcal{X} p(y|\omega)p(\omega)dx}$$

$$\propto \underbrace{p(y|\omega)}_{Likelihood\ of\ data} \underbrace{p(\omega)}_{Prior}$$

If the likelihood $p(y|\omega)$ is easy to calculate, we can estimate the distribution $p(\omega|y)$ with the Metropolis Hastings algorithm.

The transition density of Ω is denoted $q(\omega_t|\omega_{t-1})$ and if the target density $\pi(.)$ of the Metropolis–Hastings algorithm is the posterior distribution of parameters $p(\omega|y)$, the acceptance probability in Eq. (2.4) is rewritten as follows:

$$\rho(\omega_t, \omega') = \min\left\{\frac{p(\omega'|y)}{p(\omega_t|y)}\frac{q(\omega_t|\omega')}{q(\omega'|\omega_t)}, 1\right\}$$

$$= \min\left\{\frac{p(y|\omega')p(\omega')}{p(y|\omega_t)p(\omega_t)}\frac{q(\omega_t|\omega')}{q(\omega'|\omega_t)}, 1\right\}$$

Furthermore if the condition density is chosen symmetric (e.g. Normal distribution), then $\rho(\omega_t, \omega')$ becomes:

$$\rho(\omega_t, \omega') = \min\left\{\frac{p(y|\omega')p(\omega')}{p(y|\omega_t)p(\omega_t)}, 1\right\}. \tag{2.5}$$

The resulting sample of parameters (after a burn-in period), $(\omega_t)_{t=1:T}$ is next used to construct Monte Carlo approximation of the empirical distribution of $p(\omega|y)$. The parameter estimates are obtained by computing the following expectation:

$$\widehat{\Omega} = \mathbb{E}(\Omega|y) = \int_{\mathcal{X}} d\, p(\omega \mid y)\, d\omega$$

$$\approx \frac{1}{T}\sum_{t=1}^{T}\int_{\mathcal{X}} \delta_{\omega_t}(d\omega)$$

which corresponds to a collection of Dirac atoms $\delta_{\omega_t}(d\omega)$ located at ω_t with equal weights. This is also the sample mean of simulated $(\omega_t)_{t=1,\dots,T}$.

If we remember developments done in Sect. 1.9, when we assume that responses $(y_i)_{i=1,\dots,n}$ are distributed according to an exponential dispersed law, the log-likelihood of observations for a given set of parameters ω is approached by

$$\ln p(y|\omega) = \frac{1}{\phi}\sum_{i=1}^{n} v_i\,(y_i h(\widehat{y}_i) - a\,(h(\widehat{y}_i))) + \sum_{i=1}^{n}\ln c(y_i, \widehat{\phi}, v_i). \tag{2.6}$$

In Eq. (2.6), the output signal $\widehat{y}_i$ of the feed-forward network that is related to the input signal by a non-linear function $f(.)$, parameterized by Ω:

$$\widehat{y}_i = f(x_i|\Omega = \omega).$$

Whereas $\widehat{\phi}$ is an estimate of the dispersion parameter and $h(.) = \frac{\partial a}{\partial \theta}(.)$. In this case, the acceptance rate (2.5) (when $q(.|.)$ is symmetric) becomes

$$\rho(\omega_t, \omega') = \min\left\{\exp\left(\ln p(y|\omega') - \ln p(y|\omega_t)\right)\frac{p(\omega')}{p(\omega_t)}, 1\right\}. \tag{2.7}$$

As in theory $\widehat{\phi}$ is independent from Ω, the exponent in Eq. (2.7) is developed as follows

$$\ln p(y|\omega') - \ln p(y|\omega_t) =$$

$$\frac{1}{\phi}\sum_{i=1}^{n} v_i\,\Big(y_i\,\big[h(f(x_i|\omega')) - h(f(x_i|\omega_t))\big]$$

$$- \big[a\,(h(f(x_i|\omega')) - a\,(h(f(x_i|\omega_t))))\big]\Big)$$

Table 2.1 Functions $h(.)$
and $a(.)$ for the most common
exponential dispersed
distributions

	$h(y)$	$a(\theta)$	$a(h(y))$
Normal	y	$\theta^2/2$	$y^2/2$
Gamma	$-\frac{1}{y}$	$-\ln(-\theta)$	$\ln(y)$
Poisson	$\ln y$	e^θ	y
Binomial	$\ln\left(\frac{y}{1-y}\right)$	$\ln(1 + e^\theta)$	$-\ln(1 - y)$

where functions $h(.)$ and $a(.)$ depends upon the chosen exponential dispersed distribution, as reported in Table 2.1.

In numerical applications, weights are updated with a Normal transition probability

$$q(\boldsymbol{\omega}'|\boldsymbol{\omega}_t) \sim N(\boldsymbol{\omega}_t, \sigma_\omega I_m)$$

where I_m is the identity matrix of dimension m and σ_ω is a constant. This distribution being symmetric, the acceptance probability is given by Eq. (2.7). For large networks (or GLM with many covariates), updating at the same time all weights with $q(\boldsymbol{\omega}'|\boldsymbol{\omega}_t)$ leads to a high rejection rate of proposed weights $\boldsymbol{\omega}'$, if the standard deviation σ_ω is too high. Reducing this deviation improves the acceptance rate but can considerably slow down the convergence of the Metropolis–Hastings algorithm. An alternative consists to update only a subset of weights. Firstly, we partition the m-vector of weights in n_q sub-vectors of size $\frac{m}{n_q}$,

$$\boldsymbol{\omega}_t = \left(\boldsymbol{\omega}_t^1, \ldots, \boldsymbol{\omega}_t^{n_q}\right).$$

During the tth iteration of the Metropolis–Hastings algorithm, we update only the kth $= t \mod n_q$ sub-vector of weights:

$$q(\boldsymbol{\omega}^{k'}|\boldsymbol{\omega}_t^k) \sim N\left(\boldsymbol{\omega}_t^k, \sigma_\omega I_{m/n_q}\right)$$

and set the candidate vector of weights to $\boldsymbol{\omega}' = \left(\boldsymbol{\omega}_t^1, \ldots, \boldsymbol{\omega}^{k'}, \ldots, \boldsymbol{\omega}_t^{n_q}\right)$. The Algorithm 2.2 summarizes the Metropolis Hastings framework with partial update of weights that is used in the numerical illustration.

2.4 Numerical Illustration

In Sect. 1.11, we have fitted a feed-forward neural network with 4 hidden neurons and a GLM to data from the insurance company *Wasa* in order to predict the claims frequency of motorcycle drivers. The calibration was done with a gradient descent for the GLM and with the resilient back-propagation algorithm for the network. This

---

Algorithm 2.2 Metropolis Hastings algorithm with partial update of weights

---

Main procedure:

    For $t = 0$ to maximum epoch, T

        1. $k = t \mod n_q$. Simulate $\omega^{k'} \sim N\left(\omega_t^k, \sigma_\omega I_{m/n_q}\right)$ and set
$\omega' = \left(\omega_t^1, \ldots, \omega^{k'}, \ldots, \omega_t^{n_q}\right)$.

        2. Take

$$\Omega_{t+1} = \begin{cases} \omega' & with\ probability\ \rho(\omega_t, \omega') \\ \omega_t & with\ probability\ 1 - \rho(\omega_t, \omega') \end{cases}$$

    where $\rho(\omega_t, \omega')$ is the acceptance probability

$$\rho(\omega_t, \omega') = \min\left\{\exp\left(\ln p(y|\omega') - \ln p(y|\omega_t)\right) \frac{p(\omega')}{p(\omega_t)}, 1\right\} \tag{2.8}$$

    and

$$\ln p(y|\omega') - \ln p(y|\omega_t) =$$
$$\frac{1}{\phi} \sum_{i=1}^{n} v_i \left(y_i \left[h(f(x_i|\omega')) - h(f(x_i|\omega_t))\right]\right.$$
$$\left. - \left[a\left(h(f(x_i|\omega')) - a\left(h(f(x_i|\omega_t))\right)\right)\right]\right)$$

    End loop on epochs

---

motivates us to fit the same models to empirical claim frequencies with a MCMC algorithm in order to benchmark its ability of calibration.

2.4.1 Test with GLM

We first fit a GLM model by MCMC. The covariates are: the scaled owner's and vehicle ages, the gender, the geographic area and the class of the vehicle. The model counts 16 parameters. In Sect. 1.11, the deviance and AIC obtained with a gradient descent were respectively equal to 5781.66 and 7162.23. We run 40,000 iterations of the MCMC algorithm and initial parameters are drawn from a Normal random variable centered around 0 and with a unit variance. The transition probability is a normal centered on the last realization ω_t^k , with a standard deviation $\sigma_\omega = 0.10$. The left plot of Fig. 2.5 shows the evolution of the AIC at each iteration. We clearly observe a sharp decrease during the first 5000 steps whereas the convergence is next much slower. The right plot of the same figure presents the empirical statistical distribution of the AIC after a burn-in period of 30,000 iterations. It looks like a Log-Normal distribution with a mean of 7172.44 and a standard deviation of 31.10.

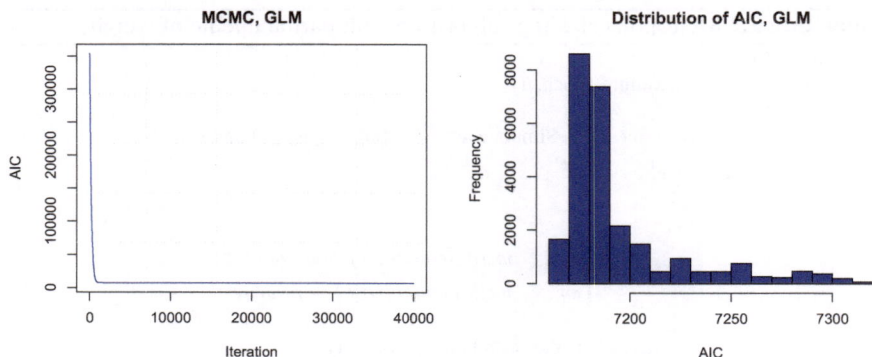

Fig. 2.5 The left plot shows the evolution of the GLM-AIC over 40,000 iterations. The right plot shows the empirical distribution of AIC, build over the last 10,000 runs

Table 2.2 Comparison of parameter estimates obtained with a gradient descent and a MCMC approach

GLM	Gradient weights	Average MCMC weights	Standard deviation MCMC weights
Intercept	−3.11	−3.08	0.23
Scaled owner's age	−4.13	−4.2	0.2
Scaled vehicle's age	−8.03	−6.11	1.82
Gender K	−0.36	−0.35	0.13
Zone 1	1.83	1.43	0.37
Zone 2	1.31	0.9	0.36
Zone 3	0.82	0.4	0.36
Zone 4	0.37	−0.06	0.37
Zone 5	0.15	−0.24	0.58
Zone 6	0.47	0.13	0.35
Class 1	−0.19	−0.03	0.28
Class 2	−0.03	0.25	0.3
Class 3	−0.62	−0.36	0.31
Class 4	−0.47	−0.26	0.27
Class 5	−0.07	0.15	0.27
Class 6	0.39	0.62	0.29

The last column report the standard deviations of estimates, computed over the last 10,000 runs

Table 2.2 compares parameter estimates obtained by the gradient method (as in Sect. 1.11) and the MCMC algorithm. The second and third columns report averages of parameters and their standard deviations, computed over the last 10,000 iterations. Most of MCMC parameters are relatively close to the gradient estimates and the AIC's are comparable. Weights for covariates "Zone 4", "Zone 5", "Class 2" and "Class 5" differ strongly but gradient estimates are located in the 95%

confidence interval of MCMC weights. This numerical exercise demonstrate the capacity of the MCMC method to estimate generalized linear models with an excellent accuracy. Furthermore, the standard deviations may help us to detect non-significant explanatory variables. In our illustration, several coefficients of modalities related to the class of vehicle or to the geographic area, have a standard deviation comparable to their estimates. It would therefore be interesting to exclude them from the model.

2.4.2 Calibration of a Neural Net with 4 Hidden Neurons

In this subsection we estimate the 69 weights of a neural network with 4 hidden neurons, similar to the one studied in Sect. 1.11. We consider all available covariates and aim to predict the claim frequency of policyholders. Initial weights are Normally distributed around 0 and with a unit variance. We run 40,000 iterations of the MCMC algorithm and update at each step, one-fifth of 69 weights ($n_q = 5$). The transition probability is a normal centered on the last realization ω_t^k, with a standard deviation $\sigma_\omega = 0.10$.

As for generalized linear models, the left plot of Fig. 2.6 shows that the AIC converges quickly to a floor after 10,000 iterations. The convergence is much slower between 10,000 to 30,000 steps and next the AIC stagnates. The right plot of this figure presents the empirical distribution of AIC after a burn-in period of 30,000 iterations. The average and standard deviation of the AIC are respectively equal to 7255.65 and 6.27 . The average AIC is higher than the one obtained with the resilient back-propagation procedure (value: 7051.43). Even if in theory parameters found by MCMC should converge to maximum log-likelihood estimators, the algorithm stay

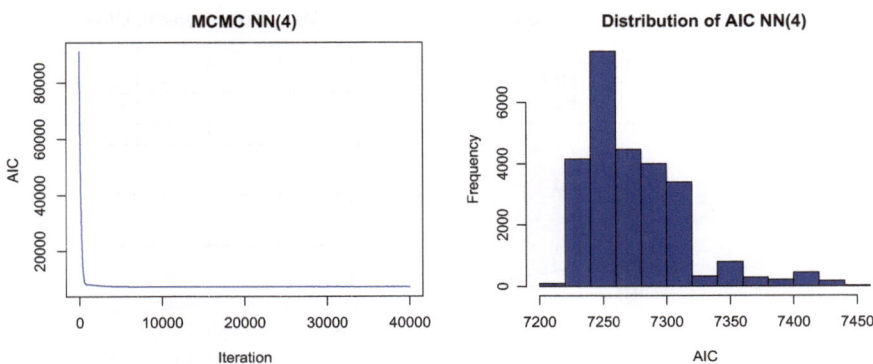

Fig. 2.6 The left plot shows the evolution of the AIC over 40,000 iterations. The right plot shows the empirical distribution of last 10,000 simulated AIC. Calibration of the NN(4) with a random starting point

trapped in a local minimum. Increasing the number of iterations could remedy to this issue but lengthens considerably the computational time.

Our tests reveal that the convergence is sensitive to the initial value of weights. In practice, better results are obtained by combining MCMC and resilient back-propagation algorithms. For example, we may first run 10,000 MCMC iterations in order to find a good set of initial values for the back-propagation algorithm. Applying this method and next running the back-propagation algorithm allows us to obtain a model with an AIC of 7107.45. This is still less good than the AIC of the model studied in Sect. 1.11. However, we remind that this model has been estimated by trial and errors with different initial weights and we only reported the best model. The combination of MCMC and back-propagation offers the advantage of being a systematic approach. Furthermore, as illustrated in Fig. 2.7, networks fitted by MCMC/back-propagation or by the model with the best AIC predict similar

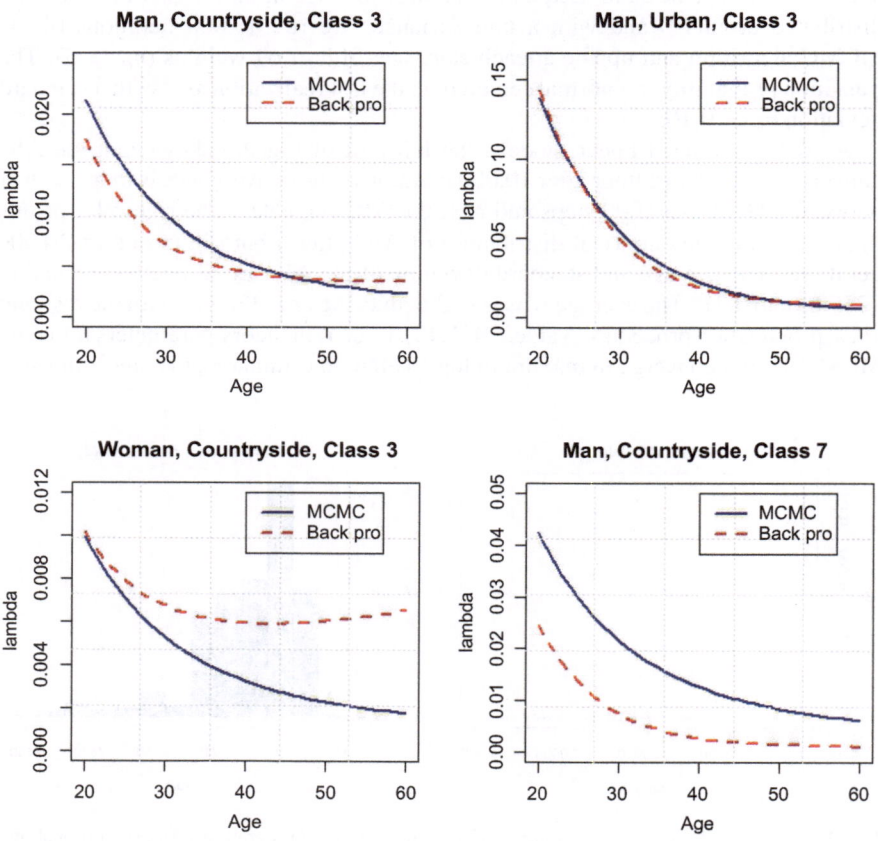

Fig. 2.7 Comparison of claim frequencies for different policyholders, forecast by a neural net with 4 hidden neurons. One network is fitted by MMCC and backpropagation and the other one is fitted by backpropagation and multiple random starting points

Table 2.3 Comparison of weights obtained with different calibration methods

				MCMC and backpropagation weights	Best AIC weights	Average MCMC weights	Standard deviation MCMC weights
1	1	1	1	−3.32	−0.78	−0.60	0.27
2	1	1	2	1.60	−6.9	−6.70	0.40
3	1	1	3	−13.86	0.35	1.40	0.40
4	1	1	4	−0.63	−1.05	−0.97	0.26
⋮	⋮	⋮	⋮	⋮	⋮	⋮	⋮
34	3	1	2	−0.57	563.42	562.23	0.33
35	3	1	3	5.45	568.56	568.97	0.62
36	3	1	4	−3.27	603.16	604.42	0.94
37	3	1	5	−7.43	45.4	46.64	0.52
38	3	1	6	−5.15	−184.84	−183.42	0.62
⋮	⋮	⋮	⋮	⋮	⋮	⋮	⋮
50	4	1	2	986.79	40.88	40.48	0.32
51	4	1	3	−793.06	2380.67	2378.37	1.24
52	4	1	4	−213.51	−4.01	−3.86	0.44
53	4	1	5	72.46	37.66	36.03	0.77
⋮	⋮	⋮	⋮	⋮	⋮	⋮	⋮

Weights in the first column are obtained by combining MCMC and back-propagation algorithms. The second column shows weights of the best AIC model. The third column reports average MCMC weights and their standard deviations, when the MCMC algorithm is initialized with best AIC parameters

claim frequencies for most of profiles of insureds. The most significant difference concerns the claim frequencies of class 7 vehicles. This frequency computed with the MCMC/back-propagation network is twice higher for drivers younger than 30 years old but after converges toward the forecast of the best AIC model. Some weights obtained by MCMC/back-propagation and these of the best AIC network are compared in Table 2.3. We do not observe any similarity between parameter estimates. This confirms that the learning of neural networks is an ill-posed problem and that several possible sets of admissible weights can exist.

In a last series of tests, we run the MCMC algorithm initialized with parameter estimates of the model with the best AIC (network studied in Sect. 1.11 fitted by back-propagation and multiple random starting points.). The left upper graph of Fig. 2.8 shows the AIC for 40,000 iterations. The average and standard deviation of the AIC are 7089.21 and 5.89. The AIC stays stable and we do not observe any improvement of the goodness of fit over the 40,000 runs. This seems to confirm that parameter estimates maximize the log-likelihood. On the other hand, we can use these simulations to study empirical distributions of all parameter estimates. The last three plots of Fig. 2.8 illustrate this. The two last columns of Table 2.3 report some average weights and their standard deviation, computed over the 40,000 runs.

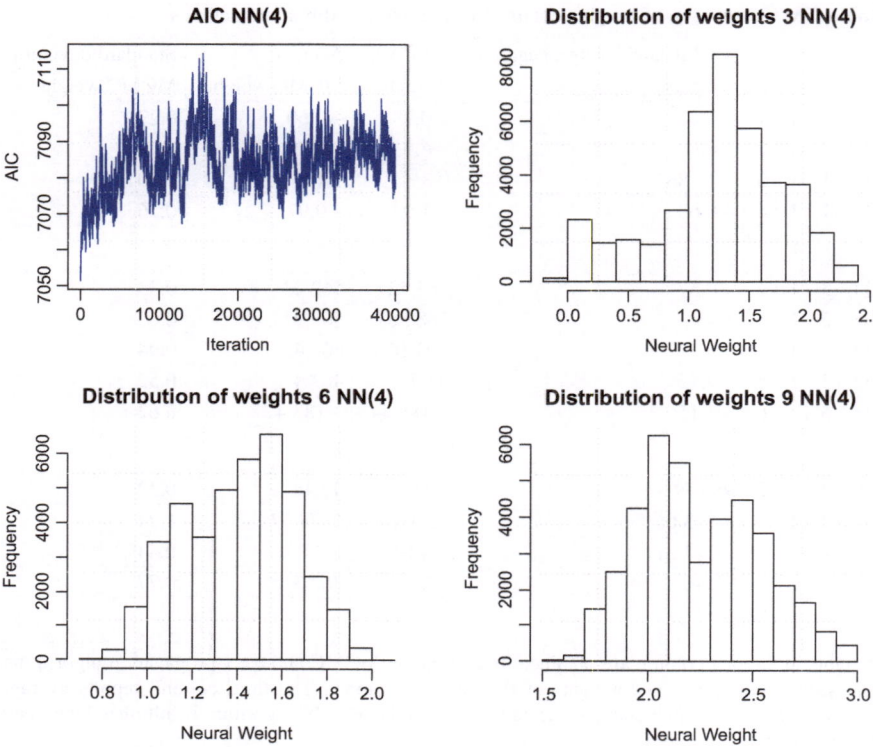

Fig. 2.8 MCMC algorithm initialized with weights for the model with the best AIC. Upper left plot: AIC for 40,000 iterations. Other plots: examples of empirical distribution of weight estimates

This information can be used to validate the significance of some neural connections and to eventually cancel nearly null weights with a high standard deviation.

2.5 Further Readings

Metropolis et al. (1953) and Hastings (1970) were the first to study Monte-Carlo Markov Chain for estimating parameters of a statistical model. In the nineties, Buntine and Weigend (1991) and Mackay (1992) showed that a principled Bayesian learning approach to neural networks is an alternative to back-propagation algorithms. Neal (1996) introduced new Bayesian simulation methods, specifically the hybrid Monte Carlo method, into the analysis of neural networks. He also proved that classes of priors whose number of hidden neurons tend to infinity, converge to Gaussian processes. de Freitas (1999) studies in his PhD thesis sequential Monte-Carlo for calibration of neural networks. The book of Robert and Casella (2004) studies the convergence and properties of MCMC algorithms. Johnson (2009) and Hastie et al. (2009) compares the performance of Bayesian neural networks with boosted trees and random forests. They conclude that Bayesian neural networks outperforms other methods. Neal and Zhang (2006) have used a Bayesian network to win in 2003, the classification competition organized during the Neural Information Processing Systems (NIPS) workshop. Results of this competition are detailed in Guyon et al. (2006). The reader can also refer to the textbook of Korb and Nicholson (2010) for more details on Bayesian artificial intelligence.

References

Buntine WL, Weigend AS (1991) Bayesian back-propagation. Complex Systems 5:603–643

de Freitas JFG (1999) Bayesian methods for neural networks. PhD Thesis, Department of Engineering, Cambridge University, Cambridge

Guyon I, Gunn S, Nikravesh M, Zadeh L (2006) Feature extraction, foundations and applications. Springer, New York

Hastie T, Tibshirani R, Friedman JH (2009) The elements of statistical learning: data mining, inference, and prediction. Springer series in statistics, 2nd edn. Springer, New York

Hastings W (1970) Monte Carlo sampling methods using Markov chains and their applications. Biometrika 57:97–109

Johnson N (2009) A study of the NIPS feature selection challenge. Standford working paper

Korb KB, Nicholson AE (2010) Bayesian artificial intelligence, 2nd edn. Chapman & Hall/CRC Computer Science & Data Analysis

Mackay DJC (1992) A practical Bayesian framework for backpropagation networks. Neural Comput 4(3):448–472

Metropolis N, Rosenbluth AW, Rosenbluth MN, Teller AH, Teller E (1953). Equations of state calculations by fast computing machines. J Chem Phys 21:1087–1092

Neal RM (1996) Bayesian learning for neural networks. Lecture notes in statistics no. 118. Springer, New York

Neal R, Zhang J (2006) High dimensional classification with Bayesian neural networks and Dirichlet diffusion trees. In: Feature extraction, foundations and applications. Springer, New-York, pp 265–296

Robert CP, Casella G (2004) Monte Carlo statistical methods. Springer, New York

Chapter 3
Deep Neural Networks

In Chap. 1, our empirical analysis was based on neural networks with a single hidden layer. These networks, called shallow, are in theory universal approximators of any continuous function. Deep neural networks use instead a cascade of multiple layers of hidden neurons. Each successive layer uses the output from the previous layer as input. As with shallow networks, many issues can arise with naively trained deep networks. Two common issues are the overfitting of the training dataset and the increase of computation time. We present the techniques for limiting the computation time and the methods of regularization for avoiding the overfitting. We next explain why deep neural networks outperform shallow networks for approximating hierarchical binary functions. This chapter is concluded by a numerical illustration.

3.1 Back-Propagation Algorithms for Deep Learning

A standard and efficient approach for fitting neural networks is the back-propagation Algorithm 1.2, introduced in Chap. 1. At iteration t, the vector of neural weights $\Omega_t = \left(\omega_1^{(t)}, \ldots, \omega_{n^{net}}^{(t)} \right)$ is updated in the opposite direction of the gradient of the loss function $\mathcal{R}(\Omega)$:

$$\Omega_{t+1} = \Omega_t - \rho \nabla \mathcal{R}(\Omega_t). \tag{3.1}$$

If the input information is contained in n vectors $(x_i)_{i=1,\ldots,n}$ of dimension p and the vector of responses is $y = (y_i)_{i=1,\ldots,n}$, this loss function is e.g. the sum of n loss functions, $\mathcal{R} = \frac{1}{n} \sum_{i=1}^{n} \mathcal{L}(y_i, \widehat{y}_i | \Omega_t)$ where $\widehat{y}_i$ is the estimate of $\mathbb{E}(y_i)$

© Springer Nature Switzerland AG 2019
M. Denuit et al., *Effective Statistical Learning Methods for Actuaries III*,
Springer Actuarial, https://doi.org/10.1007/978-3-030-25827-6_3

computed with the network. The gradient in Eq. (3.1) is computed numerically as the symmetric difference quotient:

$$
\frac{\partial \mathcal{R}(\Omega_t)}{\partial \omega_j}\Bigg|_{\omega_j=\omega_j^{(t)}} \approx \frac{\mathcal{R}\left(\ldots, \omega_j^{(t)}+h, \ldots\right) - \mathcal{R}\left(\ldots, \omega_j^{(t)}-h, \ldots\right)}{2h}
$$

$$
= \frac{1}{n}\sum_{i=1}^{n} \frac{\mathcal{L}\left(y_i, \widehat{y}_i | \ldots, \omega_j^{(t)}+h, \ldots\right) - \mathcal{L}\left(y_i, \widehat{y}_i | \ldots, \omega_j^{(t)}-h, \ldots\right)}{2h}
$$

The estimation error is equal to $-\frac{1}{6}\frac{\partial \mathcal{R}(\ldots,\omega_j^{(t)}+\epsilon,\ldots)}{\partial \omega_j}h^2$ where $\omega_j^{(t)} + \epsilon$ is a point between $\omega_j^{(t)}-h$ and $\omega_j^{(t)}+h$. Each iteration requires then $2 \times n^{net} \times n$ evaluations of the loss function. For high dimensions, dataset and deep neural networks computing the gradient is therefore extremely time consuming. Several solutions exist in practice to reduce this computation time.

3.1.1 Stochastic and Batch Gradient Descents

In the stochastic gradient approach, the true gradient $\nabla \mathcal{R}(\Omega_t)$ in the back-propagation Algorithm 1.2, is approximated by the gradient computed with a single observation (x_k, y_k), drawn randomly in the dataset:

$$
\frac{\partial \mathcal{R}(\Omega_t)}{\partial \omega_j}\Bigg|_{\omega_j=\omega_j^{(t)}} \approx \tag{3.2}
$$

$$
\frac{\mathcal{L}\left(y_k, \widehat{y}_k | \ldots, \omega_j^{(t)}+h, \ldots\right) - \mathcal{L}\left(y_k, \widehat{y}_k | \ldots, \omega_j^{(t)}-h, \ldots\right)}{2h},
$$

where $k \in \{1, \ldots, n\}$. A compromise between this approach and computing the gradient with the full data set is to compute the gradient with a small subset of observations, called a batch at each step. This can significantly improve the stochastic gradient descent and result in smoother convergence. Batches are either sequentially or either randomly selected. In both cases, the gradient is approached by:

$$
\frac{\partial \mathcal{R}(\Omega_t)}{\partial \omega_j}\Bigg|_{\omega_j=\omega_j^{(t)}} \approx \tag{3.3}
$$

$$
\sum_{k \in B_k} \frac{\mathcal{L}\left(y_k, \widehat{y}_k | \ldots, \omega_j^{(t)}+h, \ldots\right) - \mathcal{L}\left(y_k, \widehat{y}_k | \ldots, \omega_j^{(t)}-h, \ldots\right)}{2h},
$$

where B_k is a subset of n_{batch} indices drawn in $\{1, \dots, n\}$. Usually, stochastic and batch gradient descents use adaptive decreasing learning rates (e.g. $\rho_t = \rho_0 e^{-\alpha t}$ instead of ρ). The convergence of algorithms has been analyzed using the theories of convex minimization and of stochastic approximation.

On the other hand, we can add a momentum effect in the update Eq. (3.1):

$$\Omega_{t+1} = \Omega_t - \rho_{t+1} \nabla \mathcal{R}(\Omega_t) + \gamma \left(\Omega_t - \Omega_{t-1} \right).$$

Adding this momentum effect with γ set to a value close to 0.9 enables the adjustment of the coefficients to roll or move more quickly over a plateau in the error surface. Momentum and stochastic gradient methods are proposed in several software, e.g. in the neural network toolbox of Matlab or in Keras.

3.1.2   Adaptive Gradient Algorithm (Adagrad)

AdaGrad is a stochastic gradient descent with adaptive learning rates proposed by Duchi et al. (2011). Informally, the learning rate increases for more sparse data and decreases for less sparse one. This strategy often improves convergence performance over standard stochastic gradient descent in settings where data is sparse and sparse parameters are more informative. At iteration t, the learning rate ρ in Eq. (3.1) is multiplied by the elements of a vector that is the diagonal of the following sum of outer product matrix of gradient vectors. If we denote this matrix by

$$A_t = \sum_{e=1}^{t} \nabla \mathcal{R}(\Omega_e)^\top \nabla \mathcal{R}(\Omega_e),$$

and its diagonal by $\boldsymbol{a}_t = \left(a_i^{(t)} \right)_{i=1,\dots,n^{net}} = \mathrm{diag}\,(A_t)$, the neural weights in the back-propagation Algorithm 1.2 are updated as follows:

$$\Omega_{t+1} = \Omega_t - \rho \, \boldsymbol{a}_t^{-\frac{1}{2}} \otimes \nabla \mathcal{R}(\Omega_t). \tag{3.4}$$

where $\otimes$ is the element-wise product. If $g_i^{(e)} = (\nabla \mathcal{R}(\Omega_e))_i$ is the gradient of the ith parameter computed at epoch e, we have that

$$\left(a_i^{(t)} \right)^{\frac{1}{2}} = \sqrt{ \sum_{e=1}^{t} \left(g_i^{(e)} \right)^2 }$$

is the L^2 norm of the vector of previous derivatives. Therefore, neural weights significantly modified in previous iterations get dampened updates. While weights that were slightly modified in previous epochs are updated with a higher learning rate. Notice that the gradient is in practice computed on batches as detailed in the previous paragraph.

Adadelta is a more robust extension of Adagrad that adapts learning rates based on a moving window of gradient updates, instead of accumulating all past gradients. This way, Adadelta continues learning even when many updates have been done. Adadelta and Adagrad are both implemented in Keras.

3.1.3 Root Mean Square Propagation (RMSProp) and Adaptive Moment Estimation (ADAM)

RMSProp as the Adagrad algorithm is a method in which the learning rate is adapted to each neural weight. The learning rate for a weight is divided by a moving average with momentum of recent gradients. If $g_i^{(t)} = (\nabla \mathcal{R}(\Omega_t))_i$ is the gradient of the ith weight computed at epoch t, we evaluate the vector $\boldsymbol{b}_t = \left(b_i^{(t)} \right)_{i=1,\dots,n^{net}}$ defined by:

$$b_i^{(t)} = \gamma b_i^{(t-1)} + (1 - \gamma) \left(g_i^{(t)} \right)^2 .$$

where $\gamma \in (0, 1)$ is the parameter tuning the momentum effect. It can be seen as the rate at which the algorithm forgets gradients of previous iterations. Neural weights in the back-propagation Algorithm 1.2 are then updated as follows:

$$\Omega_{t+1} = \Omega_t - \rho \, \boldsymbol{b}_t^{-\frac{1}{2}} \otimes \nabla \mathcal{R}(\Omega_t) . \tag{3.5}$$

where $\otimes$ is the element-wise product. RMSProp has demonstrated excellent adaptation of learning rate in different applications. RMSProp is a generalization of the resilient back propagation Algorithm 1.3, presented in Chap. 1. It is also capable to work with batches.

The Adam algorithm developed by Kingma and Ba (2015) is based on the RMSProp optimizer. In this procedure, moving averages of first and second moments of gradients are used to adapt the learning rate. At iteration t, we evaluate the vectors $\boldsymbol{b}_t = \left(b_i^{(t)} \right)_{i=1,\dots,n^{net}}$ and $\boldsymbol{c}_t = \left(c_i^{(t)} \right)_{i=1,\dots,n^{net}}$ defined this time by:

$$c_i^{(t)} = \gamma_1 c_i^{(t-1)} + (1 - \gamma_1) g_i^{(t)}$$

$$b_i^{(t)} = \gamma_2 b_i^{(t-1)} + (1 - \gamma_2) \left(g_i^{(t)} \right)^2 .$$

where $\gamma_1, \gamma_2 \in (0, 1)$, interpretable as "forgetting rate" of past computations. Vectors $\boldsymbol{b}_t$ and $\boldsymbol{c}_t$ are next rescaled:

$$\tilde{\boldsymbol{c}}_t = \frac{\boldsymbol{c}_t}{1 - (\gamma_1)^t} \ , \ \tilde{\boldsymbol{b}}_t = \frac{\boldsymbol{b}_t}{1 - (\gamma_2)^t} \ ,$$

and weights in the back-propagation algorithm are updated by

$$\Omega_{t+1} = \Omega_t - \rho \left(\boldsymbol{b}_t^{\frac{1}{2}} + \epsilon \right)^{-1} \otimes \tilde{\boldsymbol{c}}_t . \tag{3.6}$$

In Eq. (3.6), $\epsilon \in \mathbb{R}^+$ is a small scalar to prevent division by zero. As in previous algorithm, the square root and products are computed element wise.

3.2 Regularization Techniques

A major issue of deep neural networks is the choice of their architecture. Choosing a model with only a few neurons can be sub-optimal and the standard deviation of estimators may be large even for such a network. On the other hand, a model with too many nodes often overfits data. In order to limit overfitting and to extract from a deep network a smaller structure with the best possible configuration, we introduce a penalty term in the loss function. We present L_1 and L_2 penalizations in the next two paragraphs.

L_1 Regularization: The Lasso
The Lasso (least absolute shrinkage and selection operator) is a regression tool selecting relevant variables in order to enhance the accuracy of prediction and interpretability of results. Lasso was popularized by Tibshirani (1996) for least squares regressions.

Let us consider that input information is contained in n vectors $(\boldsymbol{x}_i)_{i=1,\ldots,n}$ of dimension p and the vector of responses is $\boldsymbol{y} = (y_i)_{i=1,\ldots,n}$. To simplify future developments, we assume that a dummy constant variable is in the vector of explanatory variables: e.g. $x_{i,1} = 1$ for $i = 1, \ldots, n$. The least squares predictor $\hat{\boldsymbol{y}}$ of $\boldsymbol{y}$ is a linear function $\boldsymbol{\beta}^\top \boldsymbol{x}$ where the p vector $\boldsymbol{\beta}$ is solution of:

$$\boldsymbol{\beta} = \arg \min_{\boldsymbol{\beta}} f(\boldsymbol{\beta}) = \arg \min_{\boldsymbol{\beta}} \sum_{i=1}^{n} \left(y_i - \boldsymbol{\beta}^\top \boldsymbol{x}_i \right)^2 . \tag{3.7}$$

Let X be the matrix $\left(x_i^\top \right)_{i=1,\ldots n}$. The optimal vector of regressors $\boldsymbol{\beta}^*$ cancels the derivative of $f(\boldsymbol{\beta})$ and is then equal to $\boldsymbol{\beta}^* = \left(X^\top X \right)^{-1} X^\top \boldsymbol{y}$. However, when the number of covariates p is important compared to the sample size n and/or are highly correlated, the predictive power of this model is low due to overfitting. The Lasso

regression limits the number of covariates by adding a penalization to the objective
function (3.7). This penalty is proportional to the L_1 norm of the vector $\boldsymbol{\beta}$:

$$\boldsymbol{\beta} = \arg\min_{\boldsymbol{\beta}} f(\boldsymbol{\beta}) + \lambda \sum_{j=1}^{d} |\beta_j| \qquad (3.8)$$

where $\lambda \in \mathbb{R}^+$ is the parameter of shrinkage. The function in Eq. (3.8) corresponds
to the Lagrangian of the optimization problem:

$$\boldsymbol{\beta} = \arg\min_{\boldsymbol{\beta}} f(\boldsymbol{\beta}) \text{ subject to } .||\boldsymbol{\beta}||_1 \leq \gamma , \qquad (3.9)$$

for some upper bound $\gamma \in \mathbb{R}^+$ on the L_1 norm of $\boldsymbol{\beta}$. The Lasso regression forces the
sum of the absolute value of regression coefficients to be less than a fixed value. This
also forces some coefficients to be set to zero. To understand why certain coefficients
are cancelled, we must look to Fig. 3.1. It shows the domain of $\boldsymbol{\beta}$ satisfying the
constraint $||\boldsymbol{\beta}||_1 \leq \gamma$ and ellipses $f(\boldsymbol{\beta}) = c$ for a two dimensional problem. Without
L_1 constraint, the maximum of $f(\boldsymbol{\beta})$ is attained with $\boldsymbol{\beta}^*$. And for a given $c \leq f(\boldsymbol{\beta}^*)$,
the geometric locus of points $\boldsymbol{\beta}$ such that $f(\boldsymbol{\beta}) = c$ is an ellipse centered on $\boldsymbol{\beta}^*$.

We see that the constrained area defined by the L_1 norm is a square rotated so that
its corners lie on the axes. A convex object, like an ellipse, tangent to the constrained
boundary is likely to encounter a corner of a hypercube, for which some components
of $\boldsymbol{\beta}$ are identically zero. The L_1 penalty tends then to reduce the number of non-
zero coefficients in a regression.

In deep neural networks, the quadratic function $f(\boldsymbol{\beta})$ is replaced by a non-linear
loss function $\mathcal{R}(.)$ of neural weights $\Omega_t \in \mathbb{R}^{n_{net}}$. The geometric locus of parameters
Ω_t such that $\mathcal{R}(\Omega_t) = c$ for $c \in \mathbb{R}^+$ is no more an ellipse. But if this locus is
convex in $\mathbb{R}^{n_{net}}$, we expect that the intersection between the tangent volume and a

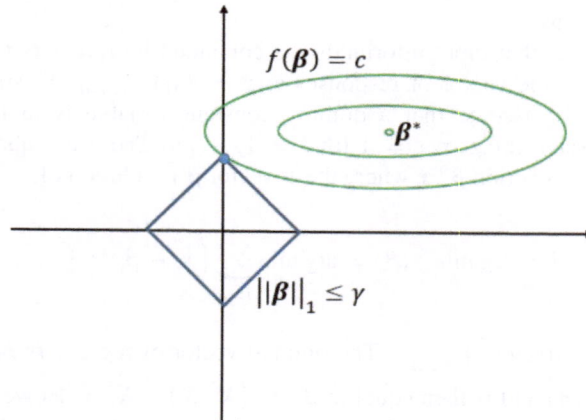

Fig. 3.1 Least squares objective function in two dimensions and domain of $\boldsymbol{\beta}$ delimited by a L_1
constraint

hypercube delimited by the constraint $||\Omega_t|| \leq \gamma$ is a high dimensional equivalent of a 2D corner. In this case, several neural weights will be set to zero.

For actuarial applications, it is recommended to use the unscaled deviance $D(y_i, \widehat{y_i})$ for the loss function. If we remember Sect. 1.7, this deviance is equal to

$$D(y_i, \widehat{y_i}) = 2\nu_i \left(y_i h(y_i) - b(h(y_i)) - y_i h(\widehat{y_i}) + b(h(\widehat{y_i}))\right),$$

where ν_i is the exposure for the ith policyholder. A Lasso loss function $\mathcal{R}^{Lasso}(\Omega)$, is defined by a L1 regularization added to the average unscaled deviance:

$$\mathcal{R}^{Lasso}(\Omega) = \frac{1}{n} \sum_{i=1}^{n} D(y_i, \widehat{y_i}) + \lambda \sum_{\omega_j \in \Omega} |\omega_j|,$$

where $\lambda \in \mathbb{R}^+$ is the parameter of shrinkage. As for the Lasso regression, a large value of λ drives the weights toward zero. This method is tested in the numerical illustration concluding this chapter.

L_2 Regularization: The Ridge Regression

Let us again consider the least squares regression of $\boldsymbol{y}$ on $\boldsymbol{x} \in \mathbb{R}^p$. The least squares predictor is the function $\boldsymbol{\beta}^{*\top}\boldsymbol{x}$ where $\boldsymbol{\beta}^* = \left(\boldsymbol{X}^\top\boldsymbol{X}\right)^{-1}\boldsymbol{X}^\top\boldsymbol{y}$. Determining $\boldsymbol{\beta}^*$ requires then to invert the $p \times p$ matrix $\left(\boldsymbol{X}^\top\boldsymbol{X}\right)^{-1}$. If p is large or if $p > n$, the matrix $\boldsymbol{X}^\top\boldsymbol{X}$ may be singular and then not invertible. In this case the least squares regression is an ill-posed problem and therefore impossible.

The most common method to solve this problem is the Thikhonov regularization. This consists to choose a Tikhonov matrix which in many cases is simply a multiple of the $p \times p$ identity matrix: $\lambda \boldsymbol{I}$, where $\lambda \in \mathbb{R}$. Since the matrix $\boldsymbol{X}^\top\boldsymbol{X} + \lambda\boldsymbol{I}$ is non-singular and invertible, the optimal coefficient of the ill-posed regression problem are approached by

$$\tilde{\boldsymbol{\beta}} = \left(\boldsymbol{X}^\top\boldsymbol{X} + \lambda\boldsymbol{I}\right)^{-1}\boldsymbol{X}^\top\boldsymbol{y}. \tag{3.10}$$

Let us define $f(\boldsymbol{\beta}) = \sum_{i=1}^{n} \left(y_i - \boldsymbol{\beta}^\top\boldsymbol{x}_i\right)^2$ and consider the following quadratic constrained optimization problem

$$\boldsymbol{\beta} = \arg\min_{\boldsymbol{\beta}} f(\boldsymbol{\beta}) \text{ subject to } ||\boldsymbol{\beta}||_2 < \gamma, \tag{3.11}$$

for some $\gamma \in \mathbb{R}^+$. For a given multiplier $\lambda \in \mathbb{R}^+$, the vector of regressors minimizing the Lagrangian of problem (3.11)

$$\boldsymbol{\beta} = \arg\min_{\boldsymbol{\beta}} f(\boldsymbol{\beta}) + \lambda||\boldsymbol{\beta}||_2, \tag{3.12}$$

is precisely $\tilde{\boldsymbol{\beta}}$ such as defined in Eq. (3.10). There is a direct link between the Thikhonov regularization for inverting singular matrix and the problem (3.11),

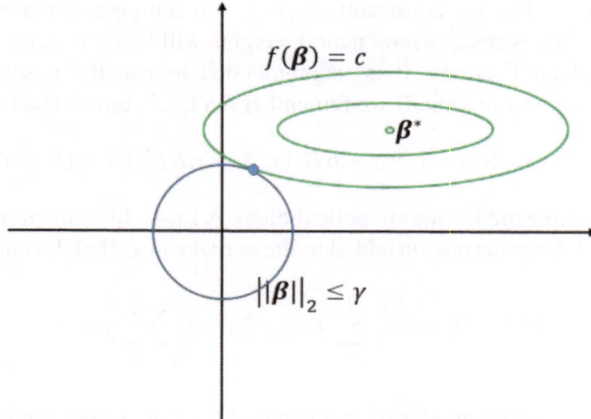

Fig. 3.2 Least squares objective function in two dimensions and domain of $\boldsymbol{\beta}$ delimited by a L_2 constraint

whose solution is called the Ridge least squares regression. The Ridge regularization allows for regressing n observations on p covariates when $n < p$. However, coefficients are not forced to be null. To understand this, let us again consider the formulation (3.11) of the Ridge regression. As shown in Fig. 3.2, the constraint $||\beta||_2 \leq \gamma$ delimits a circular area. For a given $c \leq f(\boldsymbol{\beta}^*)$, the geometric locus of points $\boldsymbol{\beta}$ such that $f(\boldsymbol{\beta}) = c$ is an ellipse centered on $\boldsymbol{\beta}^*$. in this case, the points on the boundary for which some coefficients of $\boldsymbol{\beta}$ are zero are not distinguished from the others. Furthermore, the optimal ellipse is no more likely to contact a point at which some coefficients are zero than one for which none of them are.

In deep neural networks, the quadratic function $f(\boldsymbol{\beta})$ is replaced by a general loss function. If this function is the unscaled deviance, the Ridge loss function is given by:

$$\mathcal{R}^{Ridge}(\Omega) = \frac{1}{n} \sum_{i=1}^{n} D(y_i, \widehat{y}_i) + \lambda \sum_{\omega_j \in \Omega} (\omega_j)^2 .$$

When we consider a Normal distribution for the output signal, the deviance is the quadratic function. The neural network defines a non linear function between the vector of $p-$covariates $\boldsymbol{x}_i$ and an estimated output signal $\widehat{y}_i$. If we denote this function by $\widehat{y}_i = f(\boldsymbol{x}_i \mid \Omega)$, the Ridge and Lasso loss function becomes in the Normal case:

$$\mathcal{R}^{Lasso}(\Omega) = \frac{1}{n} \sum_{i=1}^{n} v_i (y_i - f(\boldsymbol{x}_i \mid \Omega))^2 + \lambda \sum_{\omega_j \in \Omega} |\omega_j| ,$$

$$\mathcal{R}^{Ridge}(\Omega) = \frac{1}{n} \sum_{i=1}^{n} v_i (y_i - f(\boldsymbol{x}_i \mid \Omega))^2 + \lambda \sum_{\omega_j \in \Omega} (\omega_j)^2 .$$

These expressions clearly reveal the similarity with the Lasso or Ridge linear regressions, except that the linear regressor is replaced by a non linear function, $f(x_i \mid \Omega)$.

3.3 Are Deep Networks Better Than Shallow Ones?

Cybenko (1989) and Hornik et al. (1989) have shown that both shallow and deep networks are universal approximators. They can approximate arbitrarily well any continuous function. Therefore why should we work with deep neural networks? A first element answering to this question was provided in Sect. 1.13. For the same level of accuracy, the number of neurons in a shallow networks may be very large compared to a deep neural structure. Intuitively, each layer of a deep neural network composes its own level of non-linearity with the previous one. In shallow networks, non-linear relations are just summed up. However, demonstrating this statement is still an open question. But Poggio et al. (2017) provides a proof in the particular case of hierarchical binary neural networks. We present some of their results in this section.

We consider the problem that consists to regress a scalar $y \in \mathbb{R}$ on $x \in [-1, 1]^p$. The true unknown relation between x and y is a continuous function, $f(x) = y$. A neural network with n^{net} neurons is a non-linear function $f_{n^{net}}(x)$ from $[-1, 1]^p$ to $\mathbb{R}$. We wish to determine the complexity, measured by n^{net}, of a network for approximating the unknown target continuous function $f(x)$ up to a given accuracy $\epsilon > 0$. Let us denote by $V_{n^{net}}$ the set of all networks of complexity n^{net}. The accuracy is measured by the following distance:

$$d(f, V_{n^{net}}) = \inf_{f_{n^{net}}} \|f - f_{n^{net}}\|_\infty$$

$$= \inf_{f_{n^{net}}} \sup_{x \in [-1,1]^p} |f(x) - f_{n^{net}}(x)|.$$

If $d(f, V_{n^{net}}) = \mathcal{O}\left(\left(n^{net}\right)^{-\gamma}\right)$ for some $\gamma > 0$, then a network with a complexity $n^{net} = \mathcal{O}(\epsilon^{-\frac{1}{\gamma}})$ can approximate the function f with an accuracy at least equal to $\epsilon > 0$.

Let $\mathcal{S}_{n^{net}, p}$ denotes the class of shallow networks with n^{net} neurons in the hidden layer and p inputs of the form

$$f(x) = \sum_{j=1}^{n^{net}} \beta_j \phi \left(\omega_j^\top x + \omega_{j,0} \right), \tag{3.13}$$

where $\omega_j \in \mathbb{R}^p$, $\beta_j, \omega_{j,0} \in \mathbb{R}$. In Eq. (3.13), $\phi(.)$ is a measurable squashing function from $\mathbb{R}^p \to \mathbb{R}$. Let W_r^p be the set of all functions from $[-1, 1]^p$ to $\mathbb{R}$ with continuous partial derivatives up to order r such that

$$\|f\|_\infty + \sum_{1 \leq \|k\|_1 \leq r} \|D_k f\|_\infty \leq 1.$$

k is a vector of positive integers and $D_k f$ denotes here the partial derivative of f with respect to the variables x_i indexed by $i \in k$:

$$D_k f = \frac{\partial^{\|k\|_1} f}{\partial x_1^{k_1} \dots \partial x_s^{k_s}}.$$

Mhaskar (1996) shows that the following theorem is a direct consequence of results by DeVore et al. (1989).

Theorem 3.2 *Let $\phi(.) : \mathbb{R} \to \mathbb{R}$ be infinitely differentiable and not a polynomial on any subinterval of $\mathbb{R}$. For $f \in W_r^p$ the complexity of the shallow network that provides an approximation of f with an accuracy of at least ϵ is of order:*

$$n^{net} = \mathcal{O}\left(\epsilon^{-\frac{p}{r}}\right) \tag{3.14}$$

and is the best possible.

The proof is based on the fact that when $\phi(.)$ satisfies the conditions of the theorem, the algebraic polynomials in p variables of degree q are in the uniform closure of the span of $\mathcal{O}(q^p)$ functions of the form $x \to \phi\left(\omega^\top x + \omega_0\right)$.

Let us assume that $p = 2^m$ and consider hierarchical functions of the form:

$$f_{n^{net}}(x) = h_{m,1}\left(h_{m-1,1}, h_{m-1,2}\right)$$

$$\vdots$$

$$h_{2k} = h_{2k}(h_{1,(k-1)*2+1}, h_{1,k*2}) \quad k = 1, \dots, \frac{p}{4}$$

$$h_{1k} = h_{1k}(x_{(k-1)*2+1}, x_{k*2}) \quad k = 1, \dots, \frac{p}{2},$$

where the constituent functions $h_{k,j}(u, v)$ are continuous. These functions present a structure similar to a binary tree as illustrated in Fig. 3.3. We denote by $W_r^{p,2}$, the class of binary hierarchical functions from $[-1, 1]^p$ to $\mathbb{R}$ which partial derivatives up to order r and constituent functions $h_{k,j}$ in W_r^2.

The corresponding class of deep neural networks is noted $\mathcal{D}_{n^{net},2}$. This class contains all networks with n^{net} neurons and a binary tree architecture. The constituent functions are here shallow networks in $\mathcal{S}_{Q,2}$ with Q neurons, such that $n^{net} = (p - 1)Q$.

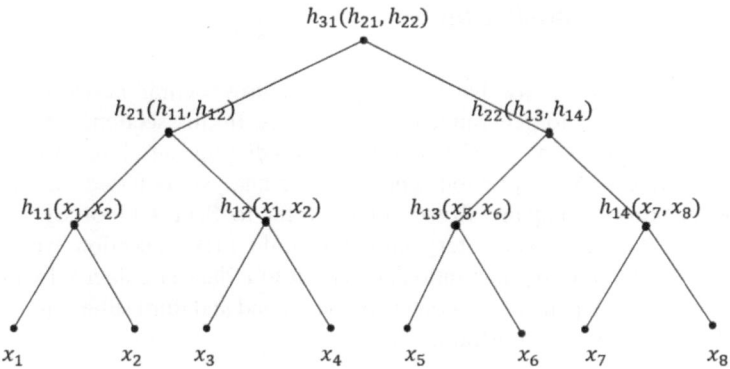

Fig. 3.3 Graph of a binomial hierarchical functions, with three levels ($m = 3$)

Theorem 3.3 *For $f \in W_r^{p,2}$ the complexity of a deep network with the same hierarchical architecture that provides an approximation of f with accuracy of at least ϵ is of order:*

$$n^{net} = \mathcal{O}\left((p-1)\epsilon^{-\frac{2}{r}}\right). \qquad (3.15)$$

Proof Since constituent functions $h_{k,j}$ of f are in W_r^2, we infer from Eq. (3.14) applied with $p = 2$ that each of these functions can be approximated from $\mathcal{S}_{n^{net},2}$ up to accuracy $\epsilon = c\left(n^{net}\right)^{-\frac{r}{2}}$. The assumption that $f \in W_r^{p,2}$ implies that each of these constituent functions is Lipschitz continuous (because its derivatives are bounded). Let us consider g, g_1, g_2 are approximations of constituent functions h, h_1, h_2, respectively within an accuracy of ϵ. Since $||g - h||_\infty \leq \epsilon$, $||g_1 - h_1||_\infty \leq \epsilon$ and $||g_2 - h_2||_\infty \leq \epsilon$ then

$$||h(h_1, h_2) - g(g_1, g_2)||_\infty$$
$$= ||h(h_1, h_2) - h(g_1, g_2) + h(g_1, g_2) - g(g_1, g_2)||_\infty$$
$$\leq ||h(h_1, h_2) - h(g_1, g_2)||_\infty + ||h(g_1, g_2) - g(g_1, g_2)||_\infty$$
$$\leq c\epsilon.$$

for some constant $c \in \mathbb{R}^+$ independent of considered functions. This result combined with the fact that there are $p - 1$ nodes allows us to prove relation (3.15). □

This result may be extended to hierarchical function in $W_r^{p,q}$, the set of functions that admit a tree representation with q branches at each node. A comparison of Eqs. (3.15) and (3.14) confirms that the same accuracy, ϵ, as a shallow network may be achieved with a deep network with less neurons (at least when ϵ tends to zero).

3.4   Numerical Illustration

In Sects. 1.11 and 2.4, we have seen that shallow neural networks offer an interesting alternative to generalized linear models. In this section, we investigate the performance of deep neural networks for predicting the claims frequency of motorcycle drivers. As in previous chapters, our analysis is based on the dataset from the insurance company *Wasa* and we refer to Sect. 1.11 of Chap. 1 for a detailed presentation of explanatory variables. In the first subsection, we fit a deep neural network without regularization constraint to a dataset enhanced with cross-occurrences between pairs of covariates. In the second and third subsections, we test L_1 and L_2 regularization constraints.

3.4.1   A Deep Neural Network for Predicting Claims Frequency

Previous models for predicting the claims frequency use as covariates, the age of the vehicle and owner's age. This information is directly sent to the network as quantitative variables. The other covariates, gender, geographic area and vehicle class are categorical variables with binary modalities. A deep network can replicate potential non-linear dependences between these input covariates. This type of network also manages a high number of input signals. To emphasize this point, we decide to extend the list of covariates in two directions. Firstly, we convert the continuous quantitative variables into categorical ones. We class policyholders by age into 16 subsets of comparable size, as detailed in Table 3.1. Whereas vehicle ages are split in 6 categories reported in Table 3.2. The last modality mainly gathers classic motorcycles, older than 20 years.

Secondly, we complete the dataset with categorical variables that indicate the cross occurrences of three pairs of covariates: gender vs. geographic zone, gender vs. vehicle class and zone vs. vehicle class. After restatement of the database, an insurance contract is described by 8 categorical variables which counts totally 107 binary modalities.

Table 3.1 Lower and upper bounds (Age Min, Age Max) of intervals delimiting the categories of owner's age

Owner's age categories							
Age Min	16	26	29	32	35	39	42	44
Age Max	25	28	31	34	38	41	43	45
Subset size	4239	4568	3952	3139	3565	3608	3859	4021
Age Min	46	48	50	51	54	57	61	66
Age Max	47	49	51	53	56	60	66	> 66
Subset size	4051	3901	3768	3505	4209	3941	3189	4921

Lines "subset size" reports the number of policyholders in each sub-class

Table 3.2 Lower and upper bounds (Age Min, Age Max) of intervals delimiting the categories of vehicle ages

Vehicle age categories						
Age Min	0	6	10	14	17	21
Age Max	5	9	13	16	20	>20
Subset size	8110	9083	10,449	11,355	8003	15,436

Lines "subset size" reports the number of contracts in each sub-class

We fit a deep network with 4 layers counting 10, 20, 10 and 1 neurons (denoted by NN(10,20,10,1)). The activation function in the input, hidden and output layers are respectively linear, sigmoidal and linear. The estimated claims frequency is the exponential of the network output signal to ensure the positivity of the network prediction. Figure 3.4 shows the structure of the neural network. The network is fed with 107 input signals and counts 1480 neural connections.

The calibration is performed with the language programming "R" and the package Keras.[1] This package is a high-level neural networks API with a focus on enabling fast experimentation. We use Tensorflow[2] as computation engine for Keras. TensorFlow is an open-source software library for machine learning applications. It is used for both research and production at Google. TensorFlow was developed by the Google Brain and first released in 2015.

The chosen loss function is the Poisson deviance. As this function is not provided in Keras, we have programmed it. The network is fitted with the root mean square propagation algorithm (RMSprop) and the gradient of the loss function is evaluated with batches of 5000 contracts. The dataset is reshuffled after each epoch to avoid training the network on same batches.

Table 3.3 reports the main statistics measuring the quality of the fit after 1500 iterations. A comparison with results of Chap. 1 reveals that the deviance obtained with a deep neural network is clearly smaller than these obtained with a shallow network (5564.86) or a GLM (5781.66). However, the AIC and BIC seriously raises due to the high number of neural weights to estimate.

In order to detect a potential overfitting, we calculate the claims frequency for different categories of policyholders. We report in Figs. 3.5 and 3.6, these forecasts by age for male and female drivers of a 4 years old motorcycle, in different geographic areas and for two classes of vehicles. As the policyholder's age is converted into a categorical variable, the predicted claims frequency is a staircase function of the age. Compared to Fig. 1.7, forecast claims frequencies have the same order of magnitude than these yield by a shallow neural network with 4 hidden neurons. The global trend is a decrease of the claims frequency with driver's age. However, the erratic behaviour of forecasts clearly reveals that the deep neural

---

[1] https://keras.rstudio.com/index.html.

[2] https://www.tensorflow.org/.

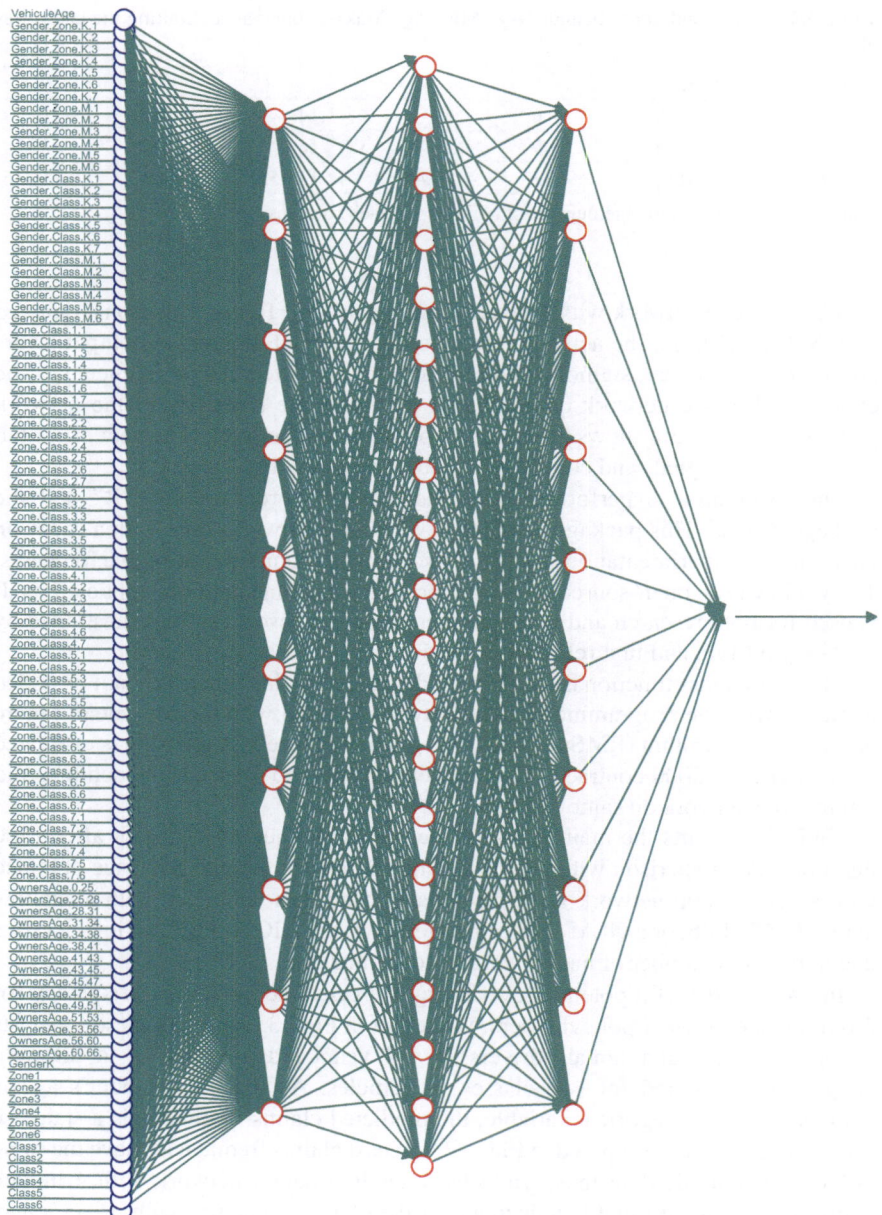

Fig. 3.4 Deep neural network with 107 entries and 3 hidden layers counting respectively 10, 20 and 10 neurons

Table 3.3 Statistics about
the calibration of a
NN(10,20,10,1) neural
networks

NN(10,20,10,1)			
Deviance	5018.17	AIC	9408.74
Log-likelihood	−3183.37	BIC	23,161.47
Number of weights	1521		

Calibration is done with 1500 steps of the RMSprop
algorithm

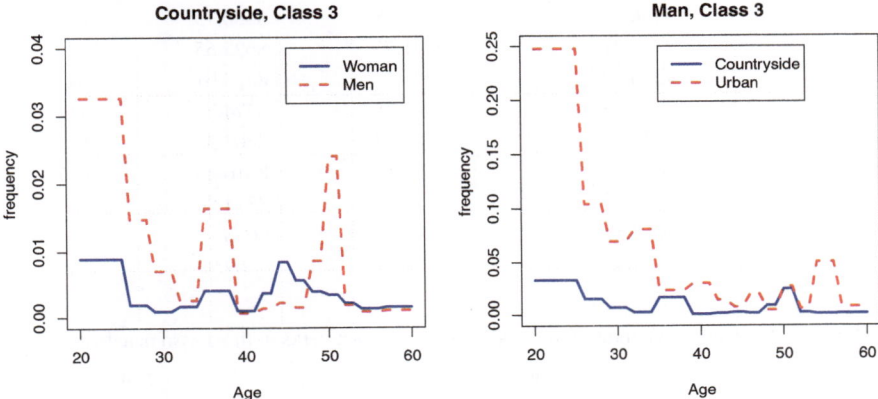

Fig. 3.5 Forecast frequencies of claims for drivers of a 4 years old vehicle, with the
NN(10,20,10,1) model

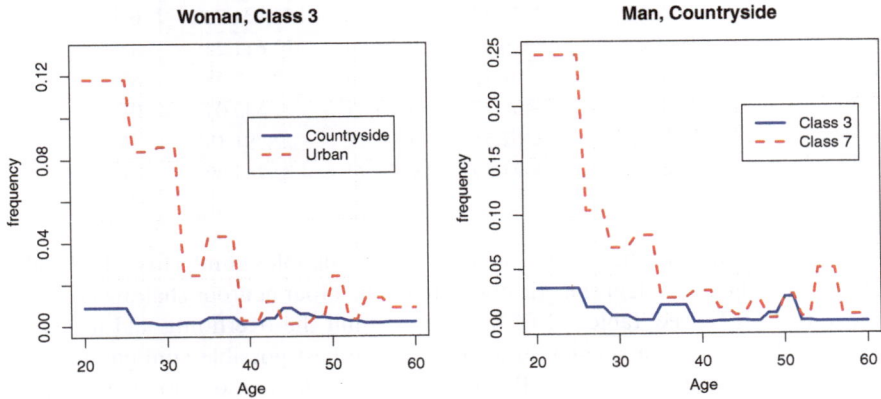

Fig. 3.6 Forecast frequencies of claims for drivers of a 4 years old vehicle, with the
NN(10,20,10,1) model

network overfits the data and prevent us to use it for constructing a real insurance
tariff.

The overfitting is confirmed by the cross validation procedure detailed in
Sect. 1.12 of Chap. 1. The results of this validation on 10 sub-samples of data are

Table 3.4 K-fold cross validation of a NN(10,20,10,1) network with a partition of the data set into 10 validation samples

K	Deviance learning set	Log-likelihood	AIC	Deviance test set
1	4332.63	−2774.6	8591.2	669.32
2	4320.74	−2772.04	8586.08	613.42
3	4410.02	−2818.99	8679.98	558.17
4	4366.03	−2804.38	8650.75	608.71
5	4344.21	−2791.77	8625.55	593.77
6	4360.12	−2791.5	8625.01	646.09
7	4330.74	−2766.35	8574.7	714.52
8	4344.16	−2780.75	8603.5	659.36
9	4367.02	−2787.18	8616.36	687.77
10	4550.13	−2886.43	8814.86	589.02
Average	4372.58	−2797.4	8636.8	634.02
St. Dev.	67.33	34.99	69.97	49.29

Table 3.5 Statistics of calibration for a NN(10,20,10,1) networks with a Lasso penalization

Lasso penalty λ	Deviance	Log-likelihood	AIC	BIC	Number of null weights
0.00	5018.17	−3183.37	9408.74	23,161.47	0
0.05	5099.27	−3223.92	9447.84	23,010.68	21
0.10	5191.7	−3270.13	9492.27	22,838.11	45
0.15	5298.11	−3323.34	9560.68	22,734.72	64
0.20	5384.42	−3366.5	9608.99	22,611.24	83
0.25	5454.88	−3401.73	9673.45	22,648.58	86
0.30	5516.98	−3432.78	9705.55	22,545.04	101
0.35	5576.45	−3462.51	9797.02	22,781.18	85
0.40	5646.2	−3497.39	9854.77	22,784.69	91

reported in Table 3.4. The average deviance on test samples climbs to 634.02 and is greater than the equivalent statistic computed with a four neurons shallow networks (value of 567.72, see Table 1.14). In order to limit the overfitting and to extract from a deep network a smaller structure with the best possible configuration, we add a Lasso penalty term to the Poisson deviance. The Lasso parameter λ varies from 0 to 0.4 by step of 0.05 and we train the networks on the full dataset. Table 3.5 summarizes the result of this procedure. As expected, the deviance increases with λ and for $\lambda = 0.4$, the deviance is close to the one obtained with a four neuron shallow network (5564.86) and still lower than the deviance of a GLM (5781.66). The last column reports the number of quasi-null weights (a weight lesser than 10^{-4} in absolute value is considered as null). Even if this number raises on average with λ, this increase is too limited for significantly reducing the AIC. The rest of our analysis focuses on the network with a Lasso penalization of $\lambda = 0.40$.

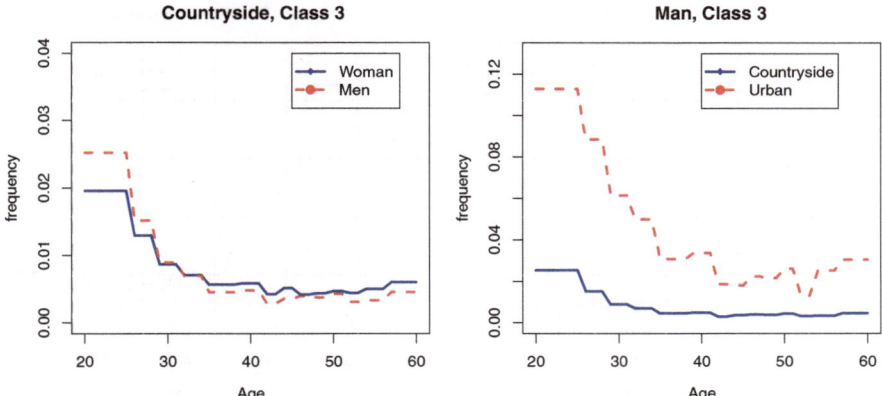

Fig. 3.7 Forecast frequency of claims for drivers of a 4 years old vehicle, with the NN(10,20,10,1) model and a Lasso penalization ($\lambda = 0.40$)

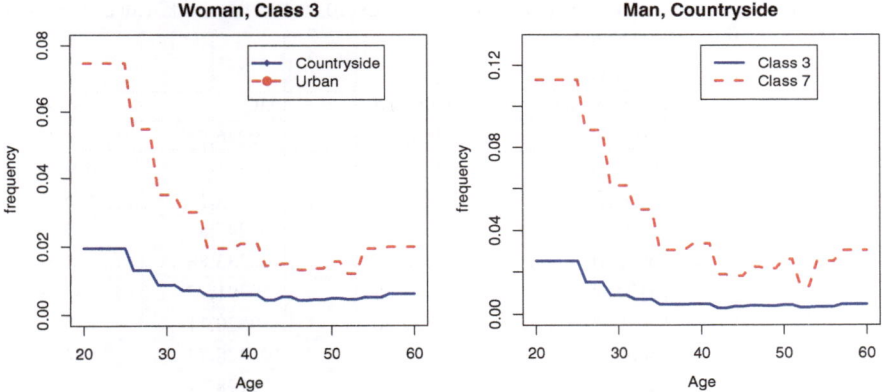

Fig. 3.8 Forecast frequency of claims for drivers of a 4 years old vehicle, with the NN(10,20,10,1) model and a Lasso penalization ($\lambda = 0.40$)

Figures 3.7, 3.8, and 3.9 show the forecast claims frequencies by age, for different types of policyholders and yield by the Lasso penalized network. A comparison with Figs. 3.5 and 3.6 clearly emphasizes that the L_1-penalization smooths the curves of predictions. The order of magnitude of predicted claims frequencies is similar to the one of forecasts computed with a GLM and the shallow networks of Chap. 1. Conclusions that we draw from these graphs appear to be consistent with previous analysis of this dataset. Whatever the policyholder's profile, the claims frequency falls with age. A driver of a class 3 vehicle can cause an accident with a higher probability in an urban area than in the countryside. The claims frequency for female policyholders is less or equal than the one of male drivers. Finally yet importantly, the power of the motorcycle is clearly a determinant of the claims frequency. Based

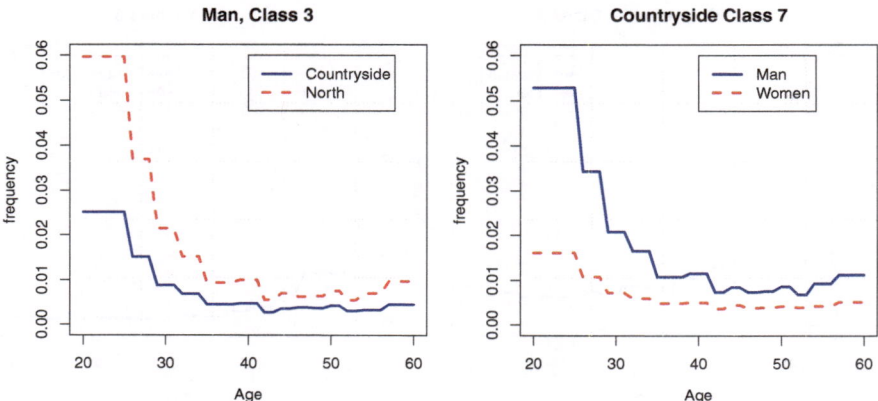

Fig. 3.9 Forecast frequency of claims for drivers of a 4 years old vehicle, with the NN(10,20,10,1) model and a Lasso penalization ($\lambda = 0.40$)

Table 3.6 K-fold cross validation of a L_1-penalized NN(10,20,10,1) network with a partition of the data set in 10 validation samples

Lasso penalty K	Deviance learning set	Log- likelihood	AIC	Deviance test set
1	5017.45	−3102.48	9246.95	704.15
2	6042.94	−3634.83	10, 311.67	604.62
3	5144.77	−3181.06	9404.11	575.88
4	5990.52	−3602.93	10, 247.87	657.04
5	5975.76	−3595.94	10, 233.88	671.84
6	5131.65	−3174.88	9391.77	585.39
7	5079.98	−3139.36	9320.71	637.64
8	6052.79	−3643.37	10, 328.74	594.98
9	4977.27	−3078.38	9198.77	746.53
10	5199.47	−3212.71	9467.42	521.05
Average	5461.26	−3336.59	9715.19	629.91
St. Dev.	481.55	246.61	493.21	66.9

The Lasso penalization is set to $\lambda = 0.40$

on these observations, the Lasso penalized deep network is an eligible model for computing real insurance tariffs.

In order to confirm that the risk of overfitting is reduced with the chosen L_1-penalized network, we perform a cross validation with 10 validation samples. Results are reported in Table 3.6. As we could expect, the average deviance on the training data (5461.26) is greater than the same statistic computed with the non-penalized deep network (4372.58). The non-penalized network fits then better the training set than its penalized version. The L_1-penalized network achieves a slightly better performance on validation samples. The average deviance (629.91)

Table 3.7 Statistics of calibration for a NN(10,20,10,1) networks with a Ridge penalization

Ridge penalty λ	Deviance	Log-likelihood	AIC	BIC	Number of nul weights	Deviance Lasso
0	4845.04	−3096.81	9235.61	22,988.34	0	5018.17
0.05	4973.49	−3161.03	9364.06	23,116.78	0	5099.27
0.1	5057.04	−3202.8	9445.61	23,189.29	1	5191.7
0.15	5144.25	−3246.41	9532.82	23,276.5	1	5298.11
0.2	5245.04	−3296.8	9635.61	23,388.33	0	5384.42
0.25	5347.65	−3348.11	9736.22	23,479.9	1	5454.88
0.3	5458.85	−3403.71	9849.42	23,602.14	0	5516.98
0.35	5546.01	−3447.29	9932.58	23,667.22	2	5576.45
0.4	5618.5	−3483.54	9997.07	23,695.55	6	5646.2

The last column recalls the deviance of Table 3.5, computed with a Lasso constraint

computed over these sets is less than the one yield by the non-penalized network (634.02). Notice also that the average deviance on test samples is still higher than the equivalent statistic computed with a four neurons shallow networks (see Table 1.14, average deviance of 567.72). This discrepancy partly comes from the fact that the ages of policyholders and vehicles are converted into categorical variables in this section. To conclude this section, we compare the impact of L_1 and L_2 penalization constraints on the goodness of fit. Table 3.7 reports the statistics of calibration for the NN(10,20,10,1) network with a Ridge penalty. The weight λ of this penalty varies from 0 to 0.4 by step of 0.05. For the same level of penalization, the deviance obtained with a Ridge penalization is systematically greater than the one computed with a Lasso constraint.[3] The second to last column reports the number of quasi-null weights (a weight lesser than 10^{-4} in absolute value is considered as null). Whatever the value of λ, most of neural coefficients are significantly different from zero. This confirm that contrary to the Lasso, a L_2 penalization does not allows us to construct a sparse neural network.

3.5 Further Readings

The convergence of stochastic gradient and batch descent has been analyzed using the theories of convex minimization and of stochastic approximation. If learning rates decrease with an appropriate rate, stochastic gradient descent converges almost surely to a global minimum when the objective function is convex or pseudo-convex, and otherwise converges almost surely to a local minimum. We refer to Bottou (1998) or Kiwiel (2001) for a study of efficiency of some algorithms presented in this chapter.

---

[3]Notice that the discrepancy between deviances for $\lambda = 0$ is explained by the random initialization of neural coefficients. This randomization leads to different deviances after 1500 iterations.

Deep neural networks have been successfully applied in many fields like speech recognition (e.g. Lample et al. 2016) or image recognition (e.g. Karpathy and Fei-Fei 2016). Deep neural networks are also used for developing artificial intelligences. Silver et al. (2016) have developed a networks for playing Go. Their network achieves a 99.8% winning rate against other Go programs and defeated the human European Go champion. Deep learning is also used for identifying patterns like in DeVries et al. (2018) who study the dynamic of aftershocks after large earthquakes. Most of these applications uses "convolution" and "recurrent" layers of neurons. We refer the reader to the book of Chollet and Allaire (2018) for a practical introduction to these concepts and implementation in Ferrario et al. (2018) provide an interesting tutorial for fitting neural networks to insurance with Keras.

References

Bottou L (1998) Online algorithms and stochastic approximations. Online learning and neural networks. Cambridge University Press, New York. ISBN 978-0-521-65263-6

Chollet F, Allaire JJ (2018) Deep learning with R. Manning Publications, New York

Cybenko G (1989) Approximation by superpositions of a sigmoidal function. Math Control Signals Syst 2:303–314

DeVore R, Howard R, Micchelli CA (1989) Optimal nonlinear approximation. Manuscripta Math 63:469–478

DeVries PMR,Viégas F, Wattenberg M, Meade BJ (2018) Deep learning of aftershock patterns following large earthquakes. Nature 560:632–634

Duchi J, Hazan E, Singer Y (2011) Adaptive subgradient methods for online learning and stochastic optimization. J Mach Learn Res 12:2121–2159

Ferrario A, Nolly A, Wuthrich MV (2018) Insights from inside neural networks. SSRN working paper. https://papers.ssrn.com/sol3/papers.cfm?abstract_id=3226852

Hornik K, Stinchcombe M, White H (1989) Multi-layer feed-forward networks are universal approximators. Neural Netw 2:359–366

Karpathy A, Fei-Fei L (2016) Deep visual-semantic alignments for generating image descriptions. IEEE Trans Pattern Anal Mach Intell 39(4):664–676

Kingma D, Ba J (2015) Adam: a method for stochastic optimization. In: 3rd international conference on learning representations, San Diego, 2015. arXiv:1412.6980

Kiwiel KC (2001) Convergence and efficiency of subgradient methods for quasiconvex minimization. Math Program 90(1):1–25

Lample G, Ballesteros M, Subramanian S, Kawakami K, Dyer C (2016) Neural architectures for named entity recognition. In: Proceedings of NAACL-HLT 2016, pp 260–270

Mhaskar HN (1996) Neural networks for optimal approximation of smooth and analytic functions. Neural Comput 8(1):164–177

Poggio T, Mhaskar H, Rosasco L, Miranda B, Liao Q (2017) Why and when can deep-but not shallow-networks avoid the curse of dimensionality: a review. Int J Autom Comput 14(5):503–519

Silver D, Huang A, Maddison CJ, Guez A, Sifre L, van den Driessche G, Schrittwieser J, Antonoglou I, Panneershelvam V, Lanctot M, Dieleman S, Grewe D, Nham J, Kalchbrenner N, Sutskever I, Lillicrap T, Leach M, Kavukcuoglu K, Graepel T, Hassabis D (2016) Mastering the game of Go with deep neural networks and tree search. Nature 529:484–489

Tibshirani R (1996) Regression shrinkage and selection via the Lasso. J R Stat Soc Ser B 58(1):267–288

Chapter 4
Dimension-Reduction with Forward Neural Nets Applied to Mortality

In this chapter, we study a particular type of neural networks that are designed for providing a representation of the input with a reduced dimensionality. These networks contains a hidden layer, called bottleneck, that contains a few nodes compared to the previous layers. The output signals of neurons in the bottleneck carry a summarized information that aggregates input signals in a non-linear way. Bottleneck networks offer an interesting alternative to principal component analysis (PCA) or non-linear PCA. In actuarial sciences, these networks can be used for understanding the evolution of longevity during the last century. We also introduce in this chapter a genetic algorithm for calibrating the neural networks. This method combined with a gradient descent speeds up the calibration.

The improvement of longevity observed over the last century is a matter of concerns for the insurance industry. This growth is explained by the reduction of mortality caused by infectious and chronic diseases at older ages. This trend calls for more advanced techniques to manage the longevity risk. A popular framework for mortality rates is the model of Lee and Carter (1992). In their approach, the log-force of mortality is the sum of a fixed age component and of an age specific function multiplied by a time component. The robustness of this approach contributes to its success among practitioners. We refer the reader to Chapter 8 in Denuit et al. (2019) where this method is comprehensively detailed in relation to generalized non-linear models (GNM).

Three approaches exist for estimating the LC model and its various extensions. The first one, pioneered by Lee and Carter (1992), counts two steps. In the first stage, age components and latent time processes are obtained by a PCA. In the second step, an autoregressive model or a random walk is fitted to forecast the time effect. The second approach of calibration is based on generalized non-linear models. Brouhns et al. (2002) use a Poisson distribution and estimate parameters of a LC model by log-likelihood maximization. Renshaw and Haberman (2006) adapt this approach for estimating a LC model with cohort effects. The third method for estimating

© Springer Nature Switzerland AG 2019
M. Denuit et al., *Effective Statistical Learning Methods for Actuaries III*,
Springer Actuarial, https://doi.org/10.1007/978-3-030-25827-6_4

parameters consists to perform the joint inference of latent time processes and age parameters, in a single step by a Markov Chain Monte-Carlo (MCMC) method.

The first approach for estimating mortality models is based on PCA. The PCA can be regarded as an extraction method that attempts to characterize lower-dimensional structure in large multivariate data sets. When the data has a nonlinear structure, as it is the case for mortality rates, it will not be detected by a PCA. In the early 1990s, a neural network based generalization of PCA to the nonlinear feature extraction problem was introduced by Kramer (1991) in the chemical engineering literature, who referred to the resulting technique as nonlinear principal component analysis (NLPCA). In this chapter inspired from Hainaut (2018), we use a feed-forward neural network with a bottleneck layer to perform a NLPCA in order to summarize the surface of mortality rates. Notice that our approach to estimate parameters is a two steps procedure. In the first stage, we fit the neural net and filter latent processes representative of time effects. In the second stage, we fit a random walk to forecast the evolution of these latent variables. In our framework, the bottleneck network defines the non linear function between latent processes and observed mortality rates.

In numerical applications, the core of our analysis focuses on mortality rates of the French population from 1946 to 2014. This sample of data is partitioned in two subsets. The first one contains observations from 1946 to 2000 and serve us to calibrate the neural analyzer. Whereas the second subset of mortality rates from 2001 to 2014 is used for validation. The neural net is benchmarked to several extensions of the Lee-Carter model. We compare it to the original and multi-factors Lee Carter model, estimated with a PCA. We also consider the Lee-Carter model, fitted by log-likelihood maximization in a GNM framework. Finally, we compare the neural analyzer to the Lee Carter model with age specific cohort effect, as proposed by Renshaw and Haberman (2006). All numerical experiments conclude that the neural analyzer has an excellent predictive power compared to the LC model. Finally, the calibration of the neural analyzer to UK and US mortality data confirms the robustness of our conclusions.

4.1 The Lee-Carter Model and Its Extensions

Lee and Carter (1992) proposed a pioneering model for mortality forecasting. They assumed that log-forces of mortality have a linear structure with respect to time and that covariates depend upon the age. This model became rapidly a standard in the industry due to its robustness and easiness of implementation. Renshaw and Haberman (2003) extended this framework by proposing a multi-factor model. Renshaw and Haberman (2006) studied a model with age-specific cohort and period effects. They also estimate the Lee-Carter model by log-likelihood maximization, under Gaussian and Poisson error structures. As we use these models as benchmark to measure the efficiency of the neural network approach, we briefly review them in this section.

Throughout the rest of this chapter, the time of decease of an individual of age x is assimilated to the first jump of a non-homogeneous Poisson process, that is denoted $(N_t^x)_{t \geq 0}$. The intensity of the mortality Poisson process is called the force of mortality or the mortality rate and depends upon time t and age x. It is denoted by $\mu(t, x)$ and may be interpreted as the instantaneous probability of death at time t, for a x year old human. The mortality rate is also related to the probability of survival till time $s \geq t$ by the following relation

$$
\begin{aligned}
{}_s p_x &:= P(\tau \geq s) \\
&= \mathbb{E}\left(\mathbf{1}_{\{N_s^x = 0\}} \mid N_t^x\right) \\
&= \exp - \int_t^s \mu(u, x + u - t) \, du \, .
\end{aligned}
$$

On the other side, the probability of dying at age x during year t, is the complementary probability of ${}_t p_x$ defined by

$$
q(t, x) := 1 - \exp\left(-\int_t^{t+1} \mu(u, x + u - t) \, du\right) .
$$

In the original Lee-Carter model, the log mortality rates are related to ages as follows:

$$
\ln \mu(t, x) = \alpha_x + \beta_x \kappa_t \, . \tag{4.1}
$$

where $\alpha_x \in \mathbb{R}$ is a constant representing the permanent impact of age on mortality. Whereas $\beta_x \in \mathbb{R}$ is a constant that quantifies the marginal effect of the latent factor κ_t on mortality at each age. In Eq. (4.1), κ_t is a latent process that describes the evolution of mortality over time. To ensure the identifiability of the model, two constraints are imposed during the calibration:

$$
\sum_x \beta_x = 1 \qquad \sum_t \kappa_t = 0 \, . \tag{4.2}
$$

In multi-factors extensions of the Lee-Carter model proposed by Renshaw and Haberman (2003), the log-force of mortality is a linear combination of d time latent factors noted $\kappa_t^{i=1,\dots,d}$, with covariates that depend on the age as follows:

$$
\ln \mu(t, x) = \alpha_x + \sum_{i=1}^{d} \beta_x^i \kappa_t^i \, . \tag{4.3}
$$

Where the $\beta_x^{i=1\dots d} \in \mathbb{R}$ are constant such that $\sum_x \beta_x^i = 1$. In Eq. (4.3), $(\kappa_t^i)_{i=1\dots d}$ are d latent processes satisfying the constraint $\sum_t \kappa_t^i = 0$ for $i = 1\dots d$, to ensure the identifiability. The Lee-Carter model and its multi-factors extension are estimated by a two-stage procedure looking first at the observation equation as a regression (ignoring the latent factor structure explicitly). In the second stage time-series models are adjusted to latent factors. In Lee and Carter (1992), this regression is performed by a singular value decomposition (SVD) and a principal component analysis (PCA). This approach being well documented in the literature, we refer to e.g. Pitacco et al. (2009) for details and just briefly sketch the procedure for the reader who is not familiar with this approach. The data set of mortality forces range from year t_{min} to t_{max} and from age x_{min} to x_{max}. The number of observations for a given year is noted $n_x = x_{max} - x_{min}$. In the first stage, we infer from the constraint $\sum_{t=t_{min}}^{t_{max}} \kappa_t^i = 0$ that the estimator of α_x is the mean of observed log-mortality rates:

$$\hat{\alpha}_x = \frac{1}{t_{max} - t_{min} + 1} \sum_{t=t_{min}}^{t_{max}} \ln \mu(t, x) \qquad x = x_{min}, \ldots, x_{max}. \qquad (4.4)$$

After removing the trend $\hat{\alpha}_x$ from observations, The next step consists to perform a PCA on the resulting matrix of residual observations, denoted by $\boldsymbol{M}$:

$$\boldsymbol{M} := \left(\ln \mu(t, x) - \hat{\alpha}_x \right)_{t=t_{min}\ldots t_{max},\, x=x_{min}\ldots x_{max}}$$

to infer its singular value decomposition (SVD):

$$\boldsymbol{M} = \sum_{i \geq 1} \sqrt{\lambda_i} \boldsymbol{v}_i^\top \boldsymbol{u}_i$$

where $\lambda_1 \geq \lambda_2 \geq \ldots \geq 0$ are the eigenvalues of $\boldsymbol{M}^\top \boldsymbol{M}$, $\boldsymbol{v}_i$ and $\boldsymbol{u}_i$ are respectively the eigenvectors and the normed eigenvectors of $\boldsymbol{M}^\top \boldsymbol{M}$. A curve of mortality rates at time t is represented by a point in a space with n_x dimensions. The SVD allows us to project these points on an hyperplan of lower dimension d, in $\mathbb{R}^{n_x}$. The coordinates of this curve in the hyperplan are contained in the d-plet $\boldsymbol{\kappa}_t := (\kappa_t^1, \kappa_t^2, \ldots, \kappa_t^d)$ calculated as follows:

$$\left(\kappa_t^i \right)_{t=t_{min}\ldots x_{max}} = \sqrt{\lambda_i} \sum_{j=x_{min}}^{x_{max}} v_{i,j} \boldsymbol{u}_i \qquad i = 1, \ldots, d \,.$$

Whereas, the age effects are the normalized projection vectors:

$$\left(\beta_x^i \right)_{x=x_{min}\ldots x_{max}} = \frac{1}{\sum_{j=x_{min}}^{x_{max}} v_{i,j}} \boldsymbol{v}_i \qquad i = 1, \ldots, r \quad .$$

Table 4.1 Summary of models to which the neural net approach is compared

	Calibration	Symbol	Log mortality
Multifactor Lee Carter	SVD	LC SVD	$\ln \mu(t, x) = \alpha_x + \sum_{i=1}^{d} \beta_x^i \kappa_t^i$
1D Lee Carter	Log-likelihood maximization	LC GLM	$\ln \mu(t, x) = \alpha_x + \beta_x \kappa_t$
Lee Carter with cohort effects	Log-likelihood maximization	LC COH	$\ln \mu(t, x) = \alpha_x + \beta_x \kappa_t + \beta_x^g \gamma_{t-x}$

The PCA method for calibrating the Lee-Carter model is robust and relatively easy to implement. This explains its popularity among practitioners. An alternative approach is proposed by Renshaw and Haberman (2006) who adapted Wilmoth (1993) and Brouhns et al. (2002) to estimate the LC model by log-likelihood maximization, in a GLM framework with a Gaussian error structure. This calibration method is out of the scope of the book and we refer the reader to the original articles for more details. However, we will compare our results to these two methods of calibration in numerical applications. The last model that we consider, adds a cohort effect in the dynamic of log-force of mortality:

$$\ln \mu(t, x) = \alpha_x + \beta_x \kappa_t + \beta_x^g \gamma_{t-x}, \qquad (4.5)$$

where $\beta_x^g \in \mathbb{R}$ represents the marginal effect of a generation factor, γ_{t-x}, on mortality. Renshaw and Haberman (2006) estimate this model by log-likelihood maximization, in a general non-linear model (GNM) framework. We refer the interested reader to their article for details about the estimation procedure. Table 4.1 summarizes the models to which our neural network analyzer is compared in the sequel. It also presents methods of calibration used for each approach.

In the second stage of the calibration procedure, a time-series model is specified for the latent processes. Most of authors use an AR(1) model or a random walk with drift. In numerical applications, we opt for the second choice and assume that increments of κ_t^i are Gaussian random variables with a mean γ_i and a variance σ_i^2:

$$\kappa_t^i - \kappa_{t-1}^i = \gamma_i + \sigma_i \epsilon_t \quad i = 1, \ldots, d \qquad (4.6)$$

where ϵ_t is a standard Normal random variable. Other dynamics, like the switching regime diffusion in Hainaut (2012) have been proposed so as to detect a change of trends in the evolution of mortality. But as the random walk model became the standard in the industry, we adopt it as reference to forecast future mortality rates by simulations. In the numerical illustration, we use a Jarque–Bera test to validate the hypothesis that increments of κ_t^i are normally distributed, at least during the most recent decades.

4.2   The Neural Net Analyzer

The main assumption underlying the LC model and its extensions is the linear dynamic of log-forces of mortality. This specification justifies to apply the PCA to fit latent stochastic processes and age effects. PCA can be regarded as an extraction method that attempts to characterize lower-dimensional structure in large multivariate data sets. If the underlying distribution is Gaussian, then PCA is an optimal feature extraction algorithm. However, if the data has a non-linear structure, as it could be the case for mortality rates, the PCA fails to detect it.

In the early 1990s, a neural-network-based generalization of PCA was introduced by Kramer (1991) in the chemical engineering literature, who referred to the resulting technique as the nonlinear principal component analysis (NLPCA). Directly inspired from the literature on neural networks, we propose here a neural net analyzer that detects the nonlinearities in the lower-dimensional structure of the log-forces of mortality.

In our data sets, the available mortality forces range from year t_{min} to t_{max} and from age x_{min} to x_{max}. The number of observations for a given year is noted $p = x_{max} - x_{min}$. Available demographic data contains the number of deaths aged x per year, $d_{x,t}$, and the initial size of the population aged x, $L_{x,t}$. The death probability is then approached by $q_x = \frac{d_{x,t}}{L_{x,t}}$. Under the assumption that the force of mortality is a stepwise constant function on $[t, t+1[\times[x, x+1[$, we calculate it as follows:

$$\mu(s, y) = -\ln(1 - q(t, x)) \quad \forall s \in [t, t+1[\ y \in [x, x+1[\ .$$

To compare our results with these yield by other models, we use as input for the neural net the centered log-forces of mortality, denoted by:

$$\boldsymbol{m}_t := \begin{pmatrix} \ln \mu(t, x_{min}) - \alpha_{x_{min}} \\ \vdots \\ \ln \mu(t, x) - \alpha_x \\ \vdots \\ \ln \mu(t, x_{max}) - \alpha_{x_{max}} \end{pmatrix} \quad t = t_{min}, \dots, t_{max}\ .$$

$\boldsymbol{m}_t = \{m_{t\,x_{min}}, \dots, m_{t\,x_{max}}\}$ is a vector of dimensions $p = x_{min} - x_{max}$ and α_x is the average log-mortality rates:

$$\alpha_x = \frac{1}{t_{max} - t_{min} + 1} \sum_{t=t_{min}}^{t_{max}} \ln \mu(t, x) \qquad x = x_{min}, \dots, x_{max}. \qquad (4.7)$$

We aim to determine two functions: an encoding and a decoding function. We denote these functions by $f^{enc} : \mathbb{R}^p \to \mathbb{R}^d$ and $f^{dec} : \mathbb{R}^d \to \mathbb{R}^p$. The encoding function, $f^{enc}(.)$, is nonlinear and projects curves of mortality rates at time t, $\boldsymbol{m}_t \in \mathbb{R}^p$ on a hyperplan of lower dimensions, in $\mathbb{R}^d$. As in the multi-factor LC model, the coordinates of the projection in $\mathbb{R}^d$ are contained in a d-plet $\kappa_t^{nn} := (\kappa_t^{nn,1}, \ldots, \kappa_t^{nn,d})$ such that

$$\kappa_t^{nn} := f^{enc}(\boldsymbol{m}_t) \quad t = t_{min}, \ldots, t_{max} .$$

The decoding function $f^{dec}(.)$ uses this summarized information to build an approximation $\hat{\boldsymbol{m}}_t \in \mathbb{R}^{n_x}$ of the initial curve of log-mortality rates:

$$\hat{\boldsymbol{m}}_t := f^{dec}\left(\kappa_t^{nn}\right) .$$

Compared to the original LC model, the product $\beta_x \kappa_t$ is replaced by a general non-linear function and log-mortality forces are ruled by the following relation:

$$\ln \mu(t, x) = \alpha_x + f^{dec}\left(x, \kappa_t^{nn}\right) \tag{4.8}$$

$$= \alpha_x + f^{dec}\left(x, f^{enc}(\boldsymbol{m}_t)\right)$$

The encoding and decoding functions are calibrated so as the minimize the sum of squared residuals between initial and reconstructed mortality curves:

$$\left(f^{enc}, f^{dec}\right) = \arg\min \sum_{t=t_{min}}^{t_{max}} \|\boldsymbol{m}_t - \hat{\boldsymbol{m}}_t\|_2^2 . \tag{4.9}$$

In the neural analyzer net, the functions $f^{enc}(.)$, $f^{dec}(.)$ are approximated by two feed-forward neural networks. A neural network is a series of parallel layers of interconnected neurons. A neuron in the ith layer receives as input, the output of neurons located in the previous layer. Let n_j be the number of neurons in layer j. The output of the ith neurons in layer j, denoted by $y_{i,j}$, is computed as follows:

$$y_{i,j} = \phi_{i,j}\left(\sum_{k=1}^{n_{j-1}} \omega_{i,k}^j y_{k,j-1}\right)$$

where $\omega_{i,k}^j$ are the weights and $\phi_{i,j}(.)$ is a transfer function. In our framework, two transfer functions are used. The first one is the hyperbolic tangent sigmoid function, $\phi_{sig}(z) : \mathbb{R} \to (-1, 1)$, defined by

$$\phi_{sig}(z) = \frac{2}{1 + \exp(-z)} - 1 .$$

The second transfer function is the identity function: $\phi_{id}(z) = z$. Cybenko (1989) and Hornik (1991) proved that a three layers neural network with n_1 input neurons, hyperbolic transfer functions in the second layer, and linear transfer functions in

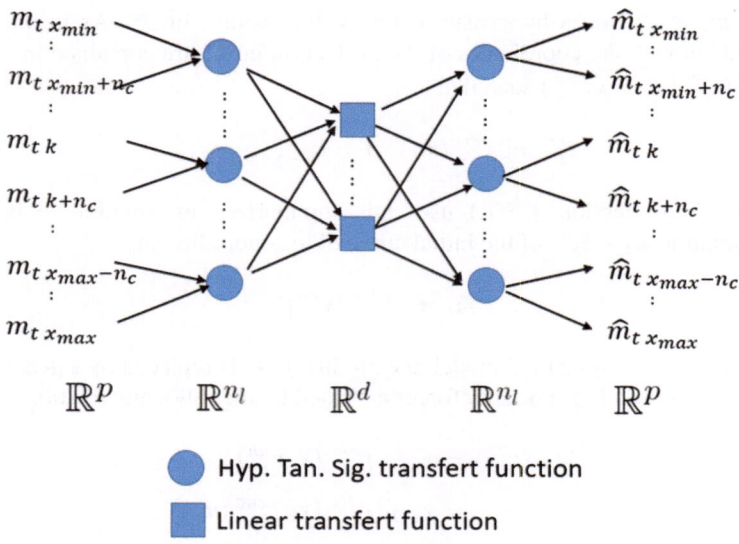

Fig. 4.1 Architecture of the neural analyzer

the third layer of n_2 neurons can approximate to arbitrary accuracy any continuous function from $\mathbb{R}^{n_1}$ to $\mathbb{R}^{n_2}$ at the condition that the number of neurons in the second layer is large enough. These fundamental results justify our approach that consists to define f^{enc} and f^{dec} by feed-forward neural networks. We test the architecture recommended in McNelis (2005) and presented in Fig. 4.1.

This architecture defines a bottleneck neural network since the intermediate hidden layer counts less neurons than others. The input and output layers count the same number of neurons, n_l, with a hyperbolic tangent sigmoid transfer function. Mortality log-forces, $\boldsymbol{m}_t$ are divided into n_l groups of $n_c = \frac{n_x}{n_l}$ elements. Each subgroup of data is sent exclusively to a single neuron of the input layer. The central layer is a bottleneck with d neurons that have a linear transfer function. The encoding phase is then summarized by the following two operations:

$$y_{i,1}(t) = \phi_{sig}\left(\sum_{k=1}^{p}\omega_{i,k}^1 m_{t\,k}\right) \quad i = 1,\ldots,n_l$$

$$\kappa_t^{nn,i} = \sum_{k=1}^{n_l}\omega_{i,k}^2 y_{k,1}(t) \quad i = 1,\ldots,n_d\,,$$

where $\omega_{i,k}^1 \neq 0$ if $x_{min} + (i-1) n_c \leq k < x_{min} + i\, n_c$ and $\omega_{i,k}^1 = 0$ otherwise. Whereas the decoding phase is given by the following two steps:

$$y_{i,3}(t) = \phi_{sig} \left(\sum_{k=1}^{n_d} \omega_{i,k}^3 \kappa_t^{nn,k} \right) \quad i = 1, \ldots, n_l$$

$$\hat{m}_{t\,i} = \sum_{k=1}^{n_l} \omega_{i,k}^4 y_{k,3}(t) \quad i = 1, \ldots, p \,,$$

where $\omega_{i,k}^4 \neq 0$ if $x_{min} + (k-1) n_c \leq i < x_{min} + k\, n_c$ and $\omega_{i,k}^4 = 0$ otherwise. The weights $\omega_{i,k}^j$ are calibrated by minimizing the quadratic spread between the input and the output, as defined in Eq. (4.9). The dimension of x_t being high, the number of parameters to calibrate is important. Applying a gradient method to adjust the network is then slow and the risk of staying eventually trapped in local minimum during the gradient descent is non negligible. For this reason, we fit the neural analyzer with a genetic algorithm that is described in the next section.

As for the LC model, we calibrate next n^d random walks each component of the d-plet $\kappa_t^{nn} := (\kappa_t^{nn,1}, \ldots, \kappa_t^{nn,d})$:

$$\kappa_t^{nn,i} - \kappa_{t-1}^{nn,i} = \gamma_i^{nn} + \sigma_i^{nn} \epsilon_t^i \quad i = 1, \ldots, d \tag{4.10}$$

where ϵ_t^i is a standard normal random variable. We will see in the numerical application that despite its relative simplicity, a random walk process provides an excellent statistical fit to the time-series of $\kappa_t^{nn,i}$.

4.3 Genetic Algorithm (GA)

As for deep neural networks, many weights must be calibrated. For this reason, estimating the parameters of the neural network with a gradient descent is time consuming. Instead, we adopt a two steps strategy. In the first stage, we start the search of optimal parameters with a genetic algorithm (GA), developed by McNelis (2005). We use next the solution found by this GA algorithm as starting point of a gradient descent. The GA algorithm is a powerful evolutionary search process that proceeds in five steps.

1. The first step consists to create a population of candidate parameters. Contrary to a gradient descent, a genetic algorithm does not start with one initial vector of parameters, but with an initial population of n^{pop} (an even number) coefficient vectors, called the first generation. Letting n^ω be the total number of coefficients to estimate, the first generation is the set of n^ω by n^{pop} random vectors: $P = \{\boldsymbol{\omega}_1, \boldsymbol{\omega}_2 \ldots, \boldsymbol{\omega}_{n^{pop}}\}$.

2. The second step is called the selection. We choose randomly two pairs of coefficients from the population, with replacement. We evaluate the goodness of fit for these four coefficient vectors, in two pair-wise combinations, according to the quadratic error function defined by Eq. (4.9). Coefficient vectors that come closer to minimizing the sum of errors receive better scores. This is a simple tournament between the two pairs of vectors: the winner of each tournament is the vector with the best scores. These two winning vectors ω_1, $\omega_2 \in P$ are retained for "breeding" purposes.

3. The third step is the crossover, in which the two parents, selected during the second stage, "breed" two children. The algorithm allows crossover to be performed on ω_1 and ω_2, with a fixed probability $p > 0$. The algorithm uses one of three different crossover operations, with each method having an equal 1/3 probability of being chosen:

    (a) Shuffle crossover. We draw n^ω random numbers from a binomial distribution. If the mth draw is equal to 1, the coefficients $\omega_{1,m}$ and $\omega_{2,m}$ are swapped; otherwise, no change is made. The two vectors resulting from these swaps are the children, denoted by c_1 and c_2.

    (b) Arithmetic crossover. A random number is chosen, $\delta \in (0, 1)$. This number is used to create two children that are linear combinations of the two parent factors, $c_1 = \delta\omega_1 + (1 - \delta)\omega_2$ and $c_2 = (1 - \delta)\omega_1 + \delta\omega_2$.

    (c) Single-point crossover. For each pair of vectors, an integer I is randomly chosen from the set $[1, n^\omega - 1]$. The two vectors are then cut at integer I and the coefficients to the right of this cut point, $\omega_{1,1:I+1}$ and $\omega_{2,1:I+1}$ are swapped to produce c_1 and c_2.

Following the crossover operation, each pair of parent vectors gives birth to two children coefficient vectors. In case of no crossover, the children are copies of their parents: $c_1 = \omega_1$ and $c_2 = \omega_2$.

4. The fourth step is the mutation of children. With some small probability p^{mut} that decreases over time, each coefficient of the two children's vectors is subjected to a mutation. We can draw a parallel between this step and the simulated annealing method. As seen in Chap. 1, simulated annealing is a probabilistic technique for approximating the global optimum of a given function. Here, the probability of each element is subject to mutation in generation $g = 1, 2, \ldots, g^*$ given by the probability $p^{mut} = 0.15 + \frac{0.33}{g}$. Here, g is the generation number, g^* is the maximum number of generations. If mutation is to be performed on a vector element, we use a non-uniform mutation operation recommended by McNelis (2005). We draw two random numbers r_1 and r_2 from the $[0, 1]$ interval and one random number s from a standard Normal distribution. The mutated coefficient $\tilde{c}_{i,k}$ for $i = 1, 2$ and $k = 1$ to n^ω is given by the following formula:

$$
\tilde{c}_{i,k} = \begin{cases} c_{i,k} + s \left(1 - r_2^{\left(1 - \frac{g}{g^*}\right)^b} \right) & if\ r_1 > 0.5 \\[2ex] c_{i,k} - s \left(1 - r_2^{\left(1 - \frac{g}{g^*}\right)^b} \right) & if\ r_1 \leq 0.5 \end{cases}
$$

where b is a parameter that governs the degree to which the mutation operation is non-uniform. We set $b = 2$. With this approach, the probability of creating via mutation a new coefficient that is far from the current coefficient value diminishes as $g \to g^*$, where g^* is the number of generations. Thus, the mutation probability itself evolves through time. McNelis (2005) mentions that the mutation operation is non-uniform since, over time, the algorithm is sampling increasingly more intensively in a neighborhood of the existing coefficient values. This more localized search allows for some fine tuning of the coefficient vector in the later stages of the search, when the vectors should be approaching close to a global optimum.

5. The fifth and last step is the election tournament. Following the mutation operation, the four members of the "family" (ω_1, ω_2, c_1, c_2) engage in a tournament. The score of children and parents is measured by their quadratic errors, as defined by Eq. (4.9). The two vectors with the best goodness of fit, whether parents or children, survive and pass to the next generation, while the two with the worst score are extinguished.

The above process is repeated, with parents returning to the population pool for possible selection again, until the next generation is populated by n^{pop} vectors.

Once the next generation is populated, we introduce elitism. It consists to evaluate all the members of new and past generations according to the score. If the best member of the older generation performs better than the best member of the new generation, this member replaces the worst member of the new generation. One continues this process for g^* generations. The literature gives us little guidance about selecting a value for g^*. Since we evaluate convergence by the score of the best member of each generation, g^* should be large enough so that we see no changes in the fitness values of the best for several generations.

4.4 Application to the French Population

This section focuses on mortality rates observed for the French population over the period 1946 to 2014. The data set is provided by the Human Mortality Database.[1] Years before 1946 are excluded from the scope of the study given the perturbations on mortality caused by the first and second world wars. The ages considered range from 20 to 109 years. The LC models and the neural networks are calibrated with mortality curves from year 1946 up to 2000. To compare the predictive capability of models, log-forces of mortality are projected by simulations over 14 years (10,000 simulations) and their average is compared with the observed mortality during the period 2001–2014.

---

[1] www.mortality.org.

Table 4.2 Goodness of fit for variants of the LC model

Model	Coef.	French population, 1946–2000			
		$\sum \|m_t - \hat{m}_t\|_2^2$	Avg.$\|m_t - \hat{m}_t\|_2$	max$(m_t - \hat{m}_t)$	min$(m_t - \hat{m}_t)$
LC SVD 1	180	152.50	0.0024	0.8567	−0.4330
LC SVD 2	270	139.35	0.0023	0.6398	−0.4352
LC SVD 3	360	134.61	0.0023	0.6364	−0.4350
LC GLM	180	29.31	0.0010	0.47017	−0.5515
LC COH	270	10.78	0.0006	0.28573	−0.2220

The first and second columns report the number of latent factors and fitted coefficients. The third and fourth columns present the sum of squared errors and the average errors. The two last columns contain the maximum and minimum errors

Table 4.3 Jarque Bera test applied to increments of latent factors over the period 1970–2000, for the LC SVD 3 model

	Jarque Bera statistics for 3D Lee Carter		
Factors	p-Value	JB statistic	Critical value 5%
$\kappa_t^1 - \kappa_{t-1}^1$	0.1679	2.0286	4.4466
$\kappa_t^2 - \kappa_{t-1}^2$	0.3555	1.2713	4.4466
$\kappa_t^3 - \kappa_{t-1}^3$	0.3355	1.3327	4.4466

Table 4.2 reports the calibration errors of LC models with one to three latent factors fitted with a SVD (LC SVD), of the LC model fitted by log-likelihood maximization (LC GLM) and of the LC model with cohort effect (LC COH). We present the sum of squared errors and the average of errors between observed and modeled log-forces of mortality. The table also provides the maximum and minimum spreads and the number of fitted parameters. An analysis of these figures reveals that calibrating the LC model with a SVD leads to a higher quadratic error than the one obtained with statistical approaches. The best fit is obtained with the model that includes a cohort effect.

We mention in Sect. 4.1 that it is common to assume that increments of latent processes κ_t^i follow a random walk with drift. This hypothesis of Normality is tested in Table 4.3 with the Jarque–Bera (JB) test, for the increments of a 3 dimensions LC model observed over the period 1970–2000.

The p-values reported in Table 4.3 show that Normality can be retained as a working assumption. However, the same test applied to the sample of increments over the whole period of calibration (1946–2000) leads to the rejection of Normality for the second latent factor. We can draw a parallel with the conclusions of Hainaut (2012) who fits a switching regime process to latent processes. This analysis clearly reveals a change of regime between 1960 and 1970. The same conclusions apply to latent processes of LC GLM and LC COH models. This change of trend may be explained by the reduction of mortality caused by coronary heart diseases, following two vast prevention campaigns launched during the sixties. For this reason, the random walks used in simulations to predict the evolution of log-forces of mortality are fitted to increments of κ_t^i observed only between 1970 and 2000.

Table 4.4 Predictive goodness of fit for variants of the LC model

Dim.	Coef.	French population, 2001–2014			
		$\sum \left\| m_t - \hat{m}_t \right\|_2^2$	Avg.$\left\| m_t - \hat{m}_t \right\|_2$	$\max \left(m_t - \hat{m}_t \right)$	$\min \left(m_t - \hat{m}_t \right)$
LC SVD 1	180	38.37	0.0049	0.7006	−0.1297
LC SVD 2	270	38.89	0.0049	0.7088	−0.1293
LC SVD 3	360	38.50	0.0049	0.6618	−0.1287
LC GLM	180	11.68	0.0027	0.5532	−0.1724
LC COH	270	17.65	0.0026	0.5329	−0.1377

The first and second columns report the number of latent factors and fitted coefficients. The third and fourth columns present the sum of squared errors and the average errors. The two last columns contain the maximum and minimum errors

The results about the predictive capability of LC models are reported in Table 4.4. An analysis of the sum of squared errors emphasizes that the performance of models fitted by SVD with one to three factors are nearly identical. The predictive capability of the model with a cohort effect is slightly less good than the one of a LC model estimated by log-likelihood maximization. These figures will be compared to these obtained with the neural net analyzer in the next paragraphs.

The neural network is fitted to the same data set of log-forces of mortality from 1946 to 2000. Several neural architectures are tested: from 3 to 8 neurons for the input/output layers and 2–3 neurons for the intermediate layer. The size of populations in the genetic algorithm is set to 100 vectors of candidate parameters randomly drawn from a standard Normal and we consider 500 generations. The time to calibrate the neural net on a personal computer varies between 5 and 15 min, depending on the processor.

The calibration errors are reported in Table 4.5. A comparison with errors presented in Table 4.2 confirms that the neural analyzer outperforms LC models fitted by SVD and provides a comparable or better fit than LC GLM and LC COH, depending upon the configuration of neurons. Increasing the number of neurons in the input/output layer improves the goodness of fit. The quadratic error obtained with a 8-3-8 neural net (8 input/output and 3 intermediate neurons) is lower than the one for the LC COH model. This confirms that the neural net approach captures age-specific cohort effects.

Figure 4.2 shows filtered latent factors by tested neural networks. For most of configurations, the latent processes $\kappa_t^{nn,i}$ exhibit a quasi-linear trend, either increasing or decreasing. As for the LC model, we assume that increments of latent factors follow a random walk with drift for the prediction. This hypothesis is checked with a Jarque Bera test for the 3-2-3 neural net, over the period 1970–2000.

Statistics of this test, reported in Table 4.6, confirm that Normality can be used as working assumption. As for LC models, the same test applied to the sample of increments over the whole period of calibration (1946–2000) rejects the Normality for the first latent factor. If we look to the evolution of this process (first graph of Fig. 4.2), we observe a change of trend between periods 1948–1960 and 1960–

Table 4.5 Goodness of fit for the neural network model

n_l	n_d	Coef.	French population, 1946–2000			
			$\sum \|m_t - \hat{m}_t\|_2^2$	Avg.$\|m_t - \hat{m}_t\|_2$	$\max(m_t - \hat{m}_t)$	$\min(m_t - \hat{m}_t)$
3	2	552	15.04	0.0008	0.2988	−0.3953
4	2	736	13.37	0.0007	0.3066	−0.3862
5	2	920	12.64	0.0007	0.3088	−0.3678
6	2	1104	12.18	0.0007	0.3047	−0.3603
7	2	1288	12.15	0.0007	0.3142	−0.3678
8	2	1472	11.97	0.0007	0.3085	−0.3648
3	3	558	14.83	0.0008	0.3027	−0.3961
4	3	744	12.64	0.0007	0.3081	−0.3894
5	3	930	11.85	0.0007	0.3072	−0.3721
6	3	1116	11.56	0.0007	0.3101	−0.3658
7	3	1302	10.68	0.0007	0.2896	−0.3336
8	**3**	**1488**	**9.71**	**0.0006**	**0.2844**	**−0.3141**

The first, second and third columns report respectively the number of input/output neurons, of latent factors and of fitted coefficients. The fourth and fifth columns present the sum of squared errors and the average errors. The two last columns contain the maximum and minimum errors. Bold values indicate the best fit

2000. As mentioned previously, This change of trend may be partly explained by the reduction of mortality caused by coronary heart diseases, following two prevention campaigns launched around the sixties.

To validate the predictive capability of the neural model, we forecast log-forces of mortality over 14 years and compare them to the real rates observed over the period 2001–2014. 10,000 simulations are performed and we consider as forecast, the yearly average of simulated log-mortality rates. Table 4.7 presents the errors of estimation. A comparison with errors of LC models confirms the excellent predictive power of the neural network: the sum of squared errors falls to 8.17, for the 3-2-3 configuration whereas the predictive error of the LC model with cohort effects has a predictive error of 17.65.

Figure 4.3 compares predicted and real log-forces of mortality for years 2001 and 2014, with this configuration of neurons. A deeper analysis of figures in Table 4.7 reveals that increasing the number of neurons deteriorates the predictive power of networks. In particular, the 8-3-8 neural net yields the highest prediction error, despite having the lowest calibration error. This phenomenon is related to the mechanism of overfitting. Overfitting occurs when the model is excessively complex, such as having too many parameters relative to the number of observations. An overfitted model has poor predictive performance and it overreacts to minor fluctuations in the training data. Overfitting may easily be avoided by choosing the neural network architecture that offers the best trade-off between calibration and prediction errors. In our case, the predictive power of the 3-2-3 configuration (3 input/output and 2 intermediate neurons) being excellent and its calibration error

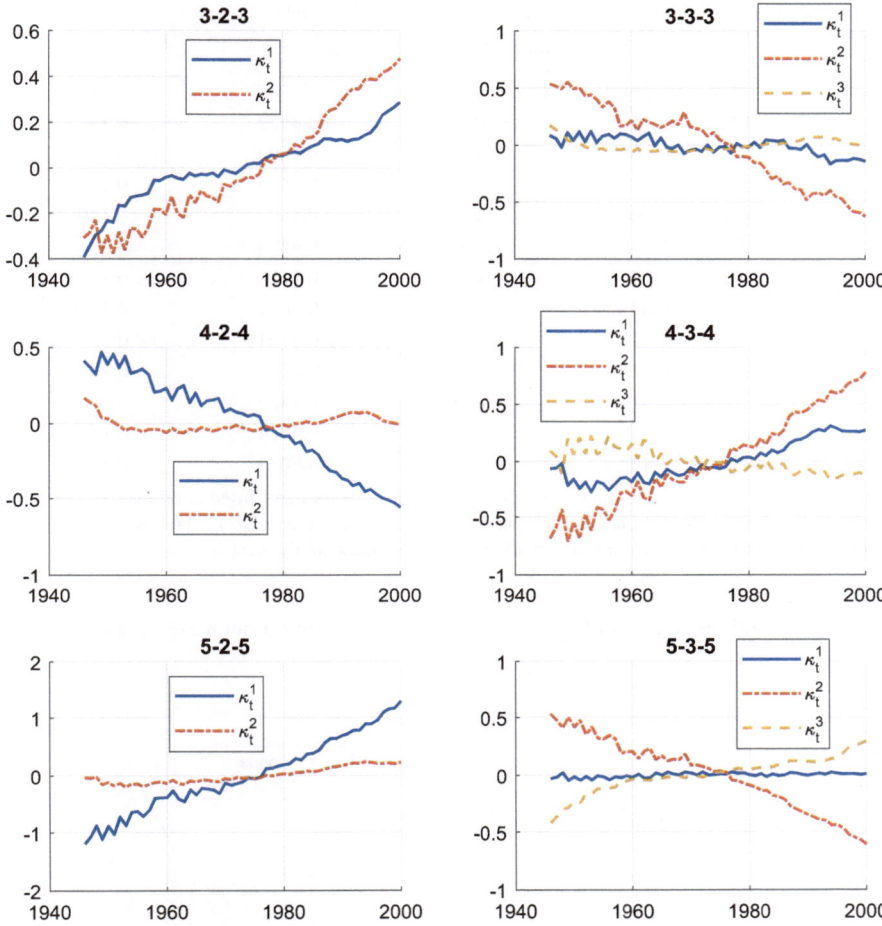

Fig. 4.2 Latent processes, $\kappa_t^{nn,i}$, filtered with different configurations of neural networks. The title of each graph reports respectively the number of neurons in the input/intermediate/output layers

Table 4.6 Jarque Bera test applied to increments of latent factors over the period 1970–2000, for neural analyzer with three neurons in input/output layers and two neurons in the intermediate layer

Factors	Jarque Bera statistics for a 3-2-3 neural net		
	p-Value	JB statistic	Critical value 5%
$\kappa_t^{nn,1} - \kappa_{t-1}^{nn,1}$	0.5000	0.0762	4.4496
$\kappa_t^{nn,2} - \kappa_{t-1}^{nn,2}$	0.3698	1.2296	4.4496

Table 4.7 Predictive goodness of fit for the neural network model

n_l	n_d	Coef.	$\sum \lVert m_t - \hat{m}_t \rVert_2^2$	Avg. $\lVert m_t - \hat{m}_t \rVert_2$	$\max\left(m_t - \hat{m}_t\right)$	$\min\left(m_t - \hat{m}_t\right)$
			French population, 2001–2014, forecast			
3	**2**	**552**	**8.17**	**0.0023**	**0.4922**	**−0.1474**
4	2	736	9.60	0.0025	0.4607	−0.1587
5	2	920	10.84	0.0026	0.4567	−0.1788
6	2	1104	14.20	0.0030	0.4549	−0.1705
7	2	1288	16.51	0.0032	0.4498	−0.1647
8	2	1472	16.62	0.0032	0.4408	−0.1754
3	3	558	9.09	0.0024	0.5021	−0.1470
4	3	744	9.71	0.0025	0.4701	−0.1546
5	3	930	10.57	0.0026	0.4603	−0.1628
6	3	1116	10.81	0.0026	0.4333	−0.1875
7	3	1302	14.24	0.0031	0.4084	−0.1902
8	**3**	**1488**	**17.43**	**0.0033**	**0.5053**	**−0.1914**

The first and second columns report the number of latent factors and fitted coefficients. The third and fourth columns present the sum of squared errors and the average errors. The two last columns contain the maximum and minimum errors. Bold values indicate the best fit

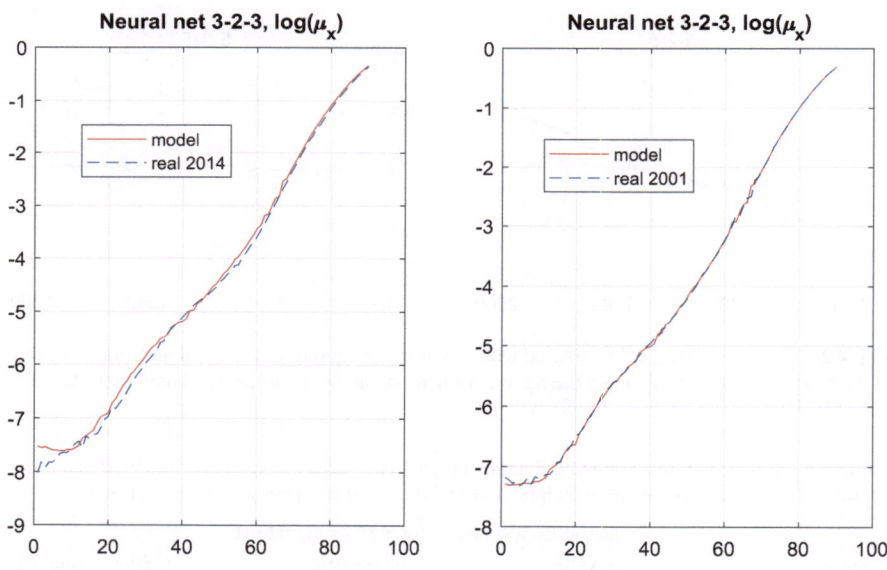

Fig. 4.3 The left and right plots compare the real log-mortality rates in 2014/2001 to the average log-forces of mortality simulated with the 3-2-3 neural analyzer. 10,000 simulations are realized

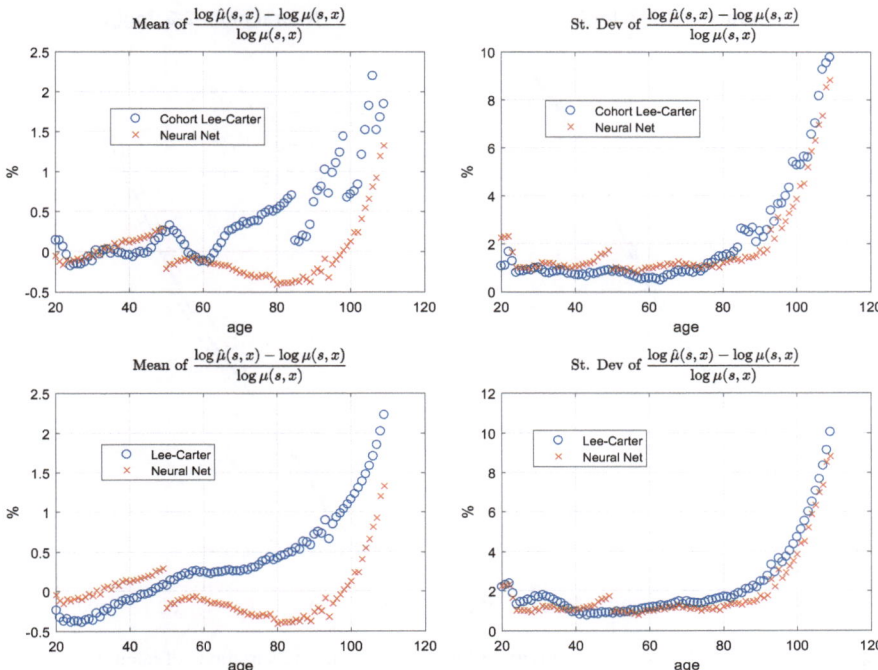

Fig. 4.4 Breakdown of means and standard deviations of relative average errors of calibration, per age. We compare the LC GLM and LC COH models to the 3-2-3 neural network

being close to the one of the LC COH model, the remainder of this section focuses on this network.

The left and right plots of Fig. 4.4 respectively present a breakdown of means and standard deviations of relative average errors of calibration, per age, and for the period 2001–2014. The 3-2-3 network is here compared to LC models with (LC GLM) and without cohort effects (LC COH). The fact that standard deviations of these errors for all models continuously increase with age, is inherent to the hypothesis of linearity of log-forces of mortality with respect to latent factors. Excepted for ages above 80 years, the deviation of relative errors for the neural net is nearly constant and may then be attributed to measurement errors, which is a desirable quality for a model. Before 50 years, the average of relative errors for the 3-2-3 net is close to zero. For the age group 50–80, average relative errors and their deviations computed with the neural net are lower in absolute value than these obtained by other approaches.

If we look to the left plot of Fig. 4.4, we observe a clear cut in the evolution of relative average errors at the age of 50 years. This cut comes from the configuration of the neural analyzer: we have three input/output neurons affected respectively and exclusively to three age groups. The first input neuron receives only information about the mortality between the ages of 20–50 years and this deteriorates the

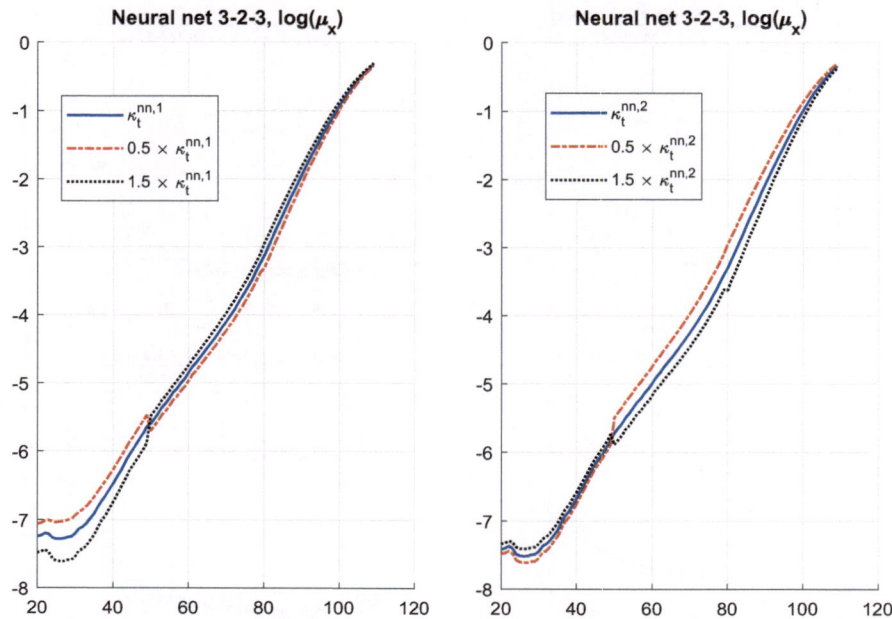

Fig. 4.5 3-2-3 neural network: sensitivity of log-mortality rates to variations of latent factors

goodness of fit around the age of 50. It is probably possible to improve the calibration by sharing some information between adjacent neurons.

Figure 4.5 illustrates the influence of latent factors filtered by a 3-2-3 neural net on the term structure of log-forces of mortality (forecast, year 2014). The left plot emphasizes that $\kappa_t^{nn,1}$ mainly influences log-mortality rates between 20 and 50 years old. Increasing $\kappa_t^{nn,1}$ reduces log-mortality rates for this age group and slightly increases log-forces of mortality for ages 50 and above. The right graph shows that the second latent factor $\kappa_t^{nn,2}$ mainly concerns individuals aged between 51 and 109 years. Increasing $\kappa_t^{nn,2}$ reduces log-mortality rates for this age range and slightly increases rates for ages before 50. Compared to LC models, latent factors yield by the neural net offer then the same ease of interpretation.

Table 4.8 compares the moments of simulated log-forces of mortality, predicted by the cohort LC model and the 3-2-3 neural analyzer. The forecasts are computed for the year 2010, with models fitted to data from 1946–2000. Figure 4.6 shows the densities of $\ln \mu(2010, 40)$ and $\ln \mu(2010, 60)$ obtained by simulations. These statistics and the graph emphasize that distributions of simulated log-forces of mortality display visible differences depending on the model. With the neural analyzer, the distribution exhibits a higher variance than for the LC model and

Table 4.8 This table compares moments of log-forces of mortality simulated by the Neural analyzer (3 input/output neurons and 2 intermediate neurons) and the cohort LC model

	3-2-3 neural analyzer			
	20 years	40 years	60 years	80 years
$\mathbb{E}(\ln\mu(t,x))$	−7.4532	−6.8255	−5.1414	−3.3868
std $(\ln\mu(t,x))$	0.0690	0.07972	0.031125	0.037625
$\mathbb{S}(\ln\mu(t,x))$	0.0714	0.07142	0.16981	0.16981
$\mathbb{K}(\ln\mu(t,x))$	2.9881	2.9881	3.1085	3.1088
	Cohort LC model (LC COH)			
	20 years	40 years	60 years	80 years
$\mathbb{E}(\ln\mu(t,x))$	−7.3365	−6.7959	−4.9346	−3.4342
std $(\ln\mu(t,x))$	0.0853	0.04085	0.0291	0.0623
$\mathbb{S}(\ln\mu(t,x))$	−0.0416	−0.0024	−0.0023	−0.0023
$\mathbb{K}(\ln\mu(t,x))$	2.9772	2.9660	2.9282	2.9791
	Historical log-mortality rates (1946–2010)			
	20 years	40 years	60 years	80 years
$\mathbb{E}(\ln\mu(t,x))$	−6.5021	−5.8614	−4.3754	−2.604
std $(\ln\mu(t,x))$	0.7993	0.6400	0.3858	0.3814
$\mathbb{S}(\ln\mu(t,x))$	1.2508	1.1690	0.4596	0.0881
$\mathbb{K}(\ln\mu(t,x))$	3.1277	2.9206	2.3744	2.0588

10,000 simulations are performed and rates are computed age 20, 40, 60 and 80, for the year 2010. The third sub-table reports the moments of observed log-forces of mortality over the period 1946–2010

right asymmetry. Whereas log-mortality rates predicted by the neural net are slightly leptokurtic, they are strictly Gaussian in the cohort LC model.[2]

It is interesting to compare simulated moments to these calculated with past mortality rates, over the period 1946–2010. We observe that the empirical historical distribution also displays a right asymmetry that is not present in the LC model. The historical variance is also much higher than the one predicted by the LC and neural models. The distribution is leptokurtic at 20 years old, and the kurtosis decreases next with age. However, these statements must be nuanced given the limited number of observations available to calculate these statistics.

We pursue our analysis of LC and neural networks by a comparison of cross-sectional lifetime expectancies predicted by models, over the period 2001–2014.

---

[2]Notice that in the LC model, the log-mortality rates are Normally distributed: their skewness and kurtosis are then respectively equal to 0 and 3. Skewness and kurtosis reported in Table 4.8 are not exactly equal to these figures because they are computed with simulated log-forces of mortality.

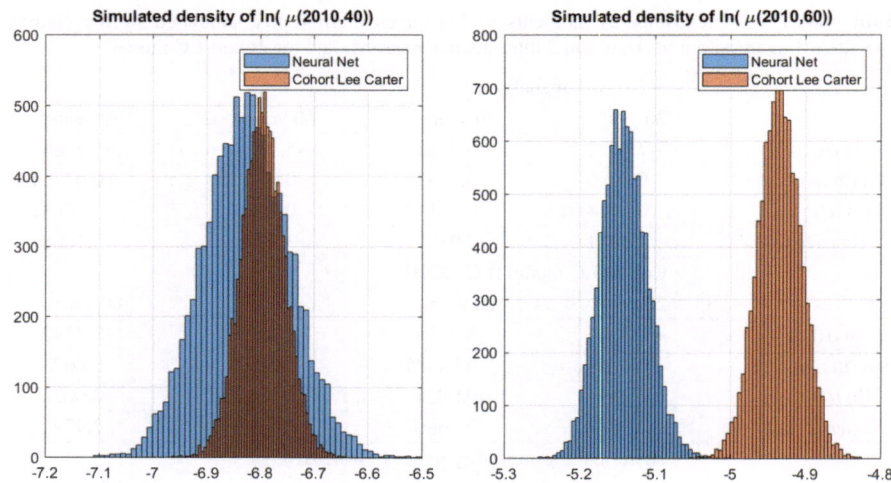

Fig. 4.6 Comparison of simulated densities for $\ln \mu(2010, 40)$ and $\ln \mu(2010, 60)$, yield by the LC with cohort effect and the 3-2-3 neural model

The lifetime expectancy for a x years old individual on year t, is defined as follows

$$e_x(t) := \sum_{s=1}^{x_{max}} {}_s p_x(t),$$

where ${}_s p_x(t)$ is the survival probability from age x to age $x + s$, calculated with cross-sectional mortality rates:

$$
{}_s p_x(t) = \exp\left(-\int_0^s \mu(t, x + u)\, du\right)
$$

$$
\approx \exp\left(-\sum_{k=0}^{s-1} \mu(t, x + k)\right)
$$

Table 4.9 presents information about cross-sectional lifetime expectancies at 20, 40, 60, 80 years old obtained with the cohort LC model (LC COH). 10,000 simulations are performed and lifetime expectancies are computed scenario per scenario. Averages of predicted expectancies are reported in the first third of the table. These figures forecast that the maximum improvement of longevity concerns the 20 years individuals who gain 2.2 years of lifetime expectancy between 2001 and 2014. This improvement is slightly lower than the real one observed over this period (2.99 years). The LC model underestimates the improvement of longevity by 0.20 years for an 80 years old person to 0.76 years for a 20 years old individual, in 2014.

Table 4.9 Cohort Lee-Carter model (LC COH): average cross-sectional lifetime expectancies and their spread with real expectancies

	$e_{20}(t)$	$e_{40}(t)$	$e_{60}(t)$	$e_{80}(t)$
2001	60.394	41.251	23.665	8.998
2005	61.111	41.899	24.312	9.382
2010	61.943	42.650	25.071	9.857
2014	62.592	43.241	25.637	10.236
	$e_{20}^{Obs}(t) - e_{20}(t)$	$e_{40}^{Obs}(t) - e_{40}(t)$	$e_{60}^{Obs}(t) - e_{60}(t)$	$e_{80}^{Obs}(t) - e_{80}(t)$
2001	−0.0115	0.0137	0.0143	0.0076
2005	0.2509	0.2087	0.0804	0.0026
2010	0.4723	0.4355	0.1648	0.1001
2014	0.7631	0.6995	0.2337	0.1959

Table 4.10 3-2-3 neural net: average cross-sectional lifetime expectancies and their spread with real expectancies

	$e_{20}^{NN}(t)$	$e_{40}^{NN}(t)$	$e_{60}^{NN}(t)$	$e_{80}^{NN}(t)$
2001	60.392	41.247	23.635	8.9842
2005	61.125	41.913	24.19	9.3227
2010	61.971	42.685	24.839	9.7315
2014	62.588	43.25	25.316	10.043
	$e_{20}^{Obs}(t) - e_{20}^{NN}(t)$	$e_{40}^{Obs}(t) - e_{40}^{NN}(t)$	$e_{60}^{Obs}(t) - e_{60}^{NN}(t)$	$e_{80}^{Obs}(t) - e_{80}^{NN}(t)$
2001	−0.0092	0.0174	0.0446	0.0217
2005	0.2367	0.1947	0.2018	0.0628
2010	0.4444	0.4012	0.3975	0.2264
2014	0.7676	0.6909	0.5548	0.3891

Table 4.10 presents information about cross-sectional lifetime expectancies at 20, 40, 60, 80 years old computed with the 3-2-3 neural analyzer. As for the LC model, the maximum improvement of longevity concerns the 20 years old generation who gains on average 2.19 years of lifetime expectancy between 2001 and 2014. In a similar way to the LC-COH model, the neural net underestimates the real improvement of longevity observed over this period. The neural analyzer predicts realistic log-mortality rates over a short period of time, following the last year of calibration. However, does it remains reliable for long term forecasting? To answer this question, we calculate the cross-sectional lifetime expectancies of a 20, 40, 60 and 80 years old individual, from 2001 to 2100. The evolution of expectancies at 20 and 60 years old are shown in Fig. 4.7. The lifetime expectancies, computed with a 3D LC model fitted by SVD grow respectively linearly from 60 to 65 years and from 23 to 27 years. The same expectancies forecast by the LC GLM model respectively increase from 60 to 74 and from 23 to 35 years. Whereas the LC model with cohort effects predicts a rise from 60 up to 74 and 23 up to 35 years. Life expectancies computed with the neural net display a concave growth. They dominate these yield by the LC model but are below the forecasts of LC GLM and LC COH models, excepted over the period 2000–2020.

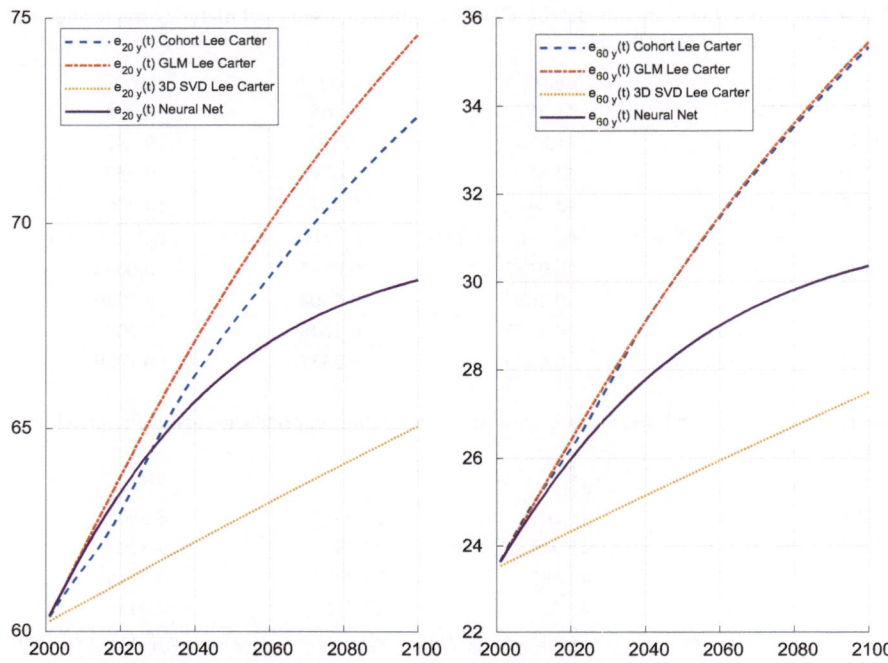

Fig. 4.7 Predicted cross-sectional lifetime expectancies with LC and neural models, over the period 2001–2100

Table 4.11 Long term predictions of cross-sectional life expectancies, computed by simulations with the cohort LC and Neural models, over the period 2001–2100

	$e_{20}^{NN}(t)$	$e_{40}^{NN}(t)$	$e_{60}^{NN}(t)$	$e_{80}^{NN}(t)$
2001	60.400	41.255	23.642	8.9887
2025	64.088	44.636	26.505	10.863
2050	66.453	46.851	28.454	12.369
2100	68.621	48.919	30.358	14.098
	$e_{20}^{LC\,COH}(t)$	$e_{40}^{LC\,COH}(t)$	$e_{60}^{LC\,COH}(t)$	$e_{80}^{LC\,COH}(t)$
2001	60.367	41.223	23.656	8.9891
2025	63.842	44.485	26.899	11.123
2050	67.542	48.068	30.33	13.232
2100	72.603	52.942	35.318	16.895

Table 4.11 compares life expectancies predicted by LC COH and Neural net models, for different ages and years. According to the neural analyzer, the average lifetime will respectively increase of 8 and 5 years for a 20–80 years old individual,

over the next century. Whereas the LC COH forecasts an increase of 12 and 8 years for persons aged 20–80 years. We cannot say which model is the most reliable for long term forecast of log-forces of mortality. However, the neural net approach predicts realistic projections.

4.5 Comparison with the UK and US Populations

In this section, we check that the neural net approach efficiently explain the US and UK mortality. Table 4.12 reports calibration errors for LC models with and without a cohort effect and for neural nets, fitted to UK log-forces of mortality, from 1946 up to 2000. As for the French population, the LC model with a cohort effect yields a low calibration error. However, neural nets achieve similar or better performances, depending upon the configuration. We also observe that increasing the number of neurons systematically reduces the calibration error. According to figures of Table 4.13, the predictive power of the LC COH over the period 2001–2013 is lower than the one of the LC model, fitted by log-likelihood maximization. For the UK population, the 3-2-3 network displays an excellent predictive capability compared to other neural configurations and to competing models. Tables 4.14 and 4.15 reports the calibration and prediction errors for models adjusted to US mortality rates. The lowest calibration errors are obtained with neural nets and three intermediate neurons. Despite that the LC COH model has an excellent explanatory power for the period 1946–2000, its predictive capability is clearly lower than these of neural nets, whatever their configuration. Contrary to French and UK cases, we don't observe any deterioration of predictive errors when we increase the number of

Table 4.12 Goodness of fit for extension of the LC model and neural nets

Model	UK population, 1946–2000			
	$\sum \|m_t - \hat{m}_t\|_2^2$	Avg.$\|m_t - \hat{m}_t\|_2$	$\max(m_t - \hat{m}_t)$	$\min(m_t - \hat{m}_t)$
LC SVD 1	131.4	0.0023	0.9390	−0.4606
LC SVD 2	114.08	0.0021	0.7845	−0.3871
LC SVD 3	111.18	0.0021	0.7509	−0.3748
LC GLM	32.85	0.0011	0.3484	−0.4444
LC COH	8.946	0.0005	0.2478	−0.1962
NN 3-2-3	15.16	0.0008	0.2104	−0.2927
NN 4-2-4	11.01	0.0007	0.2119	−0.2789
NN 5-2-5	10.09	0.0006	0.2076	−0.2655
NN 4-3-4	9.04	0.0006	0.1934	−0.2751
NN 5-3-5	7.74	0.0006	0.2153	−0.2673
NN 6-3-6	7.79	0.0006	0.2329	−0.2577

The second and third columns present the sum of squared errors and the average errors. The two last columns contain the maximum and minimum errors

Table 4.13 Predictive goodness of fit for LC models and neural networks

Model	UK population, 2001–2013			
	$\sum \|m_t - \hat{m}_t\|_2^2$	Avg. $\|m_t - \hat{m}_t\|_2$	max $(m_t - \hat{m}_t)$	min $(m_t - \hat{m}_t)$
LC SVD 1	27.43	0.0044	0.5337	−0.0974
LC SVD 2	26.90	0.0044	0.5865	−0.1022
LC SVD 3	26.68	0.0044	0.5648	−0.1035
LC GLM	11.68	0.0027	0.5532	−0.1723
LC COH	13.38	0.0026	0.3631	−0.2311
NN 3-2-3	12.83	0.0028	0.5179	−0.1410
NN 4-2-4	13.13	0.0029	0.5228	−0.1750
NN 5-2-5	14.18	0.0030	0.5409	−0.1382
NN 4-3-4	13.54	0.0029	0.5215	−0.1300
NN 5-3-5	14.38	0.0030	0.5310	−0.1043
NN 6-3-6	14.55	0.0030	0.5517	−0.1358

The second and third columns present the sum of squared errors and the average errors. The two last columns contain the maximum and minimum errors

Table 4.14 Goodness of fit for extensions of the LC model and neural nets

Model	US population, 1946–2000			
	$\sum \|m_t - \hat{m}_t\|_2^2$	Avg. $\|m_t - \hat{m}_t\|_2$	max $(m_t - \hat{m}_t)$	min $(m_t - \hat{m}_t)$
LC SVD 1	81.69	0.0018	0.4293	−0.3642
LC SVD 2	76.02	0.0017	0.3108	−0.3756
LC SVD 3	73.18	0.0017	0.3222	−0.3560
LC GLM	12.26	0.0007	0.1863	−0.2456
LC COH	6.23	0.0004	0.1403	−0.1966
NN 3-2-3	8.19	0.0006	0.1441	−0.1801
NN 4-2-4	6.55	0.0005	0.1363	−0.1675
NN 5-2-5	6.40	0.0005	0.1852	−0.1980
NN 4-3-4	5.61	0.0005	0.1452	−0.1574
NN 5-3-5	5.46	0.0005	0.1713	−0.1831
NN 6-3-6	4.75	0.0004	0.1639	−0.1796

The first and second columns report the number of latent factors and fitted coefficients. The third and fourth columns present the sum of squared errors and the average errors. The two last columns contain the maximum and minimum errors

neurons. We conclude form this analysis that the efficiency of the neural net analyzer does not depend upon the reference data set.

4.6 Conclusions and Further Readings

This chapter demonstrates that neural networks are promising for applications in life insurance. It summarizes the information carried by the surface of log-forces of mortality in a limited number of latent factors. These factors are next extrapolated

Table 4.15 Predictive goodness of fit for LC models and neural networks

Model	US population, 2000–2015			
	$\sum \|m_t - \hat{m}_t\|_2^2$	Avg.$\|m_t - \hat{m}_t\|_2$	$\max(m_t - \hat{m}_t)$	$\min(m_t - \hat{m}_t)$
LC SVD 1	15.61	0.0029	0.2998	−0.2311
LC SVD 2	19.31	0.0032	0.3082	−0.3035
LC SVD 3	21.26	0.0034	0.2945	−0.2866
LC GLM	10.73	0.0024	0.2020	−0.3567
LC COH	25.73	0.0032	0.1931	−0.6505
NN 3-2-3	11.86	0.0027	0.2152	−0.3409
NN 4-2-4	13.58	0.0029	0.2223	−0.3700
NN 5-2-5	14.75	0.0030	0.2170	−0.3615
NN 4-3-4	15.28	0.0031	0.3003	−0.3520
NN 5-3-5	16.62	0.0032	0.3223	−0.3786
NN 6-3-6	18.87	0.0034	0.2911	−0.4110

The first and second columns report the number of latent factors and fitted coefficients. The third and fourth columns present the sum of squared errors and the average errors. The two last columns contain the maximum and minimum errors

and future term structures of mortality rates are obtained by an inverse transform. Given the important number of parameters, a genetic algorithm combined to a gradient descent, is used to calibrate the network. Numerical tests performed on the French, UK and US log-forces of mortality, emphasizes that the neural analyzer outperforms LC model and its multi-factor extensions, fitted by SVD or log-likelihood maximization. The neural net approach has an explanatory power that is comparable or even better the LC model with age specific cohort effects.

As mentioned in the introduction, there exist many variants of the Lee and Carter model (1992). We refer the reader to Lee (2000), Pitacco (2004), Wong-Fupuy and Haberman (2004) or Cairns (2008) for a review of various extensions of the Lee-Carter (LC) model. The recent article of Currie (2016) provides a comprehensive survey on generalized linear and non-linear models of mortality. An alternative approach to calibrate consists to perform the joint inference of latent time processes and age parameters, in a single step by a Markov Chain Monte-Carlo (MCMC) method. Antonio et al. (2015) apply this Bayesian approach to predict the joint mortality of multiple populations. Fung et al. (2017) propose a state-space framework for mortality modelling with cohort effects. This approach is computationally intensive but remedies to the drawback of two steps procedures that are somewhat ad-hoc methods, without statistical foundations.

An alternative to the NLPCA with neural nets is provided in from Hastie and Stuetzle (1989), who named their method principal curves and surfaces (PCS). Malthouse (1998) demonstrated that NLPCA and PCS are closely related. The NLPCA based on neural networks has been applied in various field: chemical engineering (Dong and McAvoy 1996) to psychology (Fotheringhame and Baddeley 1997) or climatic mathematics (Monahan 2000).

To the best of our knowledge, only a few research articles apply neural networks to forecast mortality and in existing studies, the neural net is substituted to an econometric model or to a linear regression. For example, Atsalakis et al. (2007) propose a neural network with fuzzy logic inference. Abdulkarim and Garko (2015) fit a feed-forward neural network with a particle swarm algorithm so as to forecast the maternal mortality in a region of Nigeria. Puddu and Menotti (2009) use a multi-layer perceptron to predict the coronary heart disease mortality in seven countries. Puddu and Menotti (2012) extend this approach to predict to predict the 45-year all-cause mortality in Italian rural areas and they don't observe any difference between the performance of multi-layer perceptrons or multiple logistic regressions.

References

Abdulkarim SA, Garko AB (2015) Forecasting maternal mortality rate using particle Swarm optimization based artificial neural network. Dutse J Pure Appl Sci 1(1):55–59

Antonio K, Bardoutsos A, Ouburg W (2015) Bayesian Poisson log-bilinear models for mortality projections with multiple populations. Eur Actuar J 5:245–281

Atsalakis G, Nezis D, Matalliotakis G, Ucenic CI, Skiadas C (2007) Forecasting mortality rate using a neural network with fuzzy inference system. Working paper 0806. University of Crete

Brouhns N, Denuit M, Vermunt JK (2002) A Poisson log-bilinear regression approach to the construction of projected lifetables. Insurance Math Econom 31(3):373–393

Cairns AJC (2008) Modelling and management of mortality risk: a review. Scand Actuar J 2–3:79–113

Currie ID (2016) On fitting generalized linear and non-linear models of mortality. Scand Actuar J 4:356–383

Cybenko G (1989) Approximation by superpositions of a sigmoidal function. Math Control Signals Syst 2:303–314

Denuit M, Hainaut D, Trufin J (2019) Effective statistical learning methods for actuaries: GLMs and extensions. Springer, Berlin

Dong D, McAvoy TJ (1996) Nonlinear principal component analysis—based on principal curves and neural networks. Comp Chem Eng 20:65–78

Fotheringhame D, Baddeley R (1997) Nonlinear principal components analysis of neuronal spike train data. Biol Cybern 77:282–288

Fung MC, Peters GW, Shevchenko P (2017) A unified approach to mortality modelling using state-space framework: characterisation, identification, estimation and forecasting. Ann Actuar Sci 2(11):343–389

Hainaut D (2012) Multidimensional Lee–Carter model with switching mortality processes. Insurance Math Econom 5(2):236–246

Hainaut D (2018) A neural network analyzer for mortality forecast. ASTIN Bull 48(2):481–450

Hastie T, Stuetzle W (1989) Principal curves. J Am Stat Assoc 84(406):502–516

Hornik K (1991) Approximation capabilities of multilayer feedforward networks. Neural Netw 4(2):251–257

Kramer MA (1991) Nonlinear principal component analysis using autoassociative neural networks. AIChE J 37:233–243

Lee RD (2000) The Lee-Carter Method for forecasting mortality, with various extensions and applications. North Am Actuar J 4(1):80–91

Lee RD, Carter L (1992) Modelling and forecasting the time series of US mortality. J Am Stat Assoc 87:659–671

Malthouse EC (1998) Limitations of nonlinear PCA as performed with generic neural networks. IEEE Trans Neural Netw 9:165–173

McNelis PD (2005) Neural networks in finance: gaining predictive edge in the market. Elsevier, Amsterdam

Monahan HA (2000) Nonlinear principal component analysis by neural networks: theory and application to the Lorenz system. J Climate 13:821–835

Pitacco E (2004) Survival models in a dynamic context: a survey. Insurance Math Econom 35:279–298

Pitacco E, Denuit M, Haberman S, Olivieri A (2009) Modeling longevity dynamics for pensions and annuity business. Oxford University Press, London

Puddu PE, Menotti A (2009) Artificial neural network versus multiple logistic function to predict 25-year coronary heart disease mortality in the seven countries. Eur J Cardiovasc 16(5):583–591

Puddu PE, Menotti A (2012) Artificial neural networks versus proportional hazards Cox models to predict 45-year all-cause mortality in the Italian Rural Areas of the Seven Countries Study. BMC Med Res Methodol 12:100

Renshaw AE, Haberman S (2003) Lee-Carter mortality forecasting with age-specific enhancement. Insurance Math Econom 33:255–272

Renshaw A, Haberman S (2006) A cohort-based extension to the Lee–Carter model for mortality reduction factors. Insurance Math Econom 38:556–570

Wilmoth JR (1993) Computational methods for fitting and extrapolating the Lee-Carter model of mortality change. Technical Report; Department of Demography, University of California, Berkeley

Wong-Fupuy C, Haberman (2004) Projecting mortality trends: recent developments in the UK and the US. North Am Actuar J 8:56–83

Chapter 5
Self-organizing Maps and k-Means Clustering in Non Life Insurance

Feed-forward neural networks are algorithms with supervised learning. It means that we have to a priori identify the most relevant variables and to know the desired outputs for combinations of these variables. For example, forecasting the frequency of car accidents with a perceptron requires an a priori segmentation of some explanatory variables like the driver's age into categories, in a similar manner to Generalized Linear Models. The misspecification of these categories can induce a large bias in the forecast. On the other hand, the presence of collinearity between covariates affects the accuracy of the prediction. In this situation, the coefficient estimates of the multiple regression may change erratically in response to small changes in the model or the data. Self-organizing maps offer an elegant solution to segment explanatory variables and to detect dependence among covariates.

Self-organizing maps (SOM) are artificial neural networks that do not require any a priori information on the relevancy of variables. For this reason, they belong to the family of unsupervised algorithms. This method developed by Kohonen (1982) aims to analyze the original information by simplifying the amount of rough data, computing some basic features, and giving visual representation. SOMs can produce a low dimensional (typically two-dimensional), discretized representation of the input space of the training samples, called a map, and is therefore a method to perform reduction of dimensions.

Initially, Kohonen (1982) developed SOMs in order to analyze quantitative variables. In the first section of this chapter, we present this algorithm and apply it to an insurance data set to regroup policyholders with respect to quantitative variables. We emphasize the similarities of self-organizing maps with the k-means clustering method. In the next section, we show how SOMs and k-means algorithm may be adapted for regressing the claims frequency on quantitative variables in a Bayesian framework. Later, we introduce a measure of distance between categorical variables in order to extend SOMs to determine cluster of covariates. Finally, we modify the regressive SOM and k-means algorithm in order to include categorical covariates. This chapter is inspired from Hainaut (2019).

© Springer Nature Switzerland AG 2019
M. Denuit et al., *Effective Statistical Learning Methods for Actuaries III*,
Springer Actuarial, https://doi.org/10.1007/978-3-030-25827-6_5

5.1   The SOM for Quantitative Variables

In this section, features of insurance policies are exclusively quantitative variables. The number of insurance policies is denoted by n. Each of these policies is described by p real-valued variables. The data are arranged in a table X with n rows and p columns. The rows of table X are denoted by $x_i = X_{i,.}$ and are the inputs of the SOM algorithm.

As illustrated in Fig. 5.1, the map is represented by a two-dimensional grid, with l by l nodes or neurons, indexed by $u = 1, \ldots, l^2$. This grid is described by a $l^2 \times 2$ matrix C. Here ,C contains the coordinates of nodes in $\mathbb{R}^2$. In this chapter, the domain on which is defined the grid is the unit square $[0, 1] \times [0, 1]$. Neurons are equally spaced on this pavement. In order to define the neighbourhood of a neuron in this grid, we introduce a topological distance between neurons. Here, we use the Euclidian distance between lines u and v of the matrix C: $\left\| C_{u,.} - C_{v,.} \right\|_2$.

We associate to each neuron or node a codebook denoted by $\boldsymbol{\omega}_u = \{\omega_1^u, \ldots, \omega_p^u\}$ that is a $\mathbb{R}^p$ vector of weights for $u = 1, \ldots, l^2$. We pursue a double objective. First, we wish to associate a neuron of the grid to each insurance policy and partition the portfolio in l^2 clusters. Second, we want to determine codebook vectors $\boldsymbol{\omega}_u$ for $u = 1, \ldots, l^2$ containing the average profile of policies assigned to node u. Notice that the definition and role of a neuron in this section differs from previous ones developed in Chap. 1. Here a neuron is a mathematical cell that is activated if the data submitted to the neural map are close enough to its codebook. The vector $\boldsymbol{\omega}_u$ may be seen as the center of gravity in $\mathbb{R}^d$ of policies associated to the uth neuron. The definition of this barycenter requires the definition of a distance between the uth node and the ith insurance policy, whose respective positions in $\mathbb{R}^p$ are identified

$l \times l$ Grid of neurons

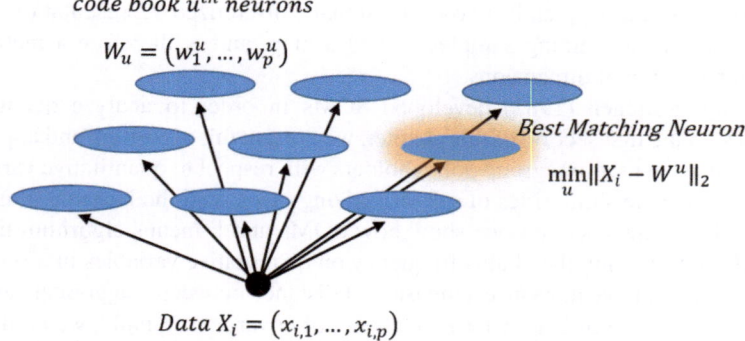

Fig. 5.1 Illustration of the grid of neurons. The data is sent to all neurons and the best matching neuron is the only one to be activated. The best matching neuron has the closest codebook to data among all neurons

by ω_u and x_i. In the case of quantitative variables, we use the Euclidian distance:

$$d(x_i, \omega_u) = \|\omega_u - x_i\|_2 \qquad u = 1, .., l^2 \ i = 1, \ldots, n$$

$$= \sqrt{\sum_{k=1}^{p} \left(\omega_k^u - x_{i,k} \right)^2}.$$

Remark that the neural map admits a double representation. One is on a pavement $[0, 1] \times [0, 1]$ and positions of neurons are fixed. The other one is in $\mathbb{R}^p$ where the positions of nodes are determined by the p-vectors ω_u for $u = 1, \ldots, l^2$. Kohonen (1982) proposed a procedure to construct this map that is recalled in Algorithm 5.1.

Algorithm 5.1 Kohonen's algorithm for quantitative variables

Initialization:
 Randomly attribute codebooks $\omega_1(0), \ldots, \omega_{l^2}(0)$.
Main procedure:
 For $e = 0$ to maximum epoch, e_{max}
  For $i = 1$ to n
   1) Find the node u matching at best the features of the ith policy

$$u = BMN(i) = \arg \min_u d(x_i, \omega_u(e)). \tag{5.1}$$

This node is the BMN (best matching node).
2) Update codebooks in the neighbourhood of the BMN
 by pulling them closer in $\mathbb{R}^p$ to the input vector:

$$\omega_v(e+1) = \omega_v(e) + \theta(u, v, e) \epsilon(e) \ (x_i - \omega_v(e)) \tag{5.2}$$

for $v = 1, \ldots, l^2$ and where

$$\epsilon(e) = \epsilon_0 \times \left(\frac{e_{max} - e}{e_{max}} \right), \tag{5.3}$$

$$\theta(u, v, e) = \theta_0 \exp \left(-\frac{(\|C_{u,.} - C_{v,.}\|_2)^2}{2\sigma(e)^2} \right), \tag{5.4}$$

$$\sigma(e) = \sigma_0 \times \left(\frac{1.2 \, e_{max} - e}{e_{max}} \right). \tag{5.5}$$

 End loop on policies, i
 3) Calculation of the total distance d^{total} between policies and BMNs:

$$d^{total} = \sum_{i=1}^{n} \|\omega_{BMN(i)} - x_i\|_2.$$

 End loop on epochs, e

The algorithm scans the portfolio and finds for each policy the neuron with the closest codebook to its features. This neuron is called the best matching node (BMN). After this step, weights of neurons in the neighbourhood of this BMN are updated in the direction of the policy features. The size of the update is proportional to the epoch of the algorithm and to the distance between the BMN and updated neurons. Notice that functions $\epsilon(e)$ and $\sigma(e)$ in Eqs. (5.3) and (5.5) may be any other decreasing functions of the epoch, e. The total distance d^{total} is the error of classification if we use the feature $\omega_{BMN(i)}$ for the ith policy, instead of the real one x_i. This distance is monitored to check the convergence of the algorithm. When it does not vary anymore, the learning of the neural net is finished.

The speed of convergence of the Kohonen's algorithm depends on initial weights $\omega_u(0)$. They should be chosen in order to reflect as much as possible the largest set of features of policies.[1] The convergence also depends on parameters, ϵ_0, θ_0 and σ_0 of Eqs. (5.3)–(5.5). If they are too high, weights of neurons oscillate during the first iterations. If ϵ_0, θ_0 and σ_0 are too small, modifications of the codebook are not enough significant. In both cases, the number of epochs must be increased to achieve convergence.

To illustrate this section, we apply the Kohonen algorithm to data from the Swedish insurance company *Wasa*, presented in Sect. 1.11. We build a map of the portfolio that regroups policyholders according to the owner's age and vehicle age rescaled on the interval $[0, 1]$ (ages are divided by their maximum values). The number of epochs is $e_{max} = 100$ and the grid of neurons counts 9 elements ($l = 9$). The initial codebooks are chosen in order to regularly cover the $[0, 1] \times [0, 1]$ pavement. The parameters of Eqs. (5.3)–(5.5) for the update of codebooks are: $\epsilon_0 = 0.01$, $\theta_0 = 1$ and $\sigma_0 = 0.10$. These values have been chosen by trials and errors in order to ensure a quick convergence.

Figure 5.2 shows the Kohonen's map in the space of variables, in which policies are identified by a dot. Each colored area represents a group of policies centered around a neuron (black dots).

The right graph of Fig. 5.3 reports the evolution of the distance $\frac{d^{total}}{n}$ between policies and BMN codebooks. This distance is nearly stable and the convergence is achieved after 60 iterations. The left graph of this figure shows the number of policies associated to each neurons. Two neurons are coupled to less than 3% of policies. The others are associated to clusters of policies representative of 7–17% of the total portfolio.

---

[1]In our approach, codebooks are randomly chosen. An alternative consists to use the initialization procedure of the k-means algorithm, presented in Sect. 5.2.

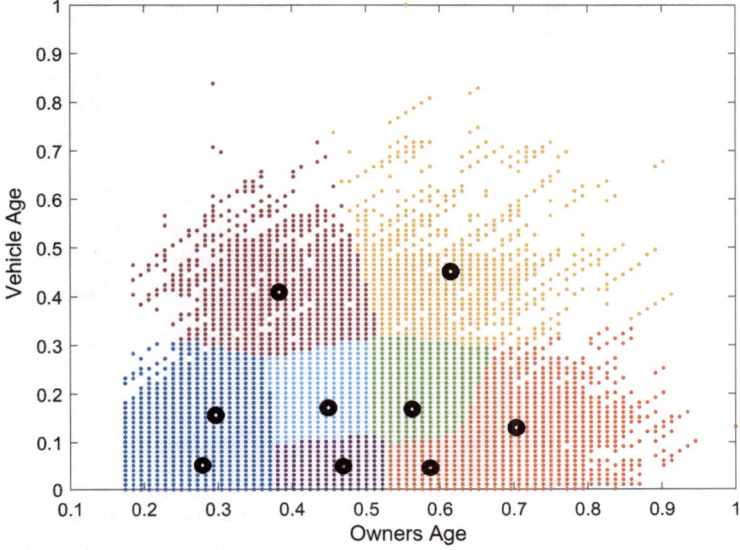

Fig. 5.2 Kohonen's map of the portfolio, with 9 neurons. The segmenting variables are the owner's and vehicle ages. Each policy is represented by a dot. Black dots point out the position of neurons. The colors identify the area of influence of neurons

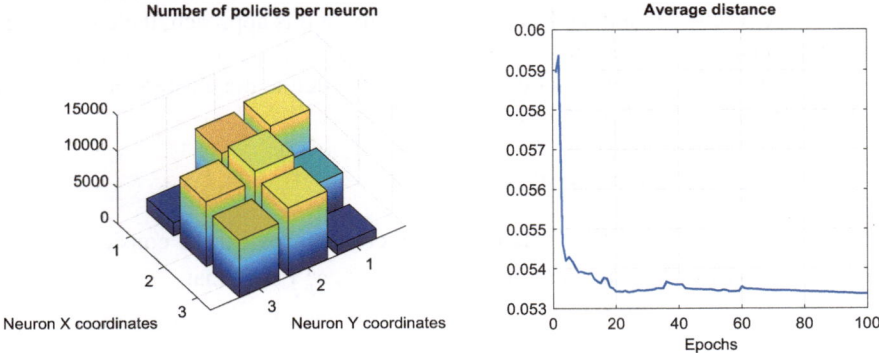

Fig. 5.3 Left panel: number of policies assigned to each neuron. Right panel: evolution of the average distance $\frac{d^{total}}{n}$, over iterations

Table 5.1 reports the codebooks of neurons. Here, $\dot{\omega}_1^u \times max(Age)$ and $\omega_1^u \times max(Veh.\ Age)$ are respectively the average owner's and vehicle ages of the uth subset of policies. An analysis of the average frequency of claims per cluster reveals that this frequency tends to decrease with the owner's age. However, the pooling of policies is based only on ages of policyholders and vehicles, not on claims frequency. In the next section, we modify the Kohonen's algorithm in order to delimit subsets of policies with homogeneous claims frequencies.

Table 5.1 Codebooks of neurons, the average frequency of claims for clusters of policies associated to each neuron, and cluster relative sizes

Neuron u	$w_1^u \times max(Age)$	$w_2^u \times max(Veh.\,Age)$	$\bar{\lambda}_u$	% of the portfolio
1	25.77	4.96	0.0413	14.96
2	27.32	15.34	0.0154	14.39
3	35.27	40.41	0.0052	2.86
4	41.36	16.85	0.0038	15.21
5	43.22	4.77	0.0096	13.59
6	51.81	16.58	0.0034	16.43
7	54.06	4.43	0.0086	12.63
8	56.66	44.60	0	2.14
9	64.78	12.77	0.0058	7.79

5.2 Comparison of SOM and k-Means Clustering

Self-organizing maps produce similar results to the method of k-means clustering. The codebook of a neuron in a SOM contains the coordinates of the center of gravity of a cluster of policies. In the method of k-means, the codebook of a neuron is called a centroid but must be interpreted in the same way. The main difference lies in the procedure for estimating these codebooks. The purpose of SOM and k-means methods is to partition a cloud of n points in $\mathbb{R}^p$ into $k = l \times l$ clusters. In this section, we briefly remind the k-means algorithm. In this approach, the coordinates of the uth centroid is contained in a vector $c_u = (c_1^u, \ldots, c_p^u)$ for $u = 1, \ldots, k$. For a given distance $d(.,.)$ and a set of k centroids, we define the clusters or classes of data S_u for $u = 1, \ldots, k$ as follows:

$$S_u = \{x_i \,:\, d(x_i, c_u) \le d(x_i, c_j) \,\forall j \in \{1, \ldots, k\}\} \quad u = 1, \ldots, k. \qquad (5.6)$$

Here, $d(.,.)$ is the Euclidian distance but other distances can be considered. The center of gravity of S_u is a p vector $g_u = \left(g_1^u, \ldots, g_p^u\right)$ such that

$$g_u = \frac{1}{|S_u|} \sum_{x_i \in S_u} x_i.$$

The center of gravity of the full dataset is denoted by $g = \frac{1}{n} \sum_{i=1}^{n} x_i$. We define the global inertia by

$$I_X = \frac{1}{n} \sum_{i=1}^{n} d(x_i, g)^2,$$

and the inertia I_u of a cluster S_u by

$$I_u = \sum_{x_i \in S_u} \frac{1}{|S_u|} d\left(x_i, g_u\right)^2 \quad u = 1, \dots, k.$$

The interclass inertia I_c is the inertia of the cloud of centers of gravity:

$$I_c = \sum_{u=1}^{k} \frac{|S_u|}{n} d\left(g_u, g\right)^2 ,$$

whereas the intraclass inertia I_a is the sum of clusters inertiae, weighted by their size:

$$I_a = \sum_{u=1}^{k} \frac{|S_u|}{n} I_u$$

$$= \frac{1}{n} \sum_{u=1}^{k} \sum_{x_i \in S_u} d\left(x_i, g_u\right)^2 .$$

According to the König-Huyghens theorem, the total inertia is the sum of the intraclass and interclass inertiae: $I_X = I_c + I_a$. An usual criterion of classification consists to seek for a partition of X minimizing the intraclass inertia I_a in order to have homogeneous clusters on average. This is equivalent to determine the partition maximizing the interclass inertia, I_c.

This problem is computationally difficult (NP-hard). However, there exist efficient heuristic procedures converging quickly to a local optimum. The most common method uses an iterative refinement technique called the k-means or Lloyd's method (1957) which is detailed in Algorithm 5.2. Given an initial set of k random centroids $c_1(0), \dots, c_k(0)$, we construct a partition $\{S_1(0), \dots, S_k(0)\}$ of the dataset according to the rule in Eq. (5.6). This partition is a set of convex polyhedrons delimited by median hyperplans of centroids as illustrated in Fig. 5.4. Next, we replace the k random centroids by the k centers of gravity $(c_u(1))_{u=1:k} = (g_u(0))_{u=1:k}$ of these classes and we iterate till convergence. At each iteration, we can prove that the intraclass inertia is reduced.

The Lloyd's algorithm proceeds by alternating between two steps. In the assignment step of the eth iteration, we associate each observation x_i to a cluster $S_u(e)$ whose centroid $c_u(e)$ has the least distance, $d(x_i, c_u(e))$. This is intuitively the nearest centroid to each observation. In the update step, we calculate the new means $g_u(e)$ to be the centroids $c_u(e+1)$ of observations in new clusters.[2] The

---

[2] A variant of this algorithm consists to recompute immediately the new position of centroids after assignment of each records of the dataset.

Algorithm 5.2 Lloyd's algorithm for k-means clustering

Initialization:

Randomly set up initial positions of centroids $c_1(0),\dots,c_k(0)$.

Main procedure:

For $e = 0$ to maximum epoch, e_{max}

Assignment step:

For $i = 1$ to n

1) Assign x_i to a cluster $S_u(e)$ where $u \in \{1,\dots,k\}$

$$S_u(e) = \{x_i \ : \ d(x_i, c_u(e)) \le d(x_i, c_j(e)) \ \forall j \in \{1,\dots,k\}\}.$$

End loop on policies, i.

Update step:

For $u = 1$ to k

2) Calculate the new centroids $c_u(e+1)$ of $S_u(e)$ as follows

$$c_u(e+1) = \frac{1}{|S_u(e)|} \sum_{x_i \in S_u(e)} x_i \,.$$

End loop on centroids, u.

3) Calculation of the total distance d^{total} between observations and closest centroids:

$$d^{total} = \sum_{u=1}^{k} \sum_{x_i \in S_u(e)} d(x_i, c_u(e+1)) \,.$$

End loop on epochs e

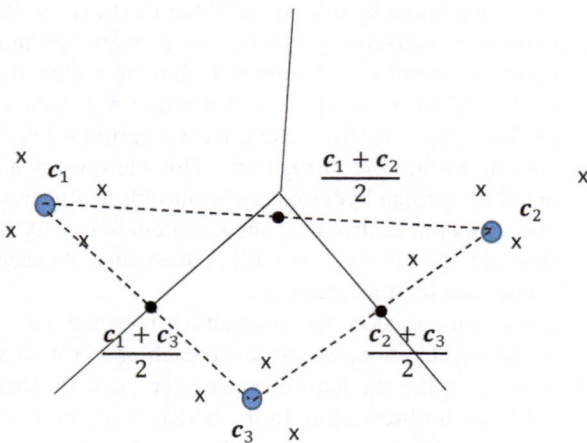

Fig. 5.4 Illustration of the partition of a dataset with the k-means algorithm

---

Algorithm 5.3 Initialization of centroids for the k-means algorithm

---

Initialization:
Select an observation uniformly at random from the data set, X. The chosen observation is the first centroid, and is denoted $c_1(0)$.
Main procedure:
    For $j = 2$ to k
        For $i = 1$ to n
            1) Calculate the distance $d(x_i, c_{j-1}(0))$ from x_i to $c_{j-1}(0)$.
        End loop on policies, i

        2) Select the next centroid, $c_j(0)$ at random from X with probability

$$\frac{d^2(x_i, c_{j-1}(0))}{\sum_{i=1}^{n} d^2(x_i, c_{j-1}(0))} \quad i = 1, \ldots, n.$$

    End loop on k

Table 5.2 Positions of centroids, the average frequency of claims for clusters of policies associated to each centroids, and cluster relative sizes

Neuron u	$c_1^u \times max(Age)$	$c_2^u \times max(Veh. Age)$	$\bar{\lambda}_u$	% of the portfolio
1	25.53	5.29	0.0404	15.70
2	28.18	15.65	0.0140	14.71
3	35.31	40.19	0.0052	2.91
4	43.01	16.65	0.0040	17.87
5	43.67	4.55	0.0092	14.47
6	53.49	16.47	0.0033	14.38
7	54.75	4.42	0.0091	11.51
8	56.62	44.63	0	2.14
9	65.97	12.44	0.0055	6.31

algorithm converges when the assignments no longer change. There is no guarantee that a global optimum is found using this algorithm. The k-means++ algorithm of Arthur and Vassilvitskii (2007) uses an heuristic to find centroid seeds for k-means clustering. The procedure to initialize the k-means heuristic is detailed in Algorithm 5.3. It improves the running time of the Lloyd's algorithm, and the quality of the final solution. It can also be used for initializing codebooks of SOM algorithms presented earlier in this chapter.

Table 5.2 reports the results of the Lloyd's algorithm applied to the dataset X of rescaled owner's and vehicle ages. A comparison with figures of Table 5.1 reveals the similarity of neural codebooks and centroids positions. However, in terms of total distance, the SOM slightly outperforms the k-means algorithm. After 100 iterations, we obtain a total distance of $d^{total} = 3331.5$ for the SOM whereas the total distance for the k-means algorithm is equal to $d^{total} = 3338.5$.

5.3   A Bayesian Regressive SOM with Quantitative Variables

We wish to construct a map of the portfolio in which policies are bundled into groups identified by a neuron. We denote by $\Omega_{u=1,\dots,l^2}$ the cluster of insurance contracts assigned to the uth neuron. We assume that the number of claims caused by a policy in Ω_u is distributed according to a Poisson law of parameter λ_u. If the total exposure is $v^{\Omega_u} = \sum_{i \in \Omega_u} v_i$, the distribution of the total number of claims in Ω_u, noted N^{Ω_u} is given by

$$P\left(N^{\Omega_u} = m\right) = \frac{\left(\lambda_u v^{\Omega_u}\right)^m}{m!} e^{-\lambda_u v^{\Omega_u}} .$$

The maximum log-likelihood estimator of λ_u is then equal to:

$$\bar{\lambda}_u = \frac{N^{\Omega_u}}{v^{\Omega_u}} = \frac{\sum_{i \in \Omega_u} N_i}{\sum_{i \in \Omega_u} v_i} .$$

The homogeneity of claims frequencies inside a subset Ω_u is measured by the standard deviation of observed claim counts N_i within Ω_u, that is,

$$d_{\Omega_u}(\bar{\lambda}_u) = \sqrt{\sum_{i \in \Omega_u} \left(N_i - \bar{\lambda}_u v_i\right)^2} ,$$

and the distance between the realized frequency of the ith policy and average frequency for the subgroup Ω_u is given by

$$d_{\Omega_u}(i, \bar{\lambda}_u) = \sqrt{\left(N_i - \bar{\lambda}_u v_i\right)^2} .$$

A first idea could be using this standard deviation to define a new distance between the ith policy and the uth, as follows:

$$d(x_i, \omega_u, \bar{\lambda}_u) := \|\omega_u - x_i\|_2 + \beta \, d_{\Omega_u}(i, \bar{\lambda}_u) \qquad u = 1, \dots, l^2 \; i = 1, \dots, n$$

$$= \sqrt{\sum_{k=1}^{p} \left(\omega_k^u - x_{i,k}\right)^2} + \beta \sqrt{\left(N_i - \bar{\lambda}_u v_i\right)^2} \tag{5.7}$$

where β is a weight adjusting the regressive feature of the map with respect to its segmentation function. This distance could be used in Eq. (5.1) of the first step of Kohonen's algorithm. Whereas $\bar{\lambda}_u$ would be updated at the end of each iteration, in the third step of Algorithm 5.1. However, this approach is not satisfactory when applied to the insurance data. Given that the database counts only 693 claims for 62,436 policies, the algorithm tends to discriminate the portfolio into two subsets: one regrouping policies with no claim and the other one gathering policies with one or more claims. To remedy to this issue, Hainaut (2019) proposes a Bayesian approach.

We consider a Gamma random variable $\Theta \sim \Gamma(\gamma, \gamma)$ and assume that conditionally to Θ, the r.v. $N^{\Omega_1}, \ldots, N^{\Omega_N}$ are independent with conditional distributions:

$$N_{u|\Theta} \sim Poi\left(\lambda_u \Theta v^{\Omega_u}\right) \quad u = 1, \ldots, l^2 \tag{5.8}$$

where $\lambda_u > 0$ and γ are the prior frequency of claims and a dispersion parameter. The choice of λ_u and γ is discussed at the end of this section. Under the Bayesian assumption, the coefficient of variation of the expected frequency is then $\gamma \frac{1}{\gamma^2} = \frac{1}{\gamma}$. We have the following standard result:

Proposition 5.6 *The posterior expected frequency on Ω_u given observations N^{Ω_u} is given by*

$$\mathbb{E}\left(\lambda_u \Theta | N^{\Omega_u}\right) = \alpha_u \bar{\lambda}_u + (1 - \alpha_u) \lambda_u, \tag{5.9}$$

where $\bar{\lambda}_u = \frac{N^{\Omega_u}}{v^{\Omega_u}} = \frac{\sum_{i \in \Omega_u} N_i}{\sum_{i \in \Omega_u} v_i}$ and with credibility weights

$$\alpha_u = \frac{v^{\Omega_u} \lambda_u}{\gamma + v^{\Omega_u} \lambda_u} \in (0, 1). \tag{5.10}$$

for $u = 1, \ldots, l^2$.

Proof Using the Bayes rule, the probability density function of $\Theta | N^{\Omega_u}$ is rewritten as

$$f_\Theta(\vartheta \mid N_{\Omega_u} = n) = \frac{f_\Theta(\vartheta) P(N_{\Omega_u} = n | \Theta = \vartheta)}{P(N_{\Omega_u} = n)}.$$

$$= \frac{\frac{\gamma^\gamma}{\Gamma(\gamma)} \vartheta^{\gamma-1} e^{-\gamma\vartheta} e^{-\lambda_u \vartheta v^{\Omega_u}} \frac{\left(\lambda_u \vartheta v^{\Omega_u}\right)^n}{n!}}{\int \frac{\gamma^\gamma}{\Gamma(\gamma)} \vartheta^{\gamma-1} e^{-\gamma\vartheta} e^{-\lambda_u \vartheta v^{\Omega_u}} \frac{\left(\lambda_u \vartheta v^{\Omega_u}\right)^n}{n!} d\vartheta}$$

$$\propto \vartheta^{\gamma+n} e^{-(\gamma+\lambda_u v^{\Omega_u})\vartheta}$$

where $\propto$ is the proportional operator. Thus, $f_\Theta(\vartheta \mid N_{\Omega_u})$ is the density of a Gamma distribution with the updated parameters:

$$\gamma' = \gamma + N_{\Omega_u},$$
$$\beta' = \gamma + \lambda_u v^{\Omega_u}.$$

As $\bar{\lambda}_u = \frac{N^{\Omega_u}}{v^{\Omega_u}} = \frac{\sum_{i \in \Omega_u} N_i}{\sum_{i \in \Omega_u} v_i}$ then

$$\mathbb{E}\left(\lambda_u \Theta | N^{\Omega_u}\right) = \lambda_u \mathbb{E}\left(\Theta | N^{\Omega_u}\right)$$

$$= \lambda_u \frac{\gamma + N^{\Omega_u}}{\gamma + \lambda_u v^{\Omega_u}}$$

$$= \lambda_u \frac{\gamma}{\gamma + \lambda_u v^{\Omega_u}} + \frac{\lambda_u v^{\Omega_u}}{\gamma + \lambda_u v^{\Omega_u}} \frac{N^{\Omega_u}}{v^{\Omega_u}}$$

$$= (1 - \alpha_u)\lambda_u + \alpha_u \bar{\lambda}_u \,.$$

□

For a given prior frequency λ_u and dispersion parameter γ, the posterior mean $\mathbb{E}\left(\lambda_u \Theta | N^{\Omega_u}\right)$, is an estimator of the expected frequency on Ω_u. Contrary to the empirical intensity $\bar{\lambda}_u$, this posterior is always strictly positive. However in practice, the prior λ_u is unknown. For this reason, we opt for an empirical version of the credibility estimator (5.8):

$$\mathbb{E}\left(\lambda_u \Theta | N^{\Omega_u}\right) = \alpha_u \bar{\lambda}_u + (1 - \alpha_u)\bar{\lambda} \,, \tag{5.11}$$

where $\bar{\lambda}_u = \frac{N^{\Omega_u}}{v^{\Omega_u}}$ and $\bar{\lambda} = \frac{\sum_{u=1}^{l^2} N^{\Omega_u}}{\sum_{u=1}^{l^2} v^{\Omega_u}}$ is the global average claims frequency. The corresponding credibility weights are equal to

$$\alpha_u = \frac{v^{\Omega_u} \bar{\lambda}}{\gamma + v^{\Omega_u} \bar{\lambda}} \,.$$

A similar assumption is done in the R package "rpart" for Poisson regression trees. In the Bayesian framework, the homogeneity of claims frequency inside a subset Ω_u is measured by $d_{\Omega_u}(\bar{\lambda}_u)$, calculated with the credibility estimator:

$$d_{\Omega_u}(\bar{\lambda}_u) = \sqrt{\sum_{i \in \Omega_u} \left(N_i - \mathbb{E}\left(\lambda_u \Theta | N^{\Omega_u}\right) v_i\right)^2} \,,$$

and the distance between the realized frequency of the ith policy and the posterior expected frequency for the subgroup Ω_u is given by

$$d_{\Omega_u}(i, \bar{\lambda}_u) = \sqrt{\left(N_i - \mathbb{E}\left(\lambda_u \Theta | N^{\Omega_u}\right) v_i\right)^2} \,,$$

whereas the distance between the ith policy and the uth neuron in the Kohonen's algorithm becomes:

$$d(\boldsymbol{x}_i, \boldsymbol{\omega}_u, \bar{\lambda}_u) := \|\boldsymbol{\omega}_u - \boldsymbol{x}_i\|_2 + \beta\, d_{\Omega_u}(i, \bar{\lambda}_u) \qquad u = 1, \ldots, l^2\, i = 1, \ldots, n$$

$$= \sqrt{\sum_{k=1}^{p} \left(\omega_k^u - x_{i,k}\right)^2 + \beta \sqrt{\left(N_i - \mathbb{E}\left(\lambda_u \Theta \,|\, N^{\Omega_u}\right) v_i\right)^2}} \qquad (5.12)$$

where β is a weight adjusting the regressive feature of the map with respect to its segmentation capacity. Algorithm 5.4 summarizes the steps of the Bayesian regressive SOM. We run the Algorithm 5.4 with 9 neurons. The weight β in the definition of the distance (5.12) is set to 10. The credibility factor γ is fixed to 4. The convergence is achieved in less than 100 iterations.

Table 5.3 provides detailed information about codebooks and estimated claims frequencies. Two neurons associated to younger insured cover 30% of the portfolio. Neurons 3 and 4 gather less 1.3% of policies and the frequency of claims associated to these risks is particularly high (71% and 12%) due to the lack of contracts in these clusters. A way to smooth frequencies of claims consists to increase the credibility parameters γ and/or to decrease the weight β. A comparison of Figs. 5.5 and 5.3 allows us to visualize the impact of the Bayesian metric on the segmentation. For example, the owners of a motorcycle older than 30 years are assigned to the 6th neuron in Table 5.3. Whereas a segmentation only based on owner's and vehicle ages divides this group into two clusters (neurons 3 and 8 in Table 5.1).

The total distance d^{total} measures the quality of the regression and of the portfolio segmentation, based on the distance in Eq. (5.12). If we aim at evaluating the goodness of fit, we may be tempted to calculate the deviance. However, using a Bayesian estimator perturbs the interpretation of this measure. The deviance is the difference between the log-likelihood of a model with as many parameters than observations and the fitted model. The model counting as many parameters than observations is called the *saturated model*. The log-likelihood of this model, denoted by $l^{saturated}$, is computed with parameters set to empirical frequency. The saturated model has no predictive power but provides the best fit from a pure statistical point of view given that it has the highest attainable log-likelihood. If we momentarily denote by l the log-likelihood for our data,[3] the deviance D^* is defined as likelihood ratio test (LRT) of the model under consideration against the saturated model:

$$D^* = 2\left(l^{saturated} - l\right).$$

---

[3]This should not be confused with the number of neurons on edges of the map.

Algorithm 5.4 Bayesian regression Kohonen's algorithm for quantitative variables

---

Initialization:

    Randomly attribute codebooks $\omega_1(0), \ldots, \omega_{l^2}(0)$.

    Set $\bar{\lambda}_u(0) = \bar{\lambda}$.

Main procedure:

    For $e = 0$ to maximum epoch, e_{max}

        For $i = 1$ to n

            1) Find the BMN (best matching node) u matching the ith policy

$$u = \arg\min_u d(x_i, \omega_u, \bar{\lambda}_u).$$

            2) Update $\Omega_u(e)$ and codebooks in the BMN neighborhood:

$$\omega_v(e+1) = \omega_v(e) + \theta(u, v, e)\,\epsilon(e)\,(x_i - \omega_v(e))$$

            for $v = 1, \ldots, l^2$, with

$$\epsilon(e) = \epsilon_0 \times \left(\frac{e_{max} - e}{e_{max}}\right),$$

$$\theta(u, v, e) = \theta_0 \exp\left(-\frac{(\|C_u - C_v\|_2)^2}{2\sigma(e)^2}\right),$$

$$\sigma(e) = \sigma_0 \times \left(\frac{1.2\, e_{max} - e}{e_{max}}\right).$$

        End loop on policies, i

        3) Update of $\bar{\lambda}_u(e+1) = \frac{\sum_{i \in \Omega^u(e)} N_i}{\sum_{i \in \Omega^u(e)} v_i}$ and $\alpha_u(e+1) = \frac{v^{\Omega_u(e)}\bar{\lambda}}{\gamma + v^{\Omega_u(e)}\bar{\lambda}}$.

        4) Calculation of the total distance d^{total} between policies and BMNs:

$$d^{total} = \sum_{i=1}^{n} \left\|\omega_{BMN(i)}(e+1) - x_i\right\|_2 + \beta\, d_{\Omega_{BMN(i)}}(i, \bar{\lambda}_{BMN(i)}(e+1))$$

    End loop on epochs, e

If $\hat{\lambda}_i$ is the estimate of the claims frequency for the ith policy forecast by the SOM, the probability of observing m claims over a period v_i is equal to

$$P(N_i = m) = \frac{\left(\hat{\lambda}_i v^i\right)^m}{m!} e^{-\hat{\lambda}_i v^i}.$$

The contribution of the ith policy to the log-likelihood l is then equal to

$$N_i \log\left(\hat{\lambda}_i v^i\right) - \hat{\lambda}_i v^i - N_i!.$$

Table 5.3 Codebooks of neurons, credibility estimators of claims frequencies and relative sizes of clusters

Neuron u	$\omega_1^u \times max(Age)$	$\omega_2^u \times max(Veh.\,Age)$	$\mathbb{E}\left(\lambda_u\Theta|N^{\Omega_u}\right)$	% of the portfolio
1	26.64	4.41	0.0019	13.89
2	26.96	15.27	0.0005	15.97
3	34.51	8.02	0.7178	0.91
4	40.50	6.53	0.1204	0.26
5	41.70	16.63	0.0003	18.13
6	45.58	44.35	0.0016	4.15
7	47.59	4.43	0.0002	20.56
8	53.27	16.68	0.003	16.65
9	60.60	9.23	0.0007	9.48

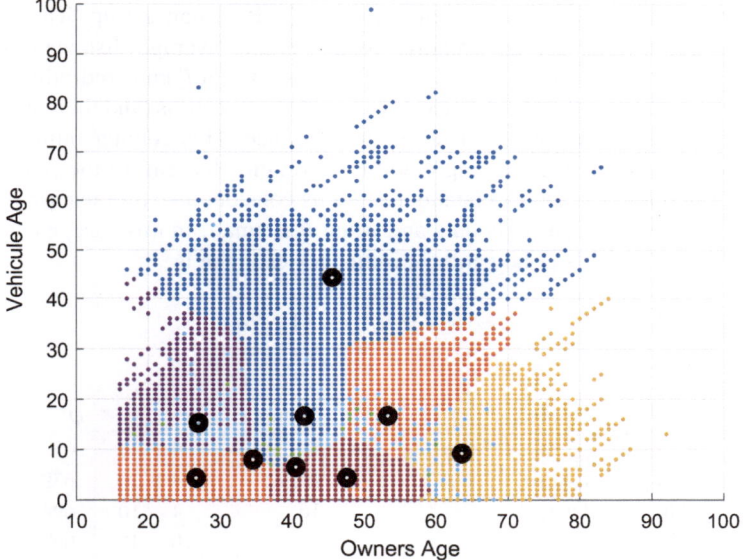

Fig. 5.5 Bayesian regressive Kohonen's map of the portfolio, with 9 neurons. The segmentation variables are the owner's age and age of the vehicle. Each policy is represented by a dot. Black dots point out the position of neurons. The colors identify the area of influence of neurons

In the saturated model, this contribution is given by

$$\begin{cases} N_i \log\left(\frac{N_i}{v^i}v^i\right) - \frac{N_i}{v^i}v^i - N_i! & if\ N_i \geq 1 \\ 0 & else\ if \end{cases}.$$

The marginal deviance for the ith policy is then equal to

$$
\begin{cases}
2N_i \left(\frac{v_i}{N_i} \hat{\lambda}_i - \log \frac{\hat{\lambda}_i v_i}{N_i} - 1 \right) & if \ N_i \geq 1 \\
2v_i \hat{\lambda}_i & if \ N_i = 0
\end{cases} .
$$

The total deviance is the sum of marginal contributions

$$
D^* = 2 \sum_{i=1}^{n} N_i \left(\frac{v_i}{N_i} \hat{\lambda}_i - \left(\log \frac{\hat{\lambda}_i v_i}{N_i} + 1 \right) I_{\{N_i \geq 1\}} \right) .
$$

The deviance is used in statistics to compare the predictive capacity of models fitted by log-likelihood maximization. This measure is however not fully adapted to compare regressive self-organizing maps in a Bayesian set-up. This point is emphasized by Table 5.4 that reports deviances and average distances between policies and best matching neurons, for different values of β and credibility weights γ. As revealed by Table 5.4, increasing the number of neurons (and then of clusters) reduces the average distance but raises the deviance. This counter-intuitive result is the direct consequence of using credibility weights that ensure the positivity of estimated frequencies. To understand this, let us consider an extreme case in which we have as many neurons than data and when β is small. In this case, each neuron is associated to a single policy and the frequency estimate is

$$
\mathbb{E} \left(\lambda_i \Theta | N_i \right) = \alpha_i \bar{\lambda}_i + (1 - \alpha_i) \bar{\lambda} ,
$$

Table 5.4 Deviances and average distances between policies and best matching neurons, for different values of β and credibility weights γ

Neuron l^2	γ	β	D^*	d^{total}
4	4	10	659	10,344
9	4	10	679	7829
16	4	10	694	7078
4	10	10	1097	11,051
9	10	10	1252	8988
16	10	10	1167	8567
4	10	1	2160	6368
9	10	1	2347	4169
16	10	1	2519	3363
9	0	0	6087	3331
GLM			6214	
Null model			6648	

We also provide the deviance of a non-regressive SOM, of a GLM model and the null model where $\bar{\lambda}_i = \frac{N_i}{v_i}$ and $\bar{\lambda} = \frac{\sum_{i=1}^{N} N_i}{\sum_{i=1}^{N} v_i}$. Given that the average intensity is low, $\bar{\lambda} =$ 0.0106, the credibility weight $\alpha_i = \left(1 + \frac{\gamma}{v_i \bar{\lambda}}\right)^{-1}$ tends to zero and $\mathbb{E}\left(\lambda_i \Theta | N_i\right) \approx \bar{\lambda}$. This model has the same log-likelihood as the one obtained with a Poisson model with a single parameter. This model, called the *null model*, has the lowest log-likelihood and the highest deviance. Then, refining the segmentation does not necessary improve the deviance if we use a credibility estimator for frequencies. This point was underlined by Wütrich and Buser (2017) who observe the same phenomenon for regression trees. On page 83, they mention that replacing the maximum likelihood estimator (MLE) by the empirical credibility estimator has the advantage that the resulting frequency estimators are always strictly positive. The disadvantage is that errors measured by the deviance may increase by adding more splits (which turns out to refine the segmentation). This is because only the MLE minimizes the corresponding deviance statistics and maximizes the log-likelihood function, respectively.

Estimated claims frequencies obtained with a credibility weight $\gamma = 4$ and reported in Table 5.3, vary a lot. Increasing the credibility parameters γ and/or to decreasing the weight β allows to smooth these frequencies. However, figures of Table 5.4 show that this smoothing deteriorates the deviance. To get an idea to which extend large deviances are admissible, we fit a generalized linear model (GLM) to our dataset. GLM is a standard approach widely used by insurers to assess their risk. We regress the claims frequency with respect to six categories of owner's age: (28;31], (31;35], (35;44], (44;51], (51;61] and (61;100]. We fit a Poisson model with a logarithmic link function. The residual deviance for this model climbs to 6214, which is quite high compared to deviances obtained with a regressive SOM. This confirms that regressive SOMs have comparable performances to GLM. Except that it is a non-parametric and non supervised approach.

In theory, it is possible to perform cross validation. In practice, we face several difficulties when we implement this method. Firstly, the number of claims being small (693 claims for 62,436 contracts), some bootstrapped data samples do not contain any claims. Secondly, the cross validation is computationally time consuming.

5.4 Comparison of Bayesian SOM and k-Means Clustering Method

This section compares the Bayesian regressive SOM to a Bayesian version of the k-means algorithm. We use the same notations as in Sect. 5.2. We partition the dataset X of n records in $\mathbb{R}^p$ into k clusters. The coordinates of the uth centroid are contained in a vector $c_u = (c_1^u, \ldots, c_p^u)$ and S_u is the cluster of policies associated to c_u for $u = 1, \ldots, k$. As for the Bayesian SOM, we replace the Euclidian distance

$d(x_i, c_u)$ in the Lloyd's algorithm by

$$d(x_i, c_u, \bar{\lambda}_u) := \|c_u - x_i\|_2 + \beta \sqrt{\left(N_i - \mathbb{E}\left(\lambda_u \Theta | N^{S_u}\right) v_i\right)^2},$$

where $\beta \in \mathbb{R}^+$. Here, N^{S_u} and v^{S_u} are respectively the number of claims and the total exposure for the uth cluster. Also, $\mathbb{E}\left(\lambda_u \Theta | N^{S_u}\right)$ is the estimate of the claims frequency for S^u where Θ is distributed as a $\Gamma(\gamma, \gamma)$ random variable. As in Sect. 5.3 $\bar{\lambda}_u = \frac{N^{S_u}}{\sum_{x_i \in S_u} v_i}$ and the observed empirical estimator of $\mathbb{E}\left(\lambda_u \Theta | N^{S_u}\right)$ is

$$\mathbb{E}\left(\lambda_u \Theta | N^{S_u}\right) = \alpha_u \bar{\lambda}_u + (1 - \alpha_u)\bar{\lambda},$$

where $\alpha_u = \frac{v^{S_u} \bar{\lambda}}{\gamma + v^{S_u} \bar{\lambda}}$. Algorithm 5.5 details a variant of the k-means algorithm including the same Bayesian regressive feature than the SOM.

Table 5.5 reports the deviances and distances obtained after 100 iterations of the adapted Lloyd's algorithm for Bayesian regression. We use the same configurations

---

Algorithm 5.5 Adapted Lloyd's algorithm for Bayesian regression

Initialization:
        Initialize positions of centroids $c_1(0), \ldots, c_k(0)$ with Algorithm 5.3.
        Set $\bar{\lambda}_u(0) = \bar{\lambda}$ for $u = 1, \ldots, k$.

Main procedure:
        For $e = 0$ to maximum epoch, e_{max}
                Assignment step:
                For $i = 1$ to n
                        1) Assign x_i to a cluster $S_u(e)$ where $u \in \{1, \ldots, k\}$

$$S_u(e) = \{x_i : d(x_i, c_u(e), \bar{\lambda}_u(e)) \le d(x_i, c_j(e), \bar{\lambda}_j(e)) \ \forall j \in \{1, \ldots, k\}\}$$

                End loop on policies, i.
                Update step:
                For $u = 1$ to k
                        2) Calculate the new centroids $c_u(e + 1)$ of $S_u(e)$ as follows

$$c_u(e + 1) = \frac{1}{|S_u(e)|} \sum_{x_i \in S_u(e)} x_i$$

                        3) Update of $\bar{\lambda}_u(e + 1) = \frac{\sum_{x_i \in S_u(e)} N_i}{\sum_{x_i \in S_u(e)} v_i}$ and $\alpha_u = \frac{v^{S_u(e)} \bar{\lambda}}{\gamma + v^{S_u(e)} \bar{\lambda}}$.
                End loop on centroids, u
                4) Calculation of the total distance d^{total} between policies and centroids:

$$d^{total} = \sum_{u=1}^{k} \sum_{x_i \in S_u(e)} d(x_i, c_u(e + 1), \bar{\lambda}_u(e + 1))$$

End loop on epochs e.

Table 5.5 Deviances and average distances between policies and centroids, for different values of β and credibility weights γ

K-means k	γ	β	D^*	d^{total}
4	4	10	660	10,333
9	4	10	659	7979
16	4	10	708	7038
4	10	10	1097	11,050
9	10	10	1121	9010
16	10	10	1179	8500
4	10	1	2131	6430
9	10	1	2366	4190
16	10	1	2500	3364
9	0	0	6095	3338

Table 5.6 Positions of centroids, the Bayesian estimator of frequency of claims for clusters of policies associated to centroids, and sub-population relative sizes

Centroid u	$c_u^1 \times max(Age)$	$c_u^2 \times max(Veh.\,Age)$	$\mathbb{E}\left(\lambda_u \Theta | N^{S_u}\right)$	% of the portfolio
1	25.94	5.70	0.0007	17.35
2	29.50	16.68	0.0005	15.63
3	34.48	8.01	0.7166	0.91
4	40.14	7.18	0.1213	0.30
5	45.94	44.38	0.0016	4.13
6	46.07	16.90	0.0002	24.33
7	47.45	4.60	0.0002	21.08
8	58.76	2.06	0.0164	1.03
9	60.98	12.62	0.0004	15.24

as for the SOMs. In terms of distance, the SOM and K-mean algorithms achieve similar performance. Table 5.6 presents the coordinates of centroids and the Bayesian claims frequency per cluster. These results are computed with $\beta = 10$, $\gamma = 4$ and 9 clusters. A comparison of centroid positions with neural codebooks in Table 5.3 confirms that both algorithms gather insurance policies in a very similar way.

5.5 Analysis of Qualitative Variables with a SOM

Many features of insurance policies are described by categorical variables that cannot be handled with a classic SOM. In this section, we modify the Kohonen algorithm in order to analyze a dataset that exclusively contains this type of variables. We first propose a procedure to study the dependencies between variables. In Sect. 5.5, this method is combined with the Bayesian SOM to regress claims frequency on quantitative and categorical variables.

First, we introduce the structure of data to which the algorithm is applied. The number of insurance policies is still denoted by n. Each of these policies is

Table 5.7 Example of a
disjunctive table for $K = 2$
variables with respectively
$m_1 = 2, m_2 = 3$ modalities

	Gender		Area		
Policy	M	F	U	S	C
1	1	0	1	0	0
2	0	1	0	0	1
$\vdots$	$\vdots$	$\vdots$	$\vdots$	$\vdots$	$\vdots$

described by K variables which have m_k binary modalities for $k = 1, \ldots, K$.
By binary, we mean that the modality j of the kth variable is identified by an
indicator variable equal to zero or one. The total number of modalities is the sum
of m_k: $m = \sum_{k=1}^{K} m_k$. In further developments, we enumerate modalities from 1
to m. The information about the portfolio may be summarized by a $n \times m$ matrix
$\boldsymbol{D} = (d_{i,j})_{i=1...n, j=1...m}$. If the ith policy presents the jth modality then $d_{i,j} = 1$
and $d_{i,j} = 0$ otherwise.

For example, let us assume that a policy is described by the gender (M=male or
F=Female) of the policyholder and by a geographic area (U=urban, S=suburban or
C=countryside). The number of variables and modalities are respectively $K = 2$,
$m_1 = 2$ and $m_2 = 3$. If the first and second policyholders are respectively a man
living in a city and a woman living in the countryside, the two first lines of the
matrix $\boldsymbol{D}$ are presented in Table 5.7.

The table $\boldsymbol{D}$ is called a disjunctive table. In order to study the dependence
between the modalities, we need to calculate the numbers $n_{i,j}$ of individuals sharing
modalities i and j, for $i, j = 1, \ldots, m$. The $m \times m$ matrix $\boldsymbol{B} = (n_{i,j})_{i,j=1,\ldots,m}$ is
a contingency table, called the Burt matrix containing this information. The Burt
matrix is directly related to the disjunctive table as follows:

$$\boldsymbol{B} = \boldsymbol{D}^\top \boldsymbol{D}.$$

This symmetric matrix is composed of $K \times K$ blocks $\boldsymbol{B}_{k,l}$ for $k, l = 1, \ldots, K$.
A block $\boldsymbol{B}_{k,l}$ is the contingency table that crosses the variables k and l. Table 5.8
shows the Burt matrix for the matrix $\boldsymbol{D}$ presented in Table 5.7. By construction, the
sum of elements of a block $\boldsymbol{B}_{k,l}$ is equal to the total number of policies, n. The sum
of $n_{i,j}$ of the same row i is equal to

$$n_{i,.} = \sum_{j=1,\ldots,m} n_{i,j} = K\, n_{i,i}.$$

The Burt matrix being symmetric, we directly infer that

$$n_{.,j} = \sum_{i=1,\ldots,m} n_{i,j} = K\, n_{j,j}.$$

Furthermore, blocks $\boldsymbol{B}_{k,k}$ for $k = 1, \ldots, K$ are diagonal matrix, whose diagonal entries are the numbers of policies who respectively present the modalities

Table 5.8 Burt matrix for the disjunctive Table 5.7

		Gender		Area		
		M	F	U	S	C
Gender	M	$n_{1,1}$	0	$n_{1,3}$	$n_{1,4}$	$n_{1,5}$
	F	0	$n_{2,2}$	$n_{2,3}$	$n_{2,4}$	$n_{2,5}$
Area	U	$n_{3,1}$	$n_{3,2}$	$n_{3,3}$	0	0
	S	$n_{4,1}$	$n_{4,2}$	0	$n_{4,4}$	0
	C	$n_{5,1}$	$n_{5,2}$	0	0	$n_{5,5}$

$1, \ldots, m_k$, for the kth variable. In our example, we have that $n_{1,1} + n_{2,2} = n$ and $n_{3,3} + n_{4,4} + n_{5,5} = n$. Here, $n_{1,1}$ and $n_{2,2}$ count the total number of men and women in the portfolio. Whereas $n_{3,3}$, $n_{4,4}$ and $n_{5,5}$ counts the number of policyholders living respectively in a urban, sub-urban or rural environment.

The self-organizing map requires the definition of a distance between categorical variables. This point is discussed in the next section.

5.6 A χ^2 Distance for Categorical Variables

Multiple correspondence Analysis (MCA) was initially developed by Burt (1950) and enhanced by Benzécri (1973), Greenacre (1984) and Lebart et al. (1984). This technique evaluates the level of dependence between categorical variables with a χ^2 distance. We first present this distance in the case of two variables and extend it later to the multivariate case.

5.6.1 The Bivariate Case

Let us first consider the case of two categorical variables, $K = 2$, with m_1 and m_2 modalities. The classical MCA studies relative frequencies of crossed modalities. The table of frequency $F = (f_{i,j})_{i,j}$ is a $m_1 \times m_2$ matrix defined as follows:

$$f_{i,j} = \frac{n_{i,j}}{n} \quad i = 1, \ldots, m_1 , \; j = 1, \ldots, m_2 .$$

The marginal frequencies are equal to

$$f_{i,.} = \sum_{j=1}^{m_2} f_{i,j} = \frac{n_{i,.}}{n} \quad i = 1, \ldots, m_1 ,$$

$$f_{.,j} = \sum_{i}^{m_1} f_{i,j} = \frac{n_{.,j}}{n} \quad j = 1, \ldots, m_2 .$$

If variables are independent, the expected number of policies with modalities i and j, is equal to $\tilde{n}_{i,j} = \frac{n_{i,.} n_{.,j}}{n}$. In this case, standardized residuals $\chi_{i,j} = \frac{n_{i,j} - \tilde{n}_{i,j}}{\sqrt{\tilde{n}_{i,j}}}$ should be approximatively $N(0, 1)$ random variables. If the two variables are independent, then the following statistics (called the inertia):

$$\chi^2 = \sum_{i=1}^{m_1} \sum_{j=1}^{m_2} \frac{\left(n_{i,j} - \tilde{n}_{i,j}\right)^2}{\tilde{n}_{i,j}}$$

$$= n \sum_{i=1}^{m_1} \sum_{j=1}^{m_2} \frac{\left(f_{i,j} - f_{i,.} f_{.,j}\right)^2}{f_{i,.} f_{.,j}}$$

$$= n \left(\sum_{i=1}^{m_1} \sum_{j=1}^{m_2} \frac{f_{i,j}^2}{f_{i,.} f_{.,j}} - 1 \right),$$

is a chi-square random variable with $(m_1 - 1)(m_2 - 1)$ degrees of freedom. This justifies to measure the distance between the modalities i and i' of the first variable by

$$\chi^2\left(i, i'\right) = \sum_{j=1}^{m_2} \frac{1}{f_{.,j}} \left(\frac{f_{i,j}}{f_{i,.}} - \frac{f_{i',j}}{f_{i',.}} \right)^2 \tag{5.13}$$

$$= \sum_{j=1}^{m_2} \frac{n}{n_{.,j}} \left(\frac{n_{i,j}}{n_{i,.}} - \frac{n_{i',j}}{n_{i',.}} \right)^2.$$

Thus, $\chi^2(i, i')$ is the distance between the rows i and i' of the matrix of frequency F. It may be checked that the dispersion of rows around their barycenter is equal to the inertia. Similarly, the chi-square distance between the modalities j and j' of the second variable is defined by

$$\chi^2\left(j, j'\right) = \sum_{i=1}^{m_1} \frac{1}{f_{i,.}} \left(\frac{f_{i,j}}{f_{.,j}} - \frac{f_{i,j'}}{f_{.,j'}} \right)^2 \tag{5.14}$$

$$= \sum_{i=1}^{m_1} \frac{n}{n_{i,.}} \left(\frac{n_{i,j}}{n_{.,j}} - \frac{n_{i,j'}}{n_{.,j'}} \right)^2.$$

This measures the distance between columns of the matrix F. In the SOM algorithm, we prefer to evaluate distances between rows and columns with the Euclidian distance. It is then convenient to replace the frequencies $f_{i,j}$ by weighted values $f_{i,j}^W$:

$$f_{i,j}^W := \frac{f_{i,j}}{\sqrt{f_{i,.} f_{.,j}}} = \frac{n_{i,j}}{\sqrt{n_{i,.} n_{.,j}}} \quad i = 1, \ldots, m_1 \; j = 1, \ldots, m_2. \tag{5.15}$$

The distances between rows (i, i') and columns (j, j') simplify as follows:

$$\chi^2\left(i, i'\right) = \sum_{j=1}^{m_2} \left(f_{i,j}^W - f_{i',j}^W\right)^2 .$$

$$\chi^2\left(j, j'\right) = \sum_{i=1}^{m_1} \left(f_{i,j}^W - f_{i,j'}^W\right)^2 .$$

Finally, notice that the matrix $\boldsymbol{F}^W = \left(f_{i,j}^W\right)_{i=1,\ldots,m_1, j=1,\ldots,m_2}$ is symmetric by construction.

5.6.2   The Multivariate Case

In this section, we extend the notion of χ^2 distance between categorical variables to the multivariate case $(K > 2)$. Remember that the Burt table, $\boldsymbol{B} = \left(n_{i,j}\right)_{i,j=1,\ldots,m}$, is a contingency table. This symmetric matrix is composed of $K \times K$ blocks $\boldsymbol{B}_{k,l}$. Where $\boldsymbol{B}_{k,l}$ is itself the contingency table crossing variables k and l. It is then natural to extend the definition of distance (5.13) between rows i and i' of the Burt matrix as follows:

$$\chi^2\left(i, i'\right) = \sum_{j=1}^m \frac{n}{n_{.,j}} \left(\frac{n_{i,j}}{n_{i,.}} - \frac{n_{i',j}}{n_{i',.}}\right)^2 \quad i, i' \in \{1, \ldots, m\}.$$

Similarly, the chi-square distance between columns j and j' of the Burt matrix is defined by

$$\chi^2\left(j, j'\right) = \sum_{i=1}^m \frac{n}{n_{i,.}} \left(\frac{n_{i,j}}{n_{.,j}} - \frac{n_{i,j'}}{n_{.,j'}}\right)^2 \quad j, j' \in \{1, \ldots, m\}.$$

As we prefer to evaluate distances with the Euclidian distance, the elements of the Burt matrix $n_{i,j}$ are replaced by weighted values $n_{i,j}^W$:

$$n_{i,j}^W := \frac{n_{i,j}}{\sqrt{n_{i,.}\, n_{.,j}}} \quad i, j = 1, \ldots, m . \tag{5.16}$$

Given that $n_{i,.} = K\, n_{i,i}$ and $n_{.,j} = K\, n_{j,j}$, we have that

$$n_{i,j}^W := \frac{n_{i,j}}{K \sqrt{n_{i,i}\, n_{j,j}}} \quad i, j = 1, \ldots, m . \tag{5.17}$$

If C is the diagonal matrix $C = diag\left(n_{11}^{-\frac{1}{2}}..n_{mm}^{-\frac{1}{2}}\right)$ then the weighted Burt matrix is denoted by B^W:

$$B^W = \frac{1}{K} C \, B \, C.$$

The distances between rows (i, i') and columns (j, j') of the Burt matrix become:

$$\chi^2\left(i, i'\right) = \sum_{j=1}^{m_2} \left(n_{i,j}^W - n_{i',j}^W\right)^2,$$

$$\chi^2\left(j, j'\right) = \sum_{i=1}^{m_1} \left(n_{i,j}^W - n_{i,j'}^W\right)^2.$$

5.6.3   Application to Insurance Data

Each modality of categorical variables is represented by a line of the weighted Burt matrix. A line of this matrix defines a point in a space with m dimensions. The level of dependence between modalities i and j is measured by the Euclidian distance between two points with coordinates contained in the ith and jth lines of B^W. Therefore, we apply the Kohonen's algorithm 5.1 directly to the weighted Burt matrix in order to study the relations between categorical variables. We consider the categorical variables described in Table 1.7 to which we add a new categorical variable representative of the owner's age. We consider three modalities for the owner's age, that are constructed according to the rule reported in Table 5.9.

We fit a 4×4 SOM to the matrix B^W. The number of epochs is set to $e_{max} = 1000$. The initial codebooks are equal to randomly drawn lines of B^W. The parameters of Eqs. (5.3)–(5.5) for the update of codebooks are: $\epsilon_0 = 0.01$, $\theta_0 = 1$ and $\sigma_0 = 0.10$. The network is trained after 850 iterations and 6 neurons are not assigned to any modality. One neuron is coupled to 4 modalities and two neurons each regroup 3 modalities. Details about the pooling of modalities are provided in Table 5.10. This reveals that the recurrent insured profile is a mature man living in the countryside and owning a vehicle of class 3. We also learn that a majority of young insureds lives in small cities whereas insured women are living in a urban

Table 5.9 Modalities for the categorical variable representative of the owner's age	Discretized age categories	
	Modality	Value
	Young	Owners age < 35 years old
	Mature	35 years old $\leq$ Owners age < 55 years old
	Old	55 years old $\leq$ Owners age

Table 5.10 Groups of modalities per neuron

Neuron u	Modalities Marginal	frequency	Neuron u	Modalities	Marginal frequency
1	Mature	0.006	9	Class 5	0.011
	Men	0.011	10	Suburban	0.016
	Countryside	0.006	12	North. village	0.006
	Class 3	0.008	13	Old	0.005
2	Young	0.027		Class 2	0.014
	Small city	0.010	15	Class 6	0.020
4	Women	0.009	16	North. city	0.006
	Urban	0.029		Gotland	0.004
	Class 4	0.008		Class 7	0.018
7	Class 1	0.009			

We also report the average claims frequencies for each of these modalities

Table 5.11 Total distances d^{total} between policies-centroids and policies-codebooks of best matching neurons

k-Means		SOM	
Number of centroids	k-Means d^{total}	Number of neurons	SOM d^{total}
2	17.25	4	15.80
4	16.63	9	11.30
6	16.01	16	10.16
8	14.89		
10	15.57		
12	15.47		
14	16.04		
16	16.73		

environment and drive a motorcycle of class 4. The older policyholders drive a vehicle of class 2 whereas people living in northern cities or Gotland mainly own a powerful motorcycle (class 7). This analysis reveals that SOMs detect the most common associations of modalities. With this information, we can draw a composite image of the average policy or define model points for further analysis.

In Sects. 5.2 and 5.3, we emphasize that SOM and k-means algorithm achieve similar performances at least when the number of features per contract is small (in our case, the owner and vehicle ages) and when the dataset is large (62,436 policies). Here, the pooling of categorical variables is done by analyzing a small dataset with a comparatively high number of modalities per data. In our case study, the Burt Matrix B^W counts only 19 observations and the same number of modalities by construction. Table 5.11 reveals that in this particular context, the k-means algorithm converges to a local minimum in term of total distance. The SOM finds a better solution whatever the configuration. Contrary to the SOM, the k-means algorithm even becomes unstable when the number of centroids increases.

5.7   A Regressive SOM with Quantitative and Categorical Variables

In this section, we exploit information contained in categorical and quantitative variables for regressing claims frequencies. The n policies are described by p quantitative and K categorical variables. The categorical variables have m_k binary modalities for $k = 1, \ldots, K$. The total number of modalities is denoted $m = \sum_{k=1}^{K} m_k$. The information about the portfolio is summarized by the $m \times m$ weighted Burt matrix $\boldsymbol{B}^W$. Whereas the feature of each policies are reported in the $n \times m$ disjunctive table $\boldsymbol{D}$. The quantitative variables stored in a table $\boldsymbol{X}$ with n rows and l columns. The neural map is a $l \times l$ array of neurons with a matrix $\boldsymbol{C}$ that contains the coordinates of nodes in $[0, 1] \times [0, 1]$. Neurons are equally spaced on this pavement. The codebook of the uth neurons is now composed of a p-vector

$$\boldsymbol{q}_u = \left(q_1^u, \ldots, q_p^u \right)$$

for quantitative variables and of a m-vector

$$\boldsymbol{\omega}_u = \left(\omega_1^u, \ldots, \omega_m^u \right)$$

for qualitative variables. The distance between the quantitative variables of the ith policy and the neuron is measured by the L_2 norm : $\left\| \boldsymbol{q}_u - \boldsymbol{x}_i \right\|_2$. The modalities of categorical variables are represented by points in a space with m dimensions and their coordinates are contained in weighted Burt matrix. It is then natural to measure the distance between the modalities of the ith policy and the uth neuron by the following norm:

$$\left\| \boldsymbol{\omega}_u - \boldsymbol{D}_{i,.} \boldsymbol{B}^W / K \right\|_2 , \tag{5.18}$$

where $\boldsymbol{D}_{i,.}$ is the ith line of the disjunctive table. The quantity (5.18) is the distance between the neuron represented by a point of coordinates $\boldsymbol{\omega}_u$ in $\mathbb{R}^m$ and the barycenter $\boldsymbol{D}_{i,.} \boldsymbol{B}^W / K$ of the cloud of points corresponding to modalities characterizing the ith policy. As in previous sections, we denote by $\Omega_{u=1,\ldots,l^2}$ the set of insurance contracts assigned to the uth neuron. The credibility estimator of the expected number of claims caused by a policy in Ω_u is equal to:

$$\mathbb{E} \left(\lambda_u \Theta | N^{\Omega_u} \right) = \alpha_u \bar{\lambda}_u + (1 - \alpha_u) \bar{\lambda} ,$$

where $\bar{\lambda}_u = \frac{N^{\Omega_u}}{v^{\Omega_u}}$ and $\bar{\lambda} = \frac{\sum_{u=1}^{l^2} N^{\Omega_u}}{\sum_{u=1}^{l^2} v^{\Omega_u}}$ is the global average claims frequency. The credibility weights are increasing with the total exposure in Ω_u:

$$\alpha_u = \frac{v^{\Omega_u} \bar{\lambda}}{\gamma + v^{\Omega_u} \bar{\lambda}} \qquad u = 1, \ldots, l^2 .$$

The homogeneity of claims frequency inside a subset Ω_u is measured by the standard deviation of observed claim counts N_i within Ω_u:

$$d_{\Omega_u}(\bar{\lambda}_u) = \sqrt{\sum_{i \in \Omega_u} \left(N_i - \mathbb{E}\left(\lambda_u \Theta | N^{\Omega_u}\right) v_i\right)^2}$$

and the distance between the number of claims reported by the ith policy and expected posterior distribution for the subgroup Ω_u is given by

$$d_{\Omega_u}(i, \bar{\lambda}_u) = \sqrt{\left(N_i - \mathbb{E}\left(\lambda_u \Theta | N^{\Omega_u}\right) v_i\right)^2}.$$

We use this standard deviation to define a new metric to measure the distance between the ith policy and the uth neuron:

$$d(x_i, D_{i,.}, q_u, \omega_u, \bar{\lambda}_u) := \left\|q_u - x_i\right\|_2 + \beta_1 \left\|\omega_u - D_{i,.} B^W / K\right\|_2 \tag{5.19}$$

$$+ \beta_2 \, d_{\Omega_u}(i, \bar{\lambda}_u) \qquad u = 1, \ldots, l^2 \; i = 1, \ldots, n$$

where β_1 and β_2 are weights adjusting the regressive feature of the map with respect to its segmentation function. Algorithm 5.6 presents the procedure to build the Bayesian regressive SOM with this distance. Notice that the initialization step can be done with Algorithm 5.3 if the convergence is too slow.

5.8 Application to Insurance Data

We fit a 5×5 regression SOM to our insurance data set. We consider the three categorical variables described in Table 1.7 and the two quantitative variables: owner's age and vehicle age. Quantitative variables are scaled on the interval $[0, 1]$ as done in Eq. (1.1) of Chap. 1. We perform 3000 iterations for 9 up to 25 neurons. Parameters for the update of codebooks are: $\epsilon_0 = 0.01$, $\theta_0 = 1$ and $\sigma_0 = 0.10$. The weights β_1 and β_2 involved in the definition of the distance (5.19) are set to one. In order to smooth the claims frequencies for different type of policies, we choose a credibility parameter $\gamma = 10$. The left plot of Fig. 5.6 presents the distribution of policies in the neural grid. As it can be seen from Table 5.13, the neuron regrouping the highest number of policies (8.2% of the dataset), regroups mainly mature men, living in the countryside and driving a class 3 vehicle. In Sect. 5.3, this set of features was identified as dominant in the portfolio. The right graph of Fig. 5.6 shows the evolution of the average distance between the features of a policy and the codebooks of the best matching neuron ($\frac{1}{n} d^{total}$). Convergence is achieved after 2000 iterations.

As mentioned in Sect. 5.3, the deviance is not necessary adapted to compare regressive self-organizing maps in a Bayesian set-up. This point is confirmed in Table 5.12: segmenting the portfolio does not improve the deviance due to the use

Algorithm 5.6 Bayesian Regression Kohonen's algorithm for quantitative and categorical variables

Initialization:

Randomly attribute codebooks $q_u(0)$ and $\omega_u(0)$, $u = 1, \ldots, l^2$.

Set $\bar{\lambda}_u = \bar{\lambda} = \frac{\sum_{u=1}^{l^2} N^{\Omega_u}}{\sum_{u=1}^{l^2} v^{\Omega_u}}$.

Main procedure:

For $e = 0$ to maximum epoch, e_{max}

   For $i = 1$ to n

      1) Find the BMN u matching the ith policy

$$u = \arg\min_u d(x_i, D_{i,.}, q_u, \omega_u, \bar{\lambda}_u).$$

      2) Update codebooks in the neighborhood of the BMN

$$q_v(e+1) = q_v(e) + \theta(u, v, e)\,\epsilon(e)\,\left(x_i - q_v(e)\right)$$

$$\omega_v(e+1) = \omega_v(e) + \theta(u, v, e)\,\epsilon(e)\,\left(D_{i,.}B^W/K - \omega_v(e)\right)$$

      where

$$\epsilon(e) = \epsilon_0 \times \left(\frac{e_{max} - e}{e_{max}}\right),$$

$$\theta(u, v, e) = \theta_0 \exp\left(-\frac{(\|C_u - C_v\|_2)^2}{2\sigma(e)^2}\right),$$

$$\sigma(e) = \sigma_0 \times \left(\frac{1.2\,e_{max} - e}{e_{max}}\right).$$

   End loop on policies, i.

   3) Calculation of the distance d^{total} between policies and BMNs:

$$d^{total} = \sum_{i=1}^n \left\| q_{BMN(i)} - x_i \right\|_2 + \beta_1 \left\| \omega_{BMN(i)} - D_{i,.}B^W/K \right\|_2$$

$$+ \beta_2\, d_{\Omega_{BMN(i)}}(i, \bar{\lambda}_{BMN(i)})$$

   4) Update of $\bar{\lambda}_u = \frac{\sum_{i \in \Omega^u} N_i}{\sum_{i \in \Omega^u} v_i}$ for $u = 1, \ldots, l^2$.

End loop on epochs, e.

of Bayesian estimator for claims frequencies. We compare the performance of the SOM to GLM by fitting a Poisson model, with a logarithmic link function. We use as covariates: the gender, the area, the class of the vehicle, six categories of owner's age ((28;31], (31;35], (35;44], (44;51], (51;61] and (61;100]) and two categories of vehicle age ((0;10] , (10;100]). The deviances for the GLM and null models are respectively equal to 5775 and 6648. In term of deviances, the SOM provides a worse fit than a GLM. However this conclusion must be nuanced given that SOMs

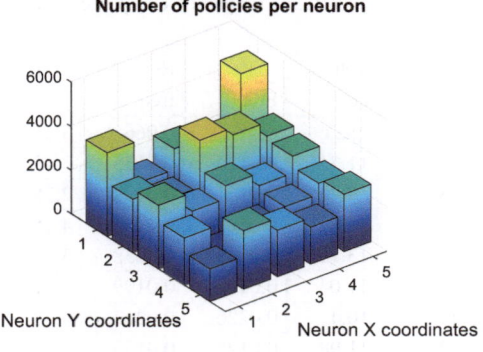
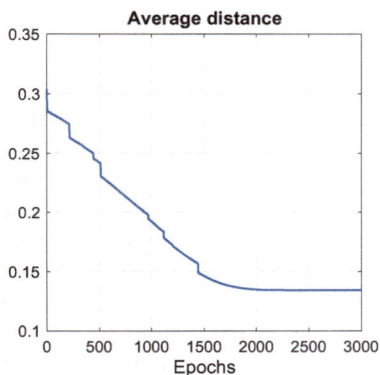

Fig. 5.6 Left panel: number of policies assigned to each neuron. Right panel: evolution of the average distance $\frac{1}{n} d^{total}$ over iterations

Table 5.12 Deviances and total distances computed with the SOM for 9, 16 and 25 neurons

Number of neurons	Deviance	d^{total}
9	5865	29,028
16	6161	25,237
25	6256	18,024

are totally unsupervised and non parametric methods. The deviance of SOM may also be adjusted by modifying the credibility parameter γ. Furthermore, as shown in Sect. 5.3, a model with a small deviance does not necessary provides a satisfactory segmentation of the portfolio.

The most interesting information for an insurer is reported in Table 5.13 that presents the dominant modalities associated to each neuron. The jth modality of the kth qualitative variable is dominant for the uth neuron if it minimizes the distance between the codebook ω_u and the lines of $\boldsymbol{B}^W$ corresponding to the m_k categories of the kth variable:

$$j = \arg \min_{v \in \{1,..,m_k\}} \left\| \omega_u - \boldsymbol{B}^W_{\sum_{d=1}^{k-1} m_d + v, \cdot} \right\|_2 . \tag{5.20}$$

For quantitative variables, the average owner's and vehicle ages of contracts assigned to the uth neuron are respectively equal to $q_1^u \times max(Age)$ and $q_2^u \times max(Veh.\,Age)$. Among the dominant profiles, we retrieve categories (Men, Countryside, Class 3), (Men, Countryside, Class 4) and (Men, Urban, Class 4). Policies assigned to these three neurons represent respectively 8.2%, 6.39% and 6.12% of the population. Only five neurons are associated to women. The reason is that the population of female drivers is under-represented in the portfolio (9843 women on 62,436 contracts).

Table 5.13 Dominant features for each neurons, estimated claims frequency and percentages of the insured population assigned to each neuron

u	Gender	Area	Type	Age	Veh. age	Freq.	Real freq	% of ptf
1	Men	Ctryside	Class3	23.87	8.26	0.0421	0.066	3.37
2	Women	Suburban	Class3	27.32	11.31	0.0123	0.0135	3.43
3	Men	Smlcity	Class6	38.94	9.76	0.0161	0.0194	2.38
4	Men	Ctryside	Class6	40.51	10.33	0.0106	0.0106	4.25
5	Women	Ctryside	Class5	40.65	13.11	0.0063	0.0038	3.27
6	Men	Suburban	Class5	41.56	11.04	0.0167	0.0199	2.97
7	Men	Urban	Class4	41.62	10.4	0.0226	0.0267	6.12
8	Men	Smlcity	Class5	42.42	11.98	0.0125	0.0133	3.34
9	Men	Suburban	Class4	42.58	11.27	0.0124	0.0134	3.05
10	Men	Urban	Class3	42.62	10.23	0.0211	0.0254	3.89
11	Men	Ctryside	Class5	43.01	12.03	0.0059	0.0052	5.88
12	Women	Ctryside	Class3	43.22	9.78	0.0076	0.0065	4.62
13	Men	Northcity	Class3	43.7	10.95	0.0075	0.0053	2.93
14	Men	Ctryside	Class4	43.87	12.71	0.0057	0.0049	6.39
15	Men	Suburban	Class6	44.04	10.22	0.0216	0.027	3.79
16	Men	Suburban	Class3	44.73	10.81	0.0091	0.0086	4.58
17	Men	Smlcity	Class4	44.88	11.24	0.0093	0.0089	4.59
18	Men	Ctryside	Class2	44.89	14.67	0.0101	0.0099	3.19
19	Men	Smlcity	Class3	45.04	11.07	0.0068	0.0057	4.86
20	Men	Ctryside	Class1	45.41	23.9	0.007	0.0054	4.17
21	Men	Ctryside	Class3	45.47	12.49	0.0038	0.0031	8.2
22	Women	Ctryside	Class6	45.67	10.12	0.0159	0.0237	1.42
23	Men	Smlcity	Class1	46.55	38.12	0.0058	0.0023	2.68
24	Women	Smlcity	Class5	47.41	11.29	0.0081	0.0058	2.37
25	Men	Northvlge	Class3	47.69	12.53	0.0059	0.0038	4.26

5.9   Comparison with the k-Means Algorithm

We use the same notations as in Sect. 5.4. We partition the dataset X of n records with p quantitative and K categorical variables, into k clusters. The information about categories is summarized by the $m \times m$ weighted Burt matrix $\boldsymbol{B}^W$ and features of policies are reported in the $n \times m$ disjunctive table $\boldsymbol{D}$. The quantitative variables stored in a table X with n rows and l columns. As in the SOM algorithm, the coordinates of the uth centroid of quantitative variables are contained in a p-vector $\boldsymbol{c}_u = (c_1^u, \ldots, c_p^u)$ and in a m-vector $\boldsymbol{v}_u = (v_1^u, \ldots, v_m^u)$ for qualitative variables. Also, S_u is the cluster of policies associated to $(\boldsymbol{c}_u, \boldsymbol{v}_u)$ for $u = 1, \ldots, k$. We replace the distance $d(\boldsymbol{x}_i, \boldsymbol{c}_u, \bar{\lambda}_u)$ in the Bayesian version of the Lloyd's algorithm by

$$d(\boldsymbol{x}_i, \boldsymbol{D}_{i,.}, \boldsymbol{c}_u, \boldsymbol{v}_u, \bar{\lambda}_u) := \|\boldsymbol{c}_u - \boldsymbol{x}_i\|_2 + \beta_1 \left\| \boldsymbol{v}_u - \boldsymbol{D}_{i,.} \boldsymbol{B}^W / K \right\|_2$$
$$+ \beta_2 \sqrt{\left(N_i - \mathbb{E}\left(\lambda_u \Theta | N^{S_u} \right) v_i \right)^2},$$

where $\beta_1, \beta_2 \in \mathbb{R}^+$ are parameters tuning the regressive feature of the map. As before, N^{S_u} and v^{S_u} are respectively the number of claims and the total exposure for the uth cluster and $\mathbb{E}\left(\lambda_u \Theta | N^{S_u} \right)$ is the Bayesian estimate of the claims frequency for S^u as defined in Sect. 5.4. Algorithm 5.7 details the variant of the k-means algorithm including the same features than the SOM.

Algorithm 5.7 Adapted Lloyd's algorithm for Bayesian regression with quantitative and categorical variables

---

Initialization:
    Initialize centroids $\boldsymbol{c}_1(0),\ldots,\boldsymbol{c}_k(0)$ and $\boldsymbol{v}_1(0),\ldots,\boldsymbol{v}_k(0)$ with Algorithm 5.3.
    Set $\bar{\lambda}_u(0) = \bar{\lambda}$ for $u = 1, \ldots, k$.

Main procedure:
    For $e = 0$ to maximum epoch, e_{max}
        Assignment step:
        For $i = 1$ to n
            1) Assign $\boldsymbol{x}_i$ to a cluster $S_u(e)$ where $u \in \{1, \ldots, k\}$

$$S_u(e) = \{ (\boldsymbol{x}_i, \boldsymbol{D}_{i,.}) \, : \, d(\boldsymbol{x}_i, \boldsymbol{D}_{i,.}, \boldsymbol{c}_u(e), \boldsymbol{v}_u(e), \bar{\lambda}_u(e))$$
$$\leq d(\boldsymbol{x}_i, \boldsymbol{D}_{i,.}, \boldsymbol{c}_j(e), \boldsymbol{v}_j(e), \bar{\lambda}_j(e)) \; \forall j \in \{1, \ldots, k\}\}$$

        End loop on policies, i.
        Update step:
        For $u = 1$ to k
            2) Calculate the new centroids $\boldsymbol{c}_u(e+1)$ and $\boldsymbol{v}_u(e+1)$

$$\boldsymbol{c}_u(e+1) = \frac{1}{|S_u(e)|} \sum_{\boldsymbol{x}_i \in S_u(e)} \boldsymbol{x}_i$$

$$\boldsymbol{v}_u(e+1) = \frac{1}{|S_u(e)|} \sum_{\boldsymbol{D}_{i,.} \in S_u(e)} \boldsymbol{D}_{i,.} \boldsymbol{B}^W / K$$

            3) Update $\bar{\lambda}_u(e+1) = \frac{\sum_{\boldsymbol{x}_i \in S_u(e)} N_i}{\sum_{\boldsymbol{x}_i \in S_u(e)} v_i}$ and $\alpha_u = \frac{v^{S_u(e)} \bar{\lambda}}{\gamma + v^{S_u(e)} \bar{\lambda}}$.

        End loop on centroids, u
        4) Calculation of the total distance d^{total} :

$$d^{total} = \sum_{u=1}^{k} \sum_{\boldsymbol{x}_i \in S_u(e)} d(\boldsymbol{x}_i, \boldsymbol{D}_{i,.}, \boldsymbol{c}_u(e+1), \boldsymbol{v}_u(e+1), \bar{\lambda}_j(e+1))$$

    End loop on epochs e.

Table 5.14 Dominant features for each centroids, estimated claims frequency and percentages of the insured population assigned to each centroids

u	Gender	Area	Type	Age	Veh. age	Freq.	Real freq	% of pop.
1	Men	Ctryside	Class4	24.91	7.88	0.0371	0.0572	3.1
2	Men	Suburban	Class6	27.75	9.41	0.0181	0.0699	2
3	Women	Ctryside	Class3	28.83	11.48	0.0091	0.0108	6.5
4	Men	Ctryside	Class3	31.75	13.22	0.0093	0.0068	3.4
5	Men	Urban	Class5	38.93	10.49	0.0371	0.038	1.83
6	Men	Smlcity	Class6	38.94	9.72	0.0067	0.0193	2.39
7	Men	Ctryside	Class6	40.53	10.31	0.0125	0.0106	4.25
8	Men	Smlcity	Class5	42.41	11.96	0.0074	0.0133	3.31
9	Men	Suburban	Class4	42.61	11.26	0.0147	0.0135	3.04
10	Men	Ctryside	Class5	43.02	12.04	0.0238	0.0052	5.88
11	Men	Smlcity	Class4	43.26	11.62	0.0097	0.0086	3.69
12	Men	Ctryside	Class5	43.44	12.61	0.0073	0.0088	1.88
13	Men	Ctryside	Class4	43.88	12.69	0.0057	0.0049	6.39
14	Men	Urban	Class3	43.99	10.43	0.0065	0.02	6.79
15	Men	Suburban	Class5	44.07	10.8	0.0059	0.0169	4.51
16	Men	Suburban	Class1	44.26	40.16	0.0078	0.0068	1.88
17	Men	Suburban	Class3	44.72	10.81	0.0028	0.0086	4.58
18	Men	Ctryside	Class2	44.73	14.73	0.0124	0.0104	3.19
19	Men	Smlcity	Class3	45.19	10.87	0.0161	0.0055	5.51
20	Men	Ctryside	Class1	45.37	23.88	0.0152	0.0058	4.17
21	Men	Smlcity	Class1	46.77	19.25	0.0105	0.0052	2.87
22	Men	Northvlge	Class3	47.35	11.95	0.0079	0.0039	2.7
23	Women	Ctryside	Class3	48.43	10.39	0.0088	0.0071	8.47
24	Men	Suburban	Class6	49.49	10.13	0.0106	0.0173	2.59
25	Men	Ctryside	Class3	55.1	12.03	0.0108	0.0017	5.08

Table 5.14 presents the dominant modalities associated to 25 centroids, calculated with Eq. (5.20) in which ω_u is replaced by v_u. Compared to dominant features identified by the SOM in Table 5.13, we retrieve similar clusters. E.g. both algorithm have neurons or centroids with dominant features corresponding to a male driver of class 4 vehicles and living in the countryside. However we observe small discrepancies between forecasts of claims frequencies for similar dominant profiles.

Number of centroids	Deviance	d^{total}
9	5848	29,204
16	6082	24,150
25	6216	18,772

Table 5.15 Deviances and total distances computed by the k-means algorithm with 9, 16 and 25 centroids

Table 5.15 reports the deviances and total distances obtained with the k-means algorithm for 9, 16 and 25 centroids. A comparison with Table 5.12 reveals that SOMs converge to solutions with a lower total distance than the k-means algorithm when we partition the dataset in 9 or 25 clusters. Whereas the k-means algorithm slightly outperforms the SOM in the configuration with 16 centroids. We also observe that the k-means algorithm converges to a solution in less than 100 iterations while the SOM needs around 2000 iterations. The SOM seems more robust than k-means but is more time consuming.

5.10   Regression with the SOM

The SOM does not only segment the portfolio, it also allows for regressing the claims frequency on explanatory variables. We create a matrix X filled with owner's and vehicle ages of policies for which we want to assess the claims frequency. Next, we construct the disjunctive table D containing the qualitative modalities of these policies. Finally, we identify the best matching neuron and report the claims frequency of this cluster. We plot in Fig. 5.7 forecast claims frequencies for different risk profiles and owner's ages. The vehicle age is constant and set to 1 year.

We draw several interesting conclusions from Fig. 5.7. Men living in a urban environment have a higher probability of accident with a class 3 motorcycle than men living in the countryside driving the same vehicle. The claims frequency for men from northern villages is higher up to 28 years old. Contrary to men, female drivers living in a city and in the countryside share the same claims frequency for ages above 33 years. Female drivers of a class 3 motorcycle in the countryside have a higher probability of reporting a claim than men of the same area. More surprising, women older than 18 years, driving the most powerful vehicle (class 7) have a smaller claims frequency than class 3 drivers. However, these results must be nuanced given that women are under-represented in the portfolio (15.76% of the data set). Male drivers of a class 7 motorcycle in the countryside have a higher claims frequency than class 3 drivers but the frequency falls above 35 years old. On average, we also expect less claims for persons living in a small city than insureds in a suburban environment. Finally, female drivers of a powerful vehicle cause less accident than male drivers, in the countryside. For men, this frequency falls around 35 years old whereas for women, the claims frequency falls around 19 years old. These results confirm that SOM may be use for segmenting policyholders and forecasting the claims frequency.

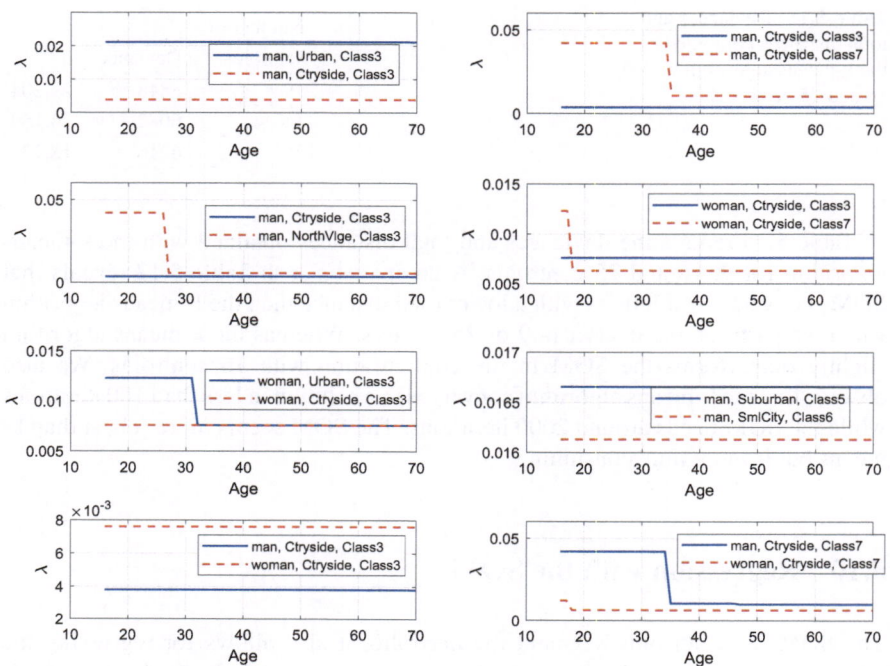

Fig. 5.7 Regressed claims frequencies for different profiles of insured

5.11   Further Readings

Self-organizing maps are artificial neural networks that do not require any a priori information on the relevancy of variables. As perceptrons, SOMs are used for fraud detection (Brockett et al. 1998) or failure prediction (Huysmans et al. 2006) but at the best of our knowledge have not been used by the actuarial community before (Hainaut 2019). Cottrell et al. (2004) proposed an extension of SOM to qualitative variables based on chi-square distance used in Multiple correspondence Analysis (MCA). The MCA was initially developed by Burt (1950) and enhanced by Benzécri (1973), Greenacre (1984) and Lebart et al. (1984). Kohonen (2013) draws a parallel between the SOM and vector quantization (VQ), which is used extensively in digital signal processing and transmission. He shows that an input information can accurately be represented by a linear mixture of a few best-matching nodes.

References

Arthur D, Vassilvitskii S (2007) K-means++: the advantages of careful seeding. In: SODA '07: proceedings of the eighteenth annual ACM-SIAM symposium on discrete algorithms, pp 1027–1035

Benzecri JP (1973) L'Analyse des données. Tome 2 : l'analyse des correspondances, Dunod, p. 619

Brockett P, Xia X, Derrig R (1998) Using Kohonen's self-organizing feature map to uncover automobile bodily injury claims fraud. J Risk Insur 65(2):245–274

Burt C (1950) The factorial analysis of qualitative data. Br J Psychol 3:166–185

Cottrell M, Ibbou S, Letrémy P (2004) SOM-based algorithms for qualitative variables. Neural Netw 17:1149–1167

Greenacre MJ (1984) Theory and applications of correspondence analysis. Academic, London

Hainaut D (2019) A self-organizing predictive map for non-life insurance. Eur Actuar J 9(1):173–207

Huysmans J, Baesens B, Vanthienen J, Van Gestel T (2006) Failure prediction with self organizing maps. Expert Syst Appl 30(3):479–487

Kohonen T (1982) Self-organized formation of topologically correct feature maps. Biol Cybern 43(1):59–69

Kohonen T (2013) Essentials of the self-organizing map. Neural Netw 37:52–65

Lebart L, Morineau A, Warwick KM (1984) Multivariate descriptive statistical analysis: correspondence analysis and related techniques for large matrices. Wiley, Chichester

Wütrich M, Buser C (2017) Data analytics for non-life insurance pricing. Lectures notes. Available on https://papers.ssrn.com/sol3/papers.cfm?abstract_id=2870308

References

Chapter 6
Ensemble of Neural Networks

The most frequent approach to data-driven modeling consists to estimate only a single strong predictive model. A different strategy is to build a bucket, or an ensemble of models for some particular learning task. One can consider building a set of weak or relatively weak models like small neural networks, which can be further combined altogether to produce a reliable prediction. The most prominent examples of such machine-learning ensemble techniques are random forests (Breiman 2001) and neural network ensembles (Hansen and Salamon 1990), which have found many successful applications in different domains. Liu et al. (2004) use this approach for predicting earthquakes. Shu and Burn (2004) forecast flood frequencies with an ensemble of networks. We start this chapter by describing the bias-variance decomposition of the prediction error. Next, we discuss how aggregated models and randomized models reduce the prediction error by decreasing the variance term in the bias-variance decomposition. Theoretical developments are inspired from the PhD thesis of Louppe (2014) on random forests.

6.1 Bias-Variance Decomposition

We consider an insurance dataset $\mathcal{D} = \{(y_k, \boldsymbol{x}_k, v_k) \, k = 1, \ldots n\}$ where $y_k \in \mathbb{R}$ is a key ratio, $\boldsymbol{x}_k$ is a real vector of dimension p and v_k is the exposure. We use this dataset for calibrating a predictor, e.g. a shallow or deep neural network parameterized by a vector of weights, Ω. This network is a non-linear mapping from $\mathbb{R}^d$ to $\mathbb{R}$ that we denote by $F_{\mathcal{D}}(\boldsymbol{x}) = \widehat{y}$. We condition this function by $\mathcal{D}$ to emphasize the dependence of this predictor to the training dataset.

Let us assume that features $(y, \boldsymbol{x})$ of the insurance contract are realizations of a random variables Y and X. The network is estimated in order to minimize an expected loss function, $\mathcal{L}(Y, F_{\mathcal{D}}(X))$. This function can be the deviance or any other functions. However the efficiency of ensemble learning is only demonstrated

© Springer Nature Switzerland AG 2019   147
M. Denuit et al., *Effective Statistical Learning Methods for Actuaries III*,
Springer Actuarial, https://doi.org/10.1007/978-3-030-25827-6_6

in the literature for a Gaussian deviance and unit exposure, $\nu = 1$. In this case, the expected loss becomes the mean square error of prediction:

$$\mathcal{L}(Y, F_{\mathcal{D}}(X)) = (Y - F_{\mathcal{D}}(X))^2 .$$

We now introduce a theoretical measure of goodness of fit:

Definition 6.17 The expected prediction error, also called generalization error of the network, $F_{\mathcal{D}}$ is defined as

$$\text{Err}(F_{\mathcal{D}}) = \mathbb{E}_{X,Y}(\mathcal{L}(Y, F_{\mathcal{D}}(X))) . \tag{6.1}$$

Equation (6.1) measures the average loss over all possible realizations of Y and X, including the observed values. Of course, this expectation cannot be evaluated in practice since the joint statistical distribution of Y and X is unknown. By conditioning on X, the generalization error becomes:

$$\mathbb{E}_{X,Y}(\mathcal{L}(Y, F_{\mathcal{D}}(X))) = \mathbb{E}_X \mathbb{E}_{Y|X}(\mathcal{L}(Y, F_{\mathcal{D}}(X))) .$$

The conditional prediction error for a given input $X = x$ is denoted by

$$\text{Err}(F_{\mathcal{D}}(x)) = \mathbb{E}_{Y|X}(\mathcal{L}(Y, F_{\mathcal{D}}(X))) . \tag{6.2}$$

The best possible model, noted $F_{\mathcal{D}}^B(X)$, minimizes the generalization error (6.1) and is such that:

$$F_{\mathcal{D}}^B(X) = \arg \min_{F_{\mathcal{D}}} \mathbb{E}_X \mathbb{E}_{Y|X}(\mathcal{L}(Y, F_{\mathcal{D}}(X))) .$$

This estimator is called for this reason the Bayes model in the literature and $\text{Err}(F_{\mathcal{D}}^B)$ is called the residual error. This is the minimal error that any supervised algorithm can achieve. This error is only caused by random deviations in the data. For this reason, a Bayes model is also defined as follows:

Definition 6.18 A neural network $F_{\mathcal{D}}^B$ is a Bayes model if for any other model $F_{\mathcal{D}}$ we have that $\text{Err}(F_{\mathcal{D}}^B) \leq \text{Err}(F_{\mathcal{D}})$.

We consider a quadratic loss function in the remainder of developments. Therefore, we assume that key ratios are distributed according to a normal law. In this case, the Bayesian estimator cancels the derivative of $\text{Err}(F_{\mathcal{D}}(x))$ with respect to $F_{\mathcal{D}}(x)$

and is then such that $\mathbb{E}_{Y|X=x}(Y) = F_{\mathcal{D}}^B(x)$. The conditional prediction error can be developed as follows:

$$\mathrm{Err}(F_{\mathcal{D}}(x)) = \mathbb{E}_{Y|X=x}\left((Y - F_{\mathcal{D}}(x))^2\right) \tag{6.3}$$

$$= \mathbb{E}_{Y|X=x}\left(\left(Y - F_{\mathcal{D}}^B(x) + F_{\mathcal{D}}^B(x) - F_{\mathcal{D}}(x)\right)^2\right)$$

$$= \mathbb{E}_{Y|X=x}\left(\left(Y - F_{\mathcal{D}}^B(x)\right)^2\right) + \mathbb{E}_{Y|X=x}\left(\left(F_{\mathcal{D}}^B(x) - F_{\mathcal{D}}(x)\right)^2\right)$$

$$+ 2\left(F_{\mathcal{D}}^B(x) - F_{\mathcal{D}}(x)\right)\mathbb{E}_{Y|X=x}\left(\left(Y - F_{\mathcal{D}}^B(x)\right)\right)$$

$$= \mathrm{Err}\left(F_{\mathcal{D}}^B(x)\right) + \left(F_{\mathcal{D}}^B(x) - F_{\mathcal{D}}(x)\right)^2.$$

The first term in this last equation is the conditional residual error at point $X = x$ and the second term is the discrepancy between the network $F_{\mathcal{D}}(x)$ and the Bayes model $F_{\mathcal{D}}^B(x)$.

We may further assume that the dataset $\mathcal{D}$ is itself a random variable. In this case, the best possible model is denoted by $F^B(x)$ and is such that:

$$F^B(x) = \arg\min_{F_{\mathcal{D}}} \mathbb{E}_{\mathcal{D}}\mathbb{E}_{Y|X=x}\left(\mathcal{L}(Y, F_{\mathcal{D}}(x))\right).$$

The conditional prediction error is therefore decomposed into three terms as stated in the next proposition proposed by Geman et al. (1992).

Proposition 6.7 *The bias-variance decomposition of the prediction error* $\mathbb{E}_{\mathcal{D}}\left(Err(F_{\mathcal{D}}(x))\right)$ *for* $X = x$ *is*

$$\mathbb{E}_{\mathcal{D}}\left(Err(F_{\mathcal{D}}(x))\right) = noise(x) + bias^2(x) + var(x) \tag{6.4}$$

where

$$noise(x) = Err\left(F^B(x)\right),$$

$$bias^2(x) = \left(F^B(x) - \mathbb{E}_{\mathcal{D}}\left(F_{\mathcal{D}}(x)\right)\right)^2,$$

$$var(x) = \mathbb{E}_{\mathcal{D}}\left((F_{\mathcal{D}}(x) - \mathbb{E}_{\mathcal{D}}\left(F_{\mathcal{D}}(x)\right))^2\right).$$

Proof According to Eq. (6.3), we have that

$$\mathbb{E}_{\mathcal{D}}\left(Err(F_{\mathcal{D}}(x))\right) = \mathbb{E}_{\mathcal{D}}\left(\mathrm{Err}\left(F^B(x)\right)\right) + \mathbb{E}_{\mathcal{D}}\left(\left(F^B(x) - F_{\mathcal{D}}(x)\right)^2\right).$$

The second term is the expected discrepancy over all possible datasets. It is equal to

$$\mathbb{E}_{\mathcal{D}}\left(\left(F^{B}(x)-F_{\mathcal{D}}(x)\right)^{2}\right)$$

$$=\mathbb{E}_{\mathcal{D}}\left(\left(F^{B}(x)-\mathbb{E}_{\mathcal{D}}\left(F_{\mathcal{D}}(x)\right)+\mathbb{E}_{\mathcal{D}}\left(F_{\mathcal{D}}(x)\right)-F_{\mathcal{D}}(x)\right)^{2}\right)$$

$$=\mathbb{E}_{\mathcal{D}}\left(\left(F^{B}(x)-\mathbb{E}_{\mathcal{D}}\left(F_{\mathcal{D}}(x)\right)\right)^{2}\right)+\mathbb{E}_{\mathcal{D}}\left(\left(F_{\mathcal{D}}(x)-\mathbb{E}_{\mathcal{D}}\left(F_{\mathcal{D}}(x)\right)\right)^{2}\right)$$

$$+2\mathbb{E}_{\mathcal{D}}\left(\left(F^{B}(x)-\mathbb{E}_{\mathcal{D}}\left(F_{\mathcal{D}}(x)\right)\right)\left(\mathbb{E}_{\mathcal{D}}\left(F_{\mathcal{D}}(x)\right)-F_{\mathcal{D}}(x)\right)\right).$$

The last term cancels since $F^{B}(x)$ and $\mathbb{E}_{\mathcal{D}}\left(F_{\mathcal{D}}(x)\right)$ are constant with respect to $\mathcal{D}$ and

$$\mathbb{E}_{\mathcal{D}}\left(\left(\mathbb{E}_{\mathcal{D}}\left(F_{\mathcal{D}}(x)\right)-F_{\mathcal{D}}(x)\right)\right)=\mathbb{E}_{\mathcal{D}}\left(F_{\mathcal{D}}(x)\right)-\mathbb{E}_{\mathcal{D}}\left(F_{\mathcal{D}}(x)\right)$$

$$=0.$$

$\square$

The noise in Eq. (6.4) is the residual error, independent from the learning set or the learning method. This is a theoretical lower bound of the prediction error. As illustrated in Fig. 6.1, the second term is the discrepancy between the average prediction obtained with a network $F_{\mathcal{D}}(x)$ and the best prediction that can achieved. The last term is the variance of the estimator and measure the sensitivity of predictions to changes in the dataset. In the next sections, we describe two approaches for reducing the prediction error: one based on bootstrapping and the other one relying on randomization.

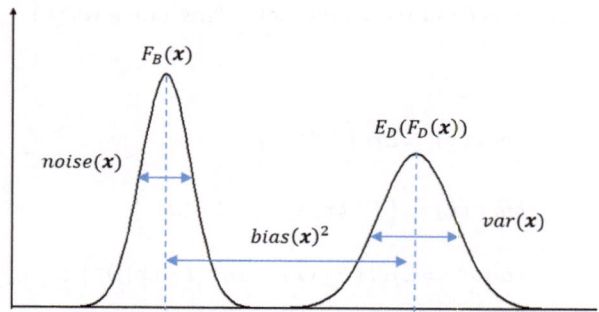

Fig. 6.1 Illustration of the bias-variance decomposition of $\mathbb{E}_{\mathcal{D}}\left(\mathrm{Err}(F_{\mathcal{D}}(x))\right)$

6.2   Bootstrap Aggregating Machine (Bagging)

The bootstrap aggregating algorithm, also called bagging, is procedure designed to improve the stability and accuracy of machine learning algorithms used in statistical classification or regression. It reduces the prediction error and partly prevents overfitting. Bagging (Bootstrap aggregating) was proposed by Leo Breiman in 1996 to improve classification by combining classifications of randomly generated training sets. Let us assume that we are given a sequence of learning datasets $(\mathcal{D}_k)_{k=1\ldots m}$, consisting of n contracts. The restriction is that we cannot aggregate these datasets to train the neural network. Instead, we have a sequence of predictors $\{F_{\mathcal{D}_k}(x)\}_{k=1,\ldots,m}$. An obvious way to improve the accuracy consists to use the average of predictors:

$$S_{\mathcal{D},m}(x) = \frac{1}{m} \sum_{k=1,\ldots,m} F_{\mathcal{D}_k}(x).$$

The $(\mathcal{D}_k)_{k=1\ldots m}$ may be seen as random arrays. In practice, independent learning sets are not available but bootstrapping allows to sample datasets $\mathcal{D}_k$ from $\mathcal{D}$. This procedure consists to generate m new learning sets $\mathcal{D}_k{}'$, each of size $n' \leq n$, by sampling data from $\mathcal{D}$ uniformly[1] and with replacement. By construction, the following equalities hold:

$$\mathbb{E}_{\mathcal{D}_i}\left(F_{\mathcal{D}_i}(x)\right) = \mathbb{E}_{\mathcal{D}}\left(F_{\mathcal{D}}(x)\right), \tag{6.5}$$

$$\mathbb{V}_{\mathcal{D}_i}\left(F_{\mathcal{D}_i}(x)\right) = \mathbb{V}_{\mathcal{D}}\left(F_{\mathcal{D}}(x)\right). \tag{6.6}$$

Since some observations may belong to several $\mathcal{D}_k$, the sample sets are then not independent and we have that:

$$\mathbb{E}_{\mathcal{D}_i,\mathcal{D}_j}\left(F_{\mathcal{D}_i}(x)F_{\mathcal{D}_j}(x)\right) \neq \mathbb{E}_{\mathcal{D}_i}\left(F_{\mathcal{D}_i}(x)\right)\mathbb{E}_{\mathcal{D}_j}\left(F_{\mathcal{D}_j}(x)\right) \quad i \neq j, \, i,j \in \{1,\ldots,m\}.$$

The bootstrap aggregating or bagging predictor is computed as the average of neural predictors calibrated on the collection of bootstrapped datasets $(\mathcal{D}_k)_{k=1\ldots m}$. The bagging machine is summarized in Algorithm 6.1. This procedure is non-parametric because no assumption is made about the distribution of observations. Instead, we use the empirical distribution of $(y_k, x_k, v_k)_{k=1,\ldots,n}$ for sampling $\mathcal{D}_j$.

To understand why bagging improves the accuracy of predictions, we consider a sample of data with an exposure equal to one ($v_k = 1$). The loss function is the

---

[1]Each contract $(y_k, x_k, v_k)_{k=1,\ldots,n}$ has a probability of $\frac{1}{n}$ to be included in the bootstrapped data sample.

Algorithm 6.1 Non-parametric bagging machine

Main procedure:
>**For** $j = 1$ to maximum epoch, m
>>1. Draw a sample $\mathcal{D}_j$ of size $n' \leq n$ from $\mathcal{D}$ uniformly and with replacement,
>>2. Calibrate a neural network $F_{\mathcal{D}_j}(x)$ on $\mathcal{D}_j$ by minimizing e.g. the Deviance,
>
>**End loop** on epochs
>Return the average of neural predictors:

$$S_{\mathcal{D},m}(x) = \frac{1}{m} \sum_{k=1,\ldots,m} F_{\mathcal{D}_k}(x).$$

---

mean square error and corresponds then to the Gaussian deviance. We also adopt the following notations for the expectation and variance of the neural network output, $F_{\mathcal{D}}(x)$:

$$\mu_{\mathcal{D}}(x) = \mathbb{E}_{\mathcal{D}}\left(F_{\mathcal{D}}(x)\right),$$
$$\sigma_{\mathcal{D}}^2(x) = \mathbb{V}_{\mathcal{D}}\left(F_{\mathcal{D}}(x)\right).$$

Under these assumptions and according to Eq. (6.4), the prediction error of the bagging predictor $\mathbb{E}_{\mathcal{D}_1,\ldots,\mathcal{D}_m}\left(S_{\mathcal{D},m}(x)\right)$, is the sum of a noise$(x)$, a bias$^2(x)$ and the variance, var(x), of $S_{\mathcal{D},m}(x)$. The noise depends only upon the residual error and is therefore independent from the learning set. The bias is the difference between the forecast of the Bayes model and the expected prediction of $S_{\mathcal{D},m}(x)$. From Eq. (6.5), we infer that

$$\mathbb{E}_{\mathcal{D}_1\ldots\mathcal{D}_m}\left(S_{\mathcal{D},m}(x)\right) = \frac{1}{m} \sum_{k=1}^{m} \mathbb{E}_{\mathcal{D}_k}\left(F_{\mathcal{D}_k}(x)\right)$$

$$= \mu_{\mathcal{D}}(x).$$

and hence the bias(x) for $S_{\mathcal{D},m}(x)$ is the same as the bias(x) for any network $F_{\mathcal{D}}(x)$ trained on a single dataset. Therefore, bootstrapping does not reduce the bias. Next, we define a coefficient of correlation $\rho_{\mathcal{D}}(x)$ between the predictions of two networks trained on bootstrapped datasets $\mathcal{D}_1$ and $\mathcal{D}_2$:

$$\rho_{\mathcal{D}}(x) = \frac{\mathbb{E}_{\mathcal{D}_1,\mathcal{D}_2}\left(\left(F_{\mathcal{D}_1}(x) - \mu_{\mathcal{D}}(x)\right)\left(F_{\mathcal{D}_2}(x) - \mu_{\mathcal{D}}(x)\right)\right)}{\sigma_{\mathcal{D}}(x)^2}. \tag{6.7}$$

The lower is this correlation, the higher is the impact of the bootstrapping on the learning process. When $\rho_{\mathcal{D}}(x)$ is close to one, the outcomes of two neural networks are highly dependent and the bootstrapping has little effect on forecasts. If $\rho_{\mathcal{D}}(x)$ is close to zero, the bootstrapped data sets are nearly independent. Notice that $\rho_{\mathcal{D}}(x)$ is positive as sample sets are bootstrapped from the same dataset. The next proposition emphasizes the role of $\rho_{\mathcal{D}}(x)$ on the variance of $S_{\mathcal{D},m}(x)$.

Proposition 6.8 *The variance term in the bias-variance decomposition of the bagging predictor error,* $\mathbb{E}_D \left(Err(S_{D,m}(x)) \right)$, *is the following sum:*

$$var(x) = \rho_D(x)\sigma_D^2(x) + \frac{1 - \rho_D(x)}{m}\sigma_D^2(x).$$

(6.8)

Proof To demonstrate this result, we first develop $var(x)$ as follows:

$$var(x) = \mathbb{E}_{D_1,\dots,D_m} \left(\left(S_{D,m}(x) - \mathbb{E}_{D_1,\dots,D_m} \left(S_{D,m}(x) \right) \right)^2 \right)$$

$$= \frac{1}{m^2} \left[\mathbb{E}_{D_1,\dots,D_m} \left(\left(\sum_{k=1}^{m} F_{D_k}(x) \right)^2 \right) \right.$$

$$\left. - \left(\mathbb{E}_{D_1,\dots,D_m} \left(\sum_{k=1}^{m} F_{D_k}(x) \right) \right)^2 \right]$$

$$= \frac{1}{m^2} \left[\mathbb{E}_{D_1,\dots,D_m} \left(\left(\sum_{k=1}^{m} F_{D_k}(x) \right)^2 \right) - m^2 \mu_D(x)^2 \right].$$

Rewriting the square of the sums as a sum of pairwise products leads to:

$$var(x) = \frac{1}{m^2} \left[\sum_{i=1}^{m} \sum_{j=1}^{m} \mathbb{E}_{D_i D_j} \left(F_{D_i}(x) F_{D_j}(x) \right) - m^2 \mu_D(x)^2 \right]$$

(6.9)

$$= \frac{1}{m^2} \left[m\mathbb{E}_D \left(F_D(x)^2 \right) + \left(m^2 - m \right) \mathbb{E}_{D_1,D_2} \left(F_{D_1}(x) F_{D_2}(x) \right) \right.$$

$$\left. - m^2 \mu_D(x)^2 \right].$$

From Eq. (6.7), we know that

$$\mathbb{E}_{D_1,D_2} \left(F_{D_1}(x) F_{D_2}(x) \right) = \rho_D(x)\sigma_D(x)^2 + \mu_D(x)^2,$$

and by definition of the variance $\sigma_D^2(x) = \mathbb{V}_D \left(F_D(x) \right)$, we have that

$$\mathbb{E}_D \left(F_D(x)^2 \right) = \sigma_D(x)^2 + \mu_D(x)^2.$$

Combining these two last relations with Eq. (6.9) give us:

$$
\begin{aligned}
\text{var}(\boldsymbol{x}) &= \frac{1}{m^2} \left[m\sigma_{\mathcal{D}}(\boldsymbol{x})^2 + m\mu_{\mathcal{D}}(\boldsymbol{x})^2 + \left(m^2 - m \right) \rho_{\mathcal{D}}(\boldsymbol{x})\sigma_{\mathcal{D}}(\boldsymbol{x})^2 \right. \\
&\quad \left. + \left(m^2 - m \right) \mu_{\mathcal{D}}(\boldsymbol{x})^2 - m^2 \mu_{\mathcal{D}}(\boldsymbol{x})^2 \right] \\
&= \frac{1}{m^2} \left[m\sigma_{\mathcal{D}}(\boldsymbol{x})^2 + \left(m^2 - m \right) \rho_{\mathcal{D}}(\boldsymbol{x})\sigma_{\mathcal{D}}(\boldsymbol{x})^2 \right],
\end{aligned}
$$

which is well the decomposition of var($\boldsymbol{x}$) in Eq. (6.8).   □

Equation (6.8) emphasizes the benefit of bagging machines: when the number of bootstrapped samples becomes arbitrarily large, i.e. as $m \to \infty$, the variance of the bagging predictor reduces to $\rho_{\mathcal{D}}(\boldsymbol{x})\sigma_{\mathcal{D}}^2(\boldsymbol{x})$ instead of $\sigma_{\mathcal{D}}^2(\boldsymbol{x})$ for a single network. As the bias and the noise remain unchanged and $\rho_{\mathcal{D}} \in [0, 1]$, the expected generalization error of a bagging predictor is lower than the expected error of a single network. If $\rho_{\mathcal{D}}(\boldsymbol{x}) \to 0$ then the variance term vanishes. If $\rho_{\mathcal{D}}(\boldsymbol{x}) \to 1$, bagging does not improve the quality of the prediction.

Theoretical developments of this section are derived when the loss function is the mean square error. Empirical tests reveals nevertheless that the bagging technique reduces the prediction errors for other types of loss function, like the Poisson deviance. Intuitively, this may be explained by the fact that the deviance of an exponential dispersed distribution can be approached by a quadratic function.[2]

6.3 Application of Bagging to the Analysis of Claims Frequency

In this section, we evaluate the ability of the bagging machine to predict the claims frequency for motorcycle insurances. As in previous chapters, our analysis is based on the dataset from the company *Wasa* and we refer to Sect. 1.11 of Chap. 1 for a detailed presentation of explanatory variables. The (scaled) age of the vehicle and owner's age are quantitative variables, directly used as input of the network. The other covariates are: gender, geographic area and vehicle class are categorical variables with binary modalities. As in Sect. 3.4.1 of Chap. 3, the dataset is completed with categorical variables that indicate the cross occurrences of three pairs of covariates: gender vs. geographic zone, gender vs. vehicle class and zone vs. vehicle class. In this numerical illustration, an insurance contract is described by 2 quantitative variables and 6 categorical variables which counts totally 89 binary modalities. We use as predictor a basic neural network with two hidden

---

[2]E.g. If we approach ln(.) in the Poisson deviance by a first order Taylor's development, we retrieve the expression of the deviance for a normal distribution.

layers with 5–10 neurons. The activation functions of the first, second and output layers are respectively linear, sigmoidal and linear. The estimated claims frequency is the exponential of the network output signal to ensure the positivity of the network prediction. Figure 6.2 shows the structure of such a neural network. The network is fed with 89 input signals and counts 521 weights. The network is fitted with the root mean square propagation algorithm (RMSprop) and the gradient of the loss function is evaluated with batches of 5000 contracts. The dataset is reshuffled after each epoch to avoid training the network on same batches.

Table 6.1 reports the main statistics about the goodness of the fit after 500 iterations and for different configurations of the bagging machine. In a first series of tests, we fit the NN(5,10,1) networks up to 50 bootstrapped samples of 10,000 contracts. In a second series, the same network is adjusted to 50 bootstrapped samples of 25,000 contracts. We next calibrate the model to the complete dataset, without bootstrapping. The number of iterations is limited to 500 in order to reduce the computation time. Whatever the configuration, bagging models achieve a better performance in term of deviance than the generalized linear model of Chap. 1 (for which we had a deviance of 5781.66). The deviance obtained with bagging estimators is slightly higher than the one computed with the model fitted to the complete dataset but this discrepancy is negligible for the machine with samples of 25,000 contracts. We also observe that averaging 50 instead of 25 predictors does not significantly improves the adjustment. Figure 6.3 shows the deviance computed with 1 up to 50 averaged models, for samples of 10,000 and 25,000 contracts. We observe that the deviance stagnates at its lowest level for bagging machines with more than 20 bootstrapped samples. This graph also emphasizes that the size of samples is a major determinant of the goodness of fit. One way to detect overfitting consists to analyze forecasted claims frequencies. Figures 6.4, 6.5, and 6.6 show these predictions by age for different types of policyholders, calculated by the bagging algorithm with 50 bootstrapped samples of 25,000 contracts. We compare these results with forecast obtained with a GLM fitted to owner's age, vehicle age, gender, geographic area and power class. We observe that the order of magnitude of frequencies computed with the bagging algorithm is similar to the one of GLM forecasts. On the other hand, whatever the policyholder's profile, the claims frequency decreases with age.

From Fig. 6.4 a driver of a class 3 vehicle can cause an accident with a higher probability in an urban area (geographic zone 1) than in the countryside (geographic zone 4). The claims frequency for female policyholders is less or equal than the one of male drivers.

The left plot of Fig. 6.5 shows that women driving a class 3 motorcycle in one of the largest Swedish cities cause on average more accidents than women driving in the countryside. For this category of drivers, the GLM and neural networks both yield very similar estimates. The right graph of the same figure reveals that the bagging algorithm fails to discriminate vehicles of different power class whereas the GLM concludes that drivers of most powerful motorcycles (class 7) are more risky than other drivers. This failure is directly related to fact that the database counts only 806 contracts insuring a class 7 vehicle (on 62,436 policies). As the bagging

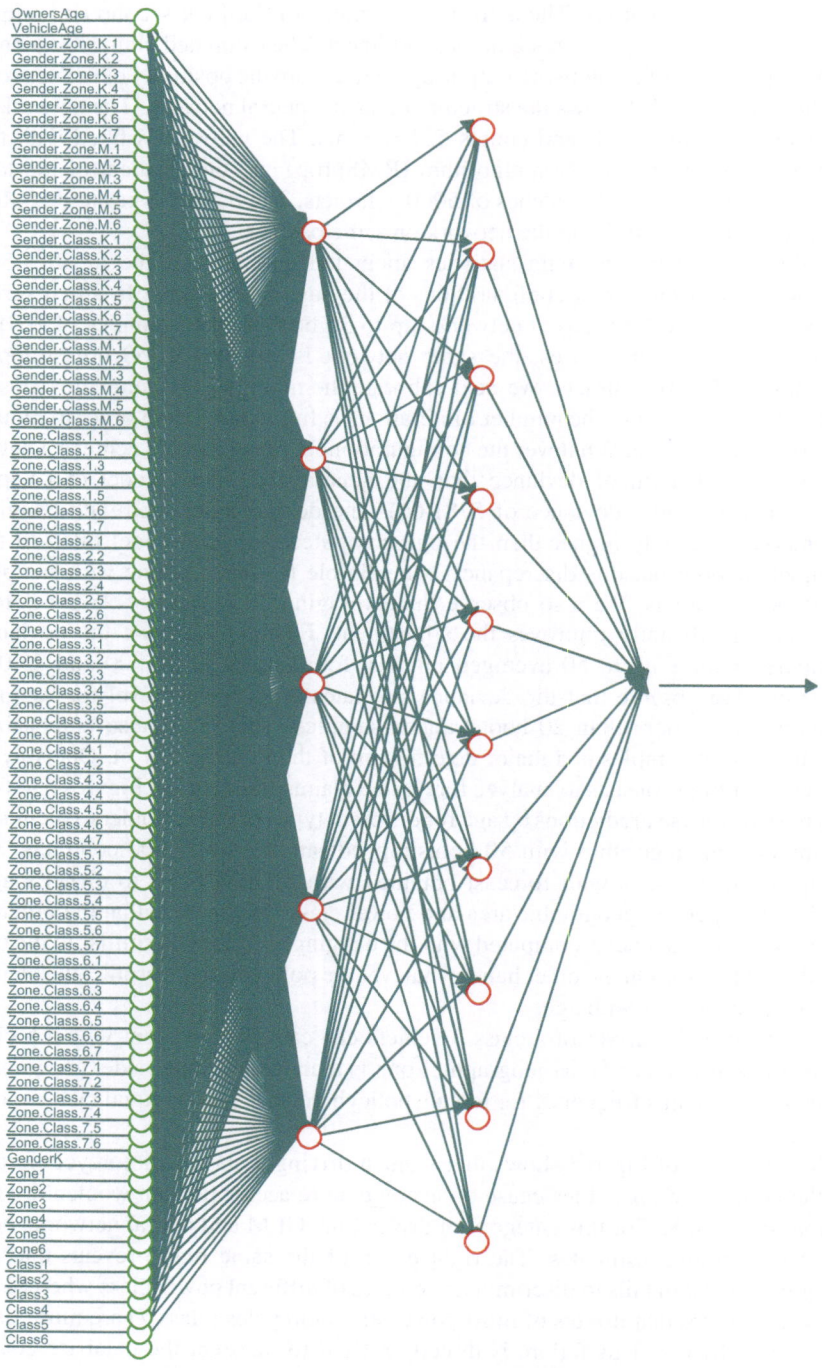

Fig. 6.2 Architecture of the neural network, NN(5,10,1) used in the bagging algorithm

Table 6.1 Statistics of calibration, bagging algorithm applied to a NN(5,10,1) network

Number of samples	1	25	50	25	50
Size of samples	All	10,000	10,000	25,000	25,000
Deviance	5593	5699	5694	5615	5610
Log. Lik.	−3471	−3523	−3521	−3482	−3479
AIC	7984	8089	8084	8006	8001
BIC	12,695	12,800	12,795	12,716	12,712

500 iterations

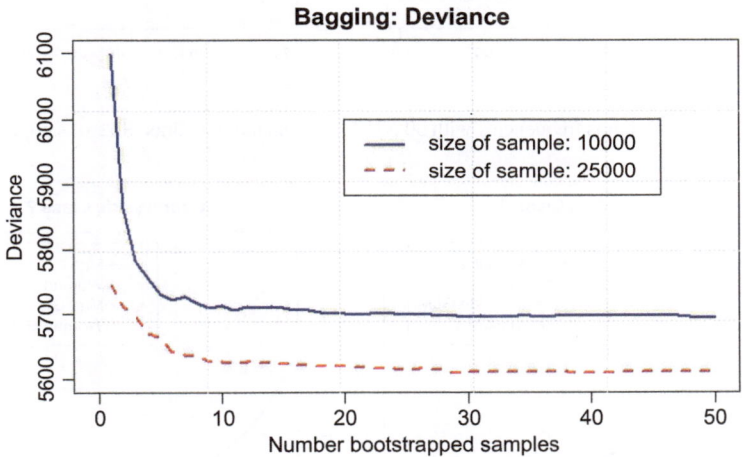

Fig. 6.3 Bagging algorithm: evolution of the deviance with the number of bootstrapped samples

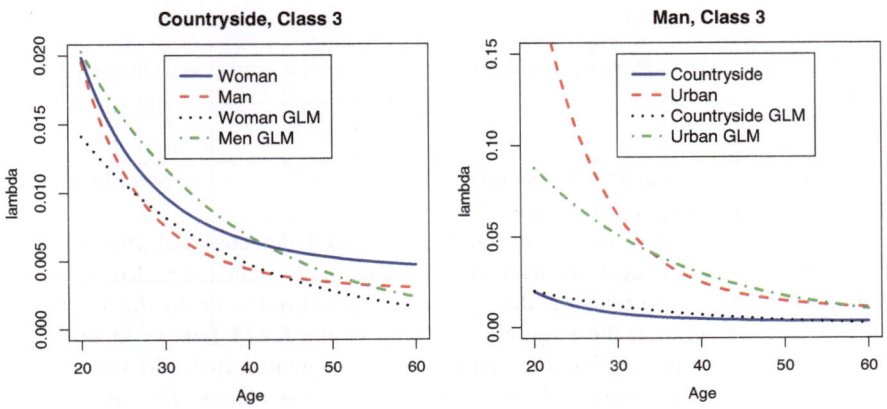

Fig. 6.4 Forecast claims frequencies with 50 NN(5,10,1) neural networks. Size of sample: 25,000 policies

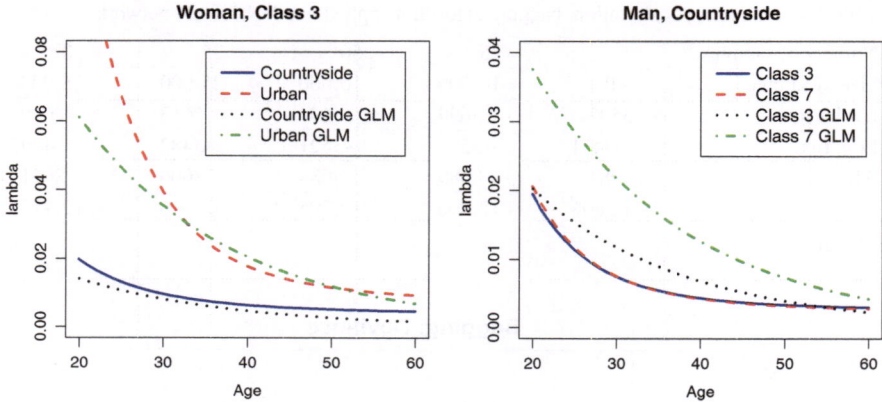

Fig. 6.5 Forecast claims frequencies with 50 NN(5,10,1) neural networks. Size of sample: 25,000 policies

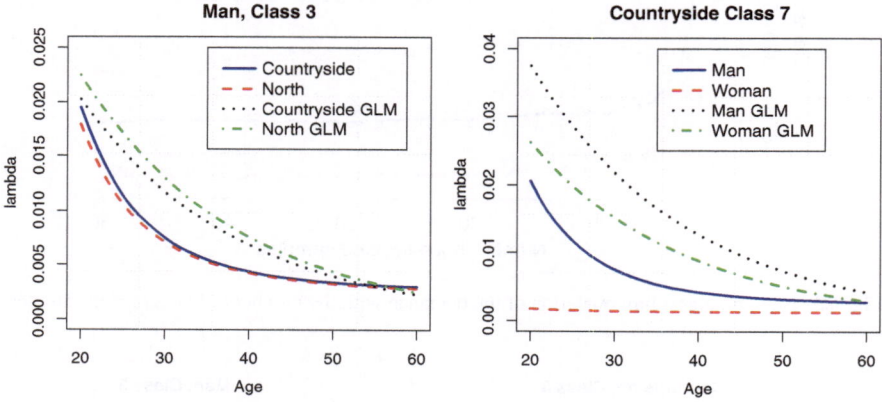

Fig. 6.6 Forecast claims frequencies with 50 NN(5,10,1) neural networks. Size of sample: 25,000 policies

algorithm runs on bootstrapped samples of 25,000 policies, some sample have no contract for this category of vehicles.

The shallow network with four neurons of Chap. 1 detected that young drivers from northern areas cause on average more accidents than other drivers in the southern countryside (zone 3), due to weather conditions. From the left graph of Fig. 6.6, we see that the bagging algorithm as the GLM fails to identify this trend. The last graph emphasizes that compared to men, female drivers of class 7 motorcycles in the countryside causes on average less claims. The GLM yields higher expected claims frequencies both for male and female drivers in this category.

The smoothness of frequencies curves confirms that the bagging machine does not overfit data. In theory, this point might be confirmed by cross-validation. In our example, this procedure is nevertheless very time consuming given that we have to calibrate 50 neural networks to bootstrapped samples drawn from each validation dataset. Estimating one bagging predictor lasts 122 min on a personal computer (processor: Intel Core I5, 2.60 GHz and 8 GB of RAM). A cross validation with 10 validation samples requires therefore 20 h of computations.

6.4 Ensemble Methods with Randomization

An alternative approach for reducing the prediction error relies on randomization of the training procedure. This approach has been applied to regression trees and is at the origin of the random forest algorithm (see Trufin et al. (2019) for details). The decomposition in Proposition 6.7 emphasizes the important role played by the variance of the predictor in the accuracy of forecasts. A elegant way for reducing this variance consists to estimate neural networks with randomly perturbed training algorithms. The ensemble predictor is obtained by combining the predictions of these networks. In this paragraph, we analyze the impact of randomization in the learning procedure on the average prediction error for a dataset with unit exposure ($\nu = 1$) and the mean square error as loss function.

In training algorithms of Chap. 1 or 3, weights are either randomly initialized and/or are either randomly updated. Therefore, we denote by Θ the set of random hyper-parameters of the training algorithm controlling the randomness of the learning procedure. Here, Θ is multivariate random variable. To emphasize the dependence of the neural network to these hyper-parameters, the prediction function is now denoted by $F_{\mathcal{D},\Theta}(.)$.

We extend the bias-variance decomposition of the conditional prediction error, in order to take into account the randomness of Θ by replacing the expectation $\mathbb{E}_{\mathcal{D}}$ by $\mathbb{E}_{\mathcal{D},\Theta}$ where $\mathbb{E}_{\mathcal{D},\Theta}$ is the expectation with respect to the distribution of $\mathcal{D}$ and Θ. The expected prediction error becomes in this case the sum of the following three terms:

$$\mathbb{E}_{\mathcal{D},\Theta}\left(\mathrm{Err}(F_{\mathcal{D},\Theta}(x))\right) = \mathrm{noise}(x) + \mathrm{bias}^2(x) + \mathrm{var}(x), \qquad (6.10)$$

where

$$\mathrm{noise}(x) = \mathrm{Err}\left(F^B(x)\right),$$

$$\mathrm{bias}^2(x) = \left(F^B(x) - \mathbb{E}_{\mathcal{D},\Theta}\left(F_{\mathcal{D},\Theta}(x)\right)\right)^2,$$

$$\mathrm{var}(x) = \mathbb{E}_{\mathcal{D},\Theta}\left(\left(F_{\mathcal{D},\Theta}(x) - \mathbb{E}_{\mathcal{D},\Theta}\left(F_{\mathcal{D},\Theta}(x)\right)\right)^2\right).$$

The variance, $\mathrm{var}(x)$ accounts for the prediction variability caused by the randomization of the learning algorithm and by the randomness of the learning dataset.

The bias is also modified by this randomization. As such, the expected prediction error of a randomized network is larger than the variance of a network trained with a deterministic approach. Randomizing a learning procedure might then appear counter-productive since it raises both bias and variance. However, we will see that combining a series of randomized models allows to improve the quality of prediction.

To demonstrate this, we consider a set of m neural networks trained on a data set $\mathcal{D}$ with randomized procedures. Their hyper-parameters are random vectors denoted by $\Theta_1, \ldots, \Theta_m$. These random variables are independent and have the same distribution as Θ. The prediction of the kth network is noted $F_{\mathcal{D}, \Theta_k}(x)$. The loss is still measured by a quadratic function. The ensemble predictor, $S_{\mathcal{D}, m}(x)$, is the average of forecasts yield by networks for $X = x$:

$$S_{\mathcal{D}, m}(x) = \frac{1}{m} \sum_{k=1}^{m} F_{\mathcal{D}, \Theta_k}(x).$$

We now study the bias-variance decomposition of this predictor. We adopt the following notations

$$\mu_{\mathcal{D}, \Theta}(x) = \mathbb{E}_{\mathcal{D}, \Theta}\left(F_{\mathcal{D}, \Theta}(x)\right),$$

$$\sigma_{\mathcal{D}, \Theta}^2(x) = \mathbb{V}_{\mathcal{D}, \Theta}\left(F_{\mathcal{D}, \Theta}(x)\right).$$

According to Eq. (6.10), the prediction error of the ensemble predictor $\mathbb{E}_{\mathcal{D}, \Theta_1, \ldots, \Theta_m}\left(S_{\mathcal{D}, m}(x)\right)$ is the sum of noise(x), bias$^2(x)$ and var(x). The noise depends only upon the residual error and is therefore independent from the learning set and method. The bias is the difference between the forecast of the Bayes model and the expected prediction of $S_{\mathcal{D}, m}(x)$. Since $(\Theta_k)_{k=1, \ldots, m}$ are independent and identically distributed, we have:

$$\mathbb{E}_{\mathcal{D}, \Theta_1, \ldots, \Theta_m}\left(S_{\mathcal{D}, m}(x)\right) = \frac{1}{m} \sum_{k=1}^{m} \mathbb{E}_{\mathcal{D}, \Theta_k}\left(F_{\mathcal{D}, \Theta_k}(x)\right)$$

$$= \mu_{\mathcal{D}, \Theta}(x).$$

The bias(x) for $S_{\mathcal{D}, m}(x)$ is the same as the bias(x) for any single network $F_{\mathcal{D}, \Theta}(x)$. Working with an ensemble predictor does not reduce the bias but we will see that it reduces the variance term. To demonstrate this, we first define a coefficient of correlation $\rho_{\Theta}(x)$ between the predictions of two randomized networks trained on the same dataset but with independent random hyper-parameters Θ_1 and Θ_2. As Θ_1 and Θ_2 are independent and identically distributed (i.i.d.), this correlation is equal to:

$$\rho_\Theta(x) = \frac{\mathbb{E}_{\mathcal{D},\Theta_1,\Theta_2}\left((F_{\mathcal{D},\Theta_1}(x) - \mu_{\mathcal{D},\Theta_1}(x))(F_{\mathcal{D},\Theta_2}(x) - \mu_{\mathcal{D},\Theta_2}(x))\right)}{\sigma_{\mathcal{D},\Theta_1}(x)\sigma_{\mathcal{D},\Theta_2}(x)}$$

$$= \frac{\mathbb{E}_{\mathcal{D},\Theta_1,\Theta_2}\left(F_{\mathcal{D},\Theta_1}(x)F_{\mathcal{D},\Theta_2}(x)\right) - \mu_{\mathcal{D},\Theta}(x)^2}{\sigma_{\mathcal{D},\Theta}(x)^2}. \tag{6.11}$$

The lower is this correlation, the higher is the impact of the randomization on the learning process. When $\rho_\Theta(x)$ is close to one, the outcomes of two neural networks are highly dependent and random perturbations have little effect on forecasts. The next proposition assesses the role of $\rho_\Theta(x)$ in the variance of the ensemble predictor.

Proposition 6.9 *The variance term in the bias-variance decomposition of the ensemble predictor error,* $\mathbb{E}_{\mathcal{D},\Theta}\left(Err(S_{\mathcal{D},m}(x))\right)$, *is the following sum:*

$$var(x) = \rho_\Theta(x)\sigma^2_{\mathcal{D},\Theta}(x) + \frac{1 - \rho_\Theta(x)}{m}\sigma^2_{\mathcal{D},\Theta}(x). \tag{6.12}$$

Proof The proof is similar to the one of Proposition 6.8 and we just sketch the reasoning. $var(x)$ can be developped as follows:

$$var(x) = \mathbb{E}_{\mathcal{D},\Theta_1,\dots,\Theta_m}\left((S_{\mathcal{D},m}(x) - \mathbb{E}_{\mathcal{D},\Theta_1,\dots,\Theta_m}(S_{\mathcal{D},m}(x)))^2\right)$$

$$= \frac{1}{m^2}\left[\mathbb{E}_{\mathcal{D},\Theta_1,\dots,\Theta_m}\left(\left(\sum_{k=1}^{m}F_{\mathcal{D},\Theta_k}(x)\right)^2\right) - m^2\,\mu_{\mathcal{D},\Theta}(x)^2\right].$$

Rewriting the square of the sums as a sum of pairwise products and using the i.i.d. feature of Θ leads to:

$$var(x) = \frac{1}{m^2}\left[m\mathbb{E}_{\mathcal{D},\Theta}\left(F_{\mathcal{D},\Theta}(x)^2\right) + \left(m^2 - m\right)\mathbb{E}_{\mathcal{D},\Theta_1,\Theta_2}\left(F_{\mathcal{D},\Theta_1}(x)F_{\mathcal{D},\Theta_2}(x)\right)\right.$$

$$\left. -m^2\,\mu_{\mathcal{D},\Theta}(x)^2\right].$$

From Eq. (6.11), and by definition of the variance $\sigma^2_{\mathcal{D},\Theta}(x)$, we can conclude. $\square$

Equation (6.12) underlines the benefit of combining randomized networks: when the size of the ensemble becomes arbitrarily large, i.e. as $m \to \infty$, the variance of the predictor reduces to $\rho_\Theta(x)\sigma^2_{\mathcal{D},\Theta}(x)$, instead of $\sigma^2_{\mathcal{D},\Theta}(x)$ for a single network. The expected generalization error of an ensemble predictor is lower than the expected error of a single network. If $\rho_\Theta(x) \to 0$ then the variance term cancels. If $\rho_\Theta(x) \to 1$, ensemble learning does not improve the accuracy of forecasts. In view of Proposition 6.9, the main principle of ensemble methods consists to introduce

---

Algorithm 6.2 Randomized back-propagation algorithm with the drop out technique

---

Initialization:

    Set initial random weights of $F_{\mathcal{D}}(x)$: $\Omega_0^{(j)}$ for $j = 1, \ldots, n^{net}$.

    Select a loss function $\mathcal{R}(.)$

Main procedure:

    For $t = 0$ to maximum epoch, T

        1. Cancel randomly a fraction $\alpha^{(j)}$ of $\Omega_t^{(j)}$ for $j = 1, \ldots, n^{net}$ (**drop out**)

        2. Calculate the gradient $\nabla \mathcal{R}(\Omega_t)$ where $\Omega_t = \left(\Omega_t^{(j)} \right)_{j=1,\ldots,n^{net}}$

        3. Adjust the step size ρ_{t+1}

        4. Update weights: $\Omega_{t+1} = \Omega_t - \rho_{t+1} \nabla \mathcal{R}(\Omega_t)$

    End loop on epochs

Return the neural predictors: $F_{\mathcal{D}}(x)$

---

random perturbations that decorrelates as much as possible forecasts of individual neural networks. This decorrelation minimizes the variance of predictions.

The randomization technique has already been implicitly used in the numerical illustration of Sect. 6.3. In order to keep under control the computational time and to avoid local minima, the gradient of the loss function is evaluated with batches of 5000 contracts from a randomly reshuffled dataset after each epoch. Measuring with accuracy the impact of this randomization in the bagging algorithm would therefore require to evaluate the gradient of loss function using the full dataset. As this operation multiplies by 5 the computational time, we skip this step.

Instead, we focus on a popular method for reducing the correlation between outcomes of an ensemble of models which consists to drop randomly a fraction of explanatory variables and/or of neurons output signals during the training. This randomization is usually combined with the bootstrapping procedure of Sect. 6.2 to maximize the decorrelation of network outputs. This approach has been successfully applied to regression trees and is called "random forests" in this context. For neural networks, this method is called the drop out technique.

Remember that our dataset is denoted by $\mathcal{D} = \{(y_k, x_k, v_k) \, k = 1, \ldots n\}$ where $y_k \in \mathbb{R}$ is a key ratio, $x_k \in \mathbb{R}^p$ is the vector of covariates and v_k is the exposure. The neural network $F_{\mathcal{D}}(x)$ that we wish to fit to this dataset has n^{net} layers and weights of layers j are denoted by $\Omega^{(j)}$ for $j = 1, \ldots, n^{net}$. The drop out technique consists in randomly setting a fraction $\alpha^{(j)}$ of input weights of the jth layer to 0 at each update, during the training time of $F_{\mathcal{D}}(x)$. This method prevents overfitting and is an alternative to Lasso or Ridge penalizations. The procedure is summarized in Algorithm 6.2 when the network is calibrated with the back-propagation algorithm. Of course, any other algorithms may be used for this step. We test this procedure in the next section.

6.5   Drop Out Method Applied to the Analysis of Claims Frequency

We test the drop out technique on the portfolio of insurance contracts from *Wasa* and focus on the analysis of the claims frequency. The covariates are the owner's age, age and class of vehicle, gender, zone, gender vs. geographic zone, gender vs. vehicle class and zone vs. vehicle class. As in Chap. 3, the continuous quantitative variables are converted into categorical ones. Policyholders are classed by age into 16 subsets of comparable size, as detailed in Table 3.1. Whereas vehicle ages are split in 6 categories reported in Table 3.2. Totally, we have 107 binary explanatory variables. In order to emphasize that the drop out technique prevents overfitting in a similar way to Lasso or Ridge penalizations, we focus on the same deep neural network as the one studied in Sect. 3.4. This networks has 4 layers counting 10, 20, 10 and 1 neurons and is denoted by NN(10,20,10,1). The analysis of forecast claims frequencies and cross-validation have revealed that this structure suffers from overfitting.

The NN(10,20,10,1) counts 1521 weights allocated as follows between layers: 1080, 220, 210 and 11 weights for layers 1, 2, 3 and 4. We calibrate this model with the root mean square propagation algorithm and for drop out rates, $\alpha^{(1)}$, varying from 0% to 40% by step of 5%. We do not drop any signal from input layers 2 to 4 $(\alpha^{(2)} = \alpha^{(3)} = \alpha^{(4)} = 0)$. Table 6.2 reports the main statistics about the goodness of the fit after 1500 iterations. Without dropping any signals, the deviance of the neural model is clearly less than the one obtained with a GLM model (5781.66, see Chap. 1) but the performance in terms of AIC and BIC is worst given the high number of parameters. We also know that this model overfits data. Increasing the drop out rate rises on average linearly the Deviance as suggested by Fig. 6.7, plotting the deviance versus the drop out rate.

Figure 6.8 presents predicted claims frequencies by age for male and female drivers of class 3 motorcycles living in the countryside and in an urban environment. These frequencies are computed with a drop out rate of 0%, 20% and 40%. Without drop out, the oscillations of forecasts per age are symptomatic of data overfitting.

Table 6.2 Statistics of calibration, about the calibration of a NN(10,20,10,1) neural networks

Drop out rate	0%	10%	20%	30%	40%
Deviance	4954	5143	5248	5314	5537
Log. Lik.	−3151	−3246	−3298	−3331	−3443
AIC	9345	9534	9639	9705	9928
BIC	23, 098	23, 287	23, 391	23, 458	23, 681

Calibration is done with 1500 steps of the RMSprop algorithm

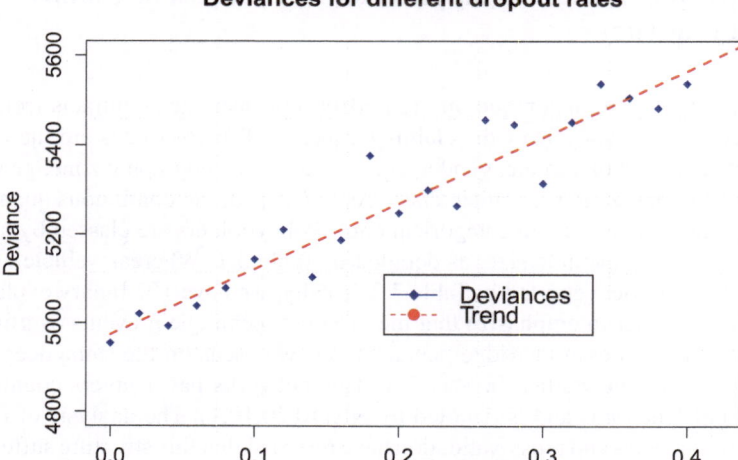

Fig. 6.7 Evolution of the NN(10,20,10,1) deviance with the drop out rate $\alpha^{(1)}$

Rising the drop out rate clearly smooths the curves of predictions and the order of magnitude of predicted claims frequencies is similar to the one of forecasts computed with a GLM. This confirms that randomization is an efficient alternative to Lasso or Ridge penalization for computing insurance premiums.

In order to confirm that the risk of overfitting is reduced by the drop out method, we perform a cross validation with 10 validation samples and with a drop out rate of 40%. Results are reported in Table 6.3 and must be compared to Table 7.2 about the validation of the same network without drop out. As we could expect, the average deviance on the training data (4992.78) is greater than the same statistic computed without dropping weights (4372.58, see Table 3.4 of Chap. 3). The network without drop out fits then better the training set than its equivalent adjusted with drop out. However, the network with drop out achieves an excellent performance on validation samples. The average deviance (582.19) computed over these sets is significantly less than the one obtained without drop out (634.02) or with a Lasso penalization (629.91, see Table 3.6 of Chap. 3). The volatility of deviances over validation tests is also lower (54.24 instead of 66.9). The cross-validation confirms therefore the efficiency of the drop out technique for reducing overfitting.

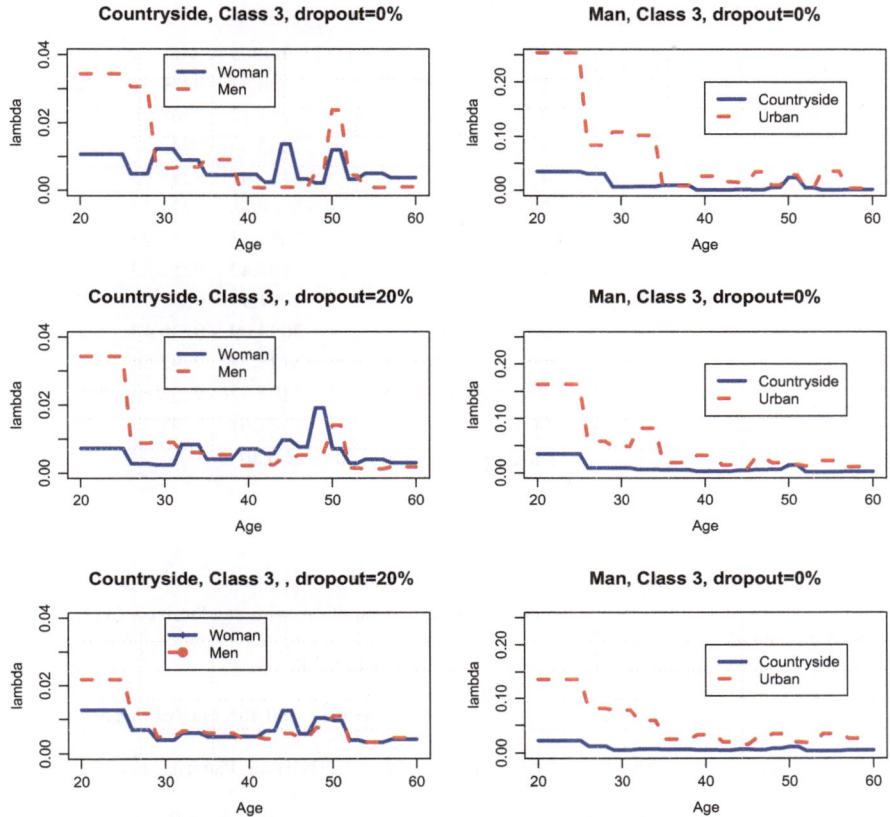

Fig. 6.8 Forecast claims frequency with the NN(10,20,10,1) neural network, for $\alpha^{(1)}$ set to 0%, 20% and 40%

6.6   Further Readings

The ensemble and randomization methods have also been successfully applied to regression trees by Tin Kam (1995). This approach, called random forests, consists to build a multitude of decision trees. It has been popularized by Breiman (1996), Amit and Geman (1997). We refer to Trufin et al. (2019) for further details. There exist many variants of ensemble methods. For example, stacking is a technique for achieving the highest generalization accuracy (Wolpert 1992). By using a meta-learner, this method tries to induce which predictors are reliable and which are not. Stacking is usually employed to combine different models. For a detailed survey on ensemble-based classifiers, we refer to Rokach (2010). Recently, the ensemble learners have showed great success in a variety of areas, including precision medicine (Luedtke and van der Laan 2016), mortality prediction (Pirracchio et al. 2015) or seismology (Shimshoni and Intrator 1998).

Table 6.3 K-fold cross validation of a NN(10,20,10,1) network with a partition of the data set in 10 validation samples

K	Deviance learning set	Log-likelihood	AIC	Deviance test set
1	5058.77	−3148.75	9339.5	506.02
2	5002.29	−3111.12	9264.25	558.99
3	5004.4	−3103.95	9249.91	642.59
4	5003.82	−3116.97	9275.93	526.69
5	4964.81	−3089.38	9220.77	610.64
6	4916.97	−3054.23	9150.47	684.34
7	4997.49	−3108.42	9258.83	589.94
8	4984.48	−3097.99	9237.98	595.38
9	4972.83	−3097.09	9236.17	566.1
10	5021.97	−3125.27	9292.54	541.27
Average	4992.78	−3105.32	9252.63	582.19
St. Dev.	37.38	24.61	49.22	54.24

The drop out rate is set to $\lambda = 0.40$

References

Amit Y, Geman D (1997) Shape quantization and recognition with randomized trees. Neural Comput 9(7):1545–1588

Breiman L (1996) Bagging predictors. Mach Learn 24(2):123–140

Breiman L (2001) Random forests. Mach Learn 45:5–32

Geman S, Bienenstock E, Doursat R (1992) Neural networks and the bias/variance dilemma. Neural Comput 4:1–58

Hansen L, Salamon P (1990) Neural network ensembles. IEEE Trans Pattern Anal Mach Intell 12:993–1001

Liu Y, Wang Y, Li Y, Zhang B, Wu G (2004) Earthquake prediction by RBF neural network ensemble. In: Yin F-L, Wang J, Guo C (eds) Advances in neural networks ISNN2004. Springer, Berlin, pp 962–969

Louppe G (2014) Understanding random forests: from theory to practice. PhD Dissertation, Faculty of Applied Sciences, Liége University

Luedtke AR, van der Laan MJ (2016) Super-learning of an optimal dynamic treatment rule. Int J Biochem 12(1):305–332

Pirracchio R, Petersen ML, Carone M, Rigon MR, Chevret S, van der Laan MJ (2015) Mortality prediction in intensive care units with the super ICU learner algorithm (SICULA): a population based study. Lancet Respir Med 3(1):42–52

Rokach L (2010) Ensemble-based classifiers. Artif Intell Rev 33:1–39

Shimshoni Y, Intrator N (1998) Classification of seismic signals by integrating ensembles of neural networks. IEEE Trans Signal Process 46(5):1194–1201

Shu C, Burn DH (2004) Artificial neural network ensembles and their application in pooled flood frequency analysis. Water Resour Res 40:1–10

Tin Kam H (1995) Random decision forests. In: Proceedings of the 3rd international conference on document analysis and recognition. IEEE, Piscataway, pp 278–282

Trufin J, Denuit M, Hainaut D (2019) Effective statistical learning methods for actuaries. From CART to GBM. Springer, Berlin

Wolpert DH (1992) Stacked generalization. Neural Netw 5:241–259

Chapter 7
Gradient Boosting with Neural Networks

Gradient boosting machines form a family of powerful machine learning techniques that have been applied with success in a wide range of practical applications. Ensemble techniques rely on simple averaging of models in the ensemble. The family of boosting methods adopts a different strategy to construct ensembles. In boosting algorithms, new models are sequentially added to the ensemble. At each iteration, a new weak base-learner is trained with respect to the error of the whole ensemble built so far.

7.1 The Gradient Boosting Machine (GBM)

As in Chap. 1, we consider an insurance portfolio of n policies. The information about policies is reported in vectors $(x_i)_{i=1,\dots,n}$ of dimension p. As before, p is the number of available descriptive variables for a contract. The vector $y = (y_i)_{i=1,\dots,n}$ contains observed key quantities for the insurer, e.g. the frequency of claims. The vector of exposure is denoted $v = (v_i)_{i=1,\dots,n}$.

We assume that Y_i are distributed according to an exponential dispersed (ED) distribution. Its density is denoted by $f_{Y_i}(y; \theta_i, \phi) = \exp\left\{\frac{y\theta_i - a(\theta_i)}{\phi/v_i}\right\} c(y, \phi, v_i)$. The cumulant function $a(\theta_i)$ is C^2 with invertible second derivative. Our objective is to build an estimate, $\widehat{y}_i$, of the expected key ratio, $\mathbb{E}(Y_i)$, based on the available information x_i. This estimator is a non-linear mapping from x_i to $\widehat{y}_i : \widehat{y}_i = F(x_i)$. The best estimator $F^*(.)$ minimizes the average of a loss function over the portfolio:

$$F^*(.) = \arg\min_F \frac{1}{n} \sum_{i=1}^{n} \mathcal{L}(y_i, F(x_i), v_i) . \tag{7.1}$$

© Springer Nature Switzerland AG 2019
M. Denuit et al., *Effective Statistical Learning Methods for Actuaries III*,
Springer Actuarial, https://doi.org/10.1007/978-3-030-25827-6_7

The loss function, $\mathcal{L}$ is proportional to the deviance. A common procedure consists to restrict $F(.)$ to belong to a parameterized class of functions. In this chapter, we assume instead that $F(.)$ is a transformation $h(.)$ of an additive form:

$$F_m(x) = h\left(\sum_{j=1}^{m} \beta_j f_j(x)\right) \tag{7.2}$$

$$= h\left(S_m(x)\right)$$

where $m \in \mathbb{N}$ is the order of the expansion and $h(.)$ is a C^2 invertible monotone function. The function $f_j(x)$ is here the response of a feedforward neural networks with n_j weights contained in a vector $\Omega_j = \left(\omega_k^j\right)_{k=1,\dots,n_j}$. The function $f_j(x)$ will be a relatively basic network, e.g. like a shallow network, and is for this reason called a base learner. The function $h(.)$ ensures that the output of the predictor is well located in the domain of Y_i. We will e.g. choose $h(x) = \exp(x)$ or $h(x) = \mathrm{logit}(x)$ if Y_i has a Poisson or a binomial distribution. The quantity $S_m(x) = \sum_{j=1}^{m} \beta_j f_j(x)$ is the linear combination of base learners. Under the assumption (7.2), we move then to another optimization problem that takes place in a functions space. Under the assumption (7.2), the optimal estimator of order m is solution of:

$$S_m^*(.) = \arg\min_{S_m(.)} \frac{1}{n}\sum_{i=1}^{n} \mathcal{L}\left(y_i, h\left(S_m(x_i)\right), v_i\right) .$$

Finding a global solution to this optimization problem is in practice a very hard task. Instead, we proceed iteratively with a steepest descent algorithm. Let us assume that we have already performed m iterations of this algorithm and found a series of optimal base learners $\left(f_1^*, \dots, f_m^*\right)$ such that $F_m^*(x) = h\left(\sum_{j=1}^{m} \beta_j^* f_j^*(x)\right)$. In order to find the optimal estimator of order $m + 1$, we first calculate the negative gradient vector of the aggregated loss. Let us recall that the contribution of the ith contract to the aggregated deviance is proportional to the exposure v_i (see Table 1.3 of Chap. 1). The gradient is hence also a multiple of the exposure. For this reason, the negative gradient vector is a vector: $(v_i g_i)_{i=1,\dots,n}$ such that

$$v_i g_i = -\left.\frac{\partial}{\partial S_m(x_i)}\left(\frac{1}{n}\sum_{k=1}^{n} \mathcal{L}\left(y_k, h\left(S_m(x_k)\right), v_k\right)\right)\right|_{S_m(x_i)=S_m^*(x_i)}$$

$$= v_i \underbrace{\left(-\frac{1}{n}\frac{\partial \mathcal{L}\left(y_i, F_m(x_i), 1\right)}{\partial F_m(x_i)}\left.\frac{\partial h\left(S_m(x_i)\right)}{\partial S_m(x_i)}\right|_{S_m(x_i)=S_m^*(x_i)}\right)}_{g_i} .$$

A first-order Taylor's expansion of the loss function leads to the following approximation for any other function $S(.)$:

$$\mathcal{L}\left(y_i, h\left(S(\boldsymbol{x}_i)\right), v_i\right) \approx \mathcal{L}\left(y_i, h\left(S_m^*(\boldsymbol{x}_i)\right), v_i\right) - v_i g_i \left(S(\boldsymbol{x}_i) - S_m^*(\boldsymbol{x}_i)\right). \quad (7.3)$$

Therefore, for a small enough step of size $\rho_m > 0$, the following estimator of order $m + 1$

$$S_{m+1}(\boldsymbol{x}_i) = S_m^*(\boldsymbol{x}_i) + \rho_{m+1}\, g_i\, , \quad (7.4)$$

locally decreases the loss since

$$\mathcal{L}\left(y_i, h\left(S_{m+1}(\boldsymbol{x}_i)\right), v_i\right) = \mathcal{L}\left(y_i, h\left(S_m^*(\boldsymbol{x}_i)\right), v_i\right) - \rho_{m+1} v_i g_i^2$$
$$< \mathcal{L}\left(y_i, h\left(S_m^*(\boldsymbol{x}_i)\right), v_i\right) .$$

The optimal step size ρ_{m+1}^* is obtained by minimizing the sum of right terms in Eq. (7.3):

$$\rho_{m+1}^* = \arg\min_{\rho} \sum_{i=1}^{n} \left(\mathcal{L}\left(y_i h\left(S_m^*(\boldsymbol{x}_i)\right), v_i\right) - \rho g_i^2\right) .$$

The optimal base learner of order $m + 1$ takes values $f_{m+1}^*(\boldsymbol{x}_i)$ at points $(\boldsymbol{x}_i)_{i,...,n}$ but we have no idea about values of the estimator $F(\boldsymbol{x}) = h\left(S_{m+1}^*(\boldsymbol{x})\right)$ for an insurance contract with a combination of specifications $\boldsymbol{x}$ that is not recorded in the training dataset. To solve this issue, we can fit the base learner $f_{m+1}^*(\boldsymbol{x})$ regressing the $\rho_{m+1}\, g_i$ on $\boldsymbol{x}_i$. Instead, a common practice consists first to calibrate the network $f_{m+1}^*(\boldsymbol{x})$ that regresses the adjusted gradient vector $\boldsymbol{g} = (g_i)_{i=1,...,n}$ on portfolio characteristics $(\boldsymbol{x}_i)_{i=1,...,n}$. The neural network is fitted by minimizing the mean squared error between the gradient and the output of the base learner. We detail this point in the following subsections. The optimal weight β_m^* is next found by minimizing the total loss:

$$\beta_{m+1}^* = \arg\min_{\beta} \frac{1}{n} \sum_{i=1}^{n} \left(\mathcal{L}\left(y_i, h\left(S_m^*(\boldsymbol{x}_i) + \beta\, f_{m+1}^*(\boldsymbol{x}_i)\right), v_i\right)\right) . \quad (7.5)$$

The gradient boosting procedure, is summarized in Algorithm 7.1. The next four subsections detail the expressions of the negative gradient $\boldsymbol{g}$, of β_0 and the solution of Eq. (7.5) when Y_i has a Normal, Poisson, Gamma and Binomial distribution.

Algorithm 7.1 Gradient boosting machine (GBM)

Initialization:

Choose an architecture for neural base learners.

Initialize the estimator $F_0^*(x) = h(f_0^*(x))$:

$$f_0^*(x) = \beta_0^* = \arg\min_\beta \frac{1}{n} \sum_{i=1}^n \mathcal{L}(y_i, h(\beta), v_i).\tag{7.6}$$

Main procedure:

For $m = 0$ to maximum epoch, M

1. Calculate the negative gradient $\boldsymbol{g} = (g_i)_{i=1,\dots,n}$ of the aggregated loss:

$$g_i = -\frac{1}{n} \frac{\partial \mathcal{L}(y_i, F_m(x_i), 1)}{\partial F_m(x_i)} \frac{\partial h(S_m(x_i))}{\partial S_m(x_i)}\bigg|_{S_m(x_i)=S_m^*(x_i)}$$

2. Estimate the weights Ω_{m+1} of a neural network $f_{m+1}^*(x)$, regressing $(g_i)_{i=1,\dots,n}$ on $(x_i)_{i=1,\dots,n}$.

3. Compute β_{m+1}^* such that

$$\beta_{m+1}^* = \arg\min_\beta \frac{1}{n} \sum_{i=1}^n \left(\mathcal{L}\left(y_i, h\left(S_m^*(x_i) + \beta f_{m+1}^*(x_i)\right), v_i\right)\right).\tag{7.7}$$

4. Update the estimator $F_{m+1}^*(x)$ as follows:

$$F_{m+1}^*(x) = h\left(S_m^*(x) + \beta_{m+1}^* f_{m+1}^*(x)\right)$$

End loop on epochs

7.1.1 The Gaussian Case

In this paragraph, we focus on the prediction of key quantities that are Normally distributed. The domain of Y_i is in this case equal to $\mathbb{R}$ and the function matching this domain with the one of the base learner is the identity function, $h(x) = x$. We use the unscaled deviance as loss function, which is in this case the mean square error weighted by the exposure:

$$\mathcal{L}(y_i, , F_m(x_i), v_i) = v_i (y_i - F_m(x_i))^2.$$

A direct calculation, leads to the following expression for the negative gradient in the mth iteration of the gradient boosting machine:

$$\begin{aligned}
g_i &= -\frac{1}{n} \frac{\partial \mathcal{L}(y_i, F_m(x_i), 1)}{\partial F_m(x_i)} \frac{\partial h(S_m(x_i))}{\partial S_m(x_i)}\bigg|_{F_m(x_i)=F_m^*(x_i)} \\
&= -\frac{2}{n}\left(F_m^*(x_i) - y_i\right) \\
&= -\frac{2}{n}\left(\widehat{y}_i^{(m)} - y_i\right) i = 1, \dots, n.
\end{aligned}\tag{7.8}$$

where $\widehat{y}_i^{(m)} = F_m^*(x_i)$ is the best estimate of y_i at iteration m. The gradient is therefore proportional to errors of prediction, weighted by the exposure. The base learner $f_m^*(x)$ in Algorithm 7.1 is fitted to negative gradients $\boldsymbol{g} = (g_i)_{i=1,\dots,n}$. Let us recall that $f_m(.)$ is a neural network with n_m weights contained in a vector $\Omega_m = (\omega_k^m)_{k=1,\dots,n_j}$. The optimal neural weights Ω_m^* of $f_m^*(x)$ are found by minimizing the squared errors between negative gradients and outputs of $f_m(.)$. In order to take into account the exposure, squared errors are weighted by v_i:

$$\Omega_m^* = \arg\min_{\Omega_m} \frac{1}{n} \sum_{i=1}^n v_i \left(g_i - f_m(x_i)\right)^2 . \tag{7.9}$$

If we refer to Table 1.3 of Chap. 1, this objective function (7.9) is also the deviance of a Gaussian distribution. This means that we implicitly assume that g_i are realizations of a Normal random variable. Here, this assumption is well satisfied since the gradient (7.8) is directly proportional to Y_i that has a Gaussian distribution.

Next we estimate β_m^* by solving the problem (7.5). The mechanism of the GBM is clear in this case. The algorithm recursively fit base learners to errors of prediction till the required accuracy is reached. The following proposition details the GBM initialization.

Proposition 7.10 *The parameter β_0^* in Eq. (7.6) used for initializing the GBM procedure is equal to*

$$\beta_0^* = \frac{\sum_{i=1}^n v_i y_i}{\sum_{i=1}^n v_i} . \tag{7.10}$$

Proof The parameter β_0^* in Eq. (7.6) is in this case solution of the following optimization problem:

$$\beta_0^* = \arg\min_{\beta} \frac{1}{n} \sum_{i=1}^n v_i \left(y_i - \beta\right)^2 .$$

Cancelling the derivative of this objective with respect to β allows us to infer Eq. (7.10). □

In the Gaussian case, finding the β_m^* that minimizes the distance between observations and the predictors does not require any numerical procedure:

Proposition 7.11 *The parameter β_m^* solution of Eq. (7.7) is equal to*

$$\beta_m^* = \frac{\sum_{i=1}^n v_i f_m^*(x_i) \left(y_i - \widehat{y}_i^{(m-1)}\right)}{\sum_{i=1}^n v_i f_m^*(x_i)} \tag{7.11}$$

Proof Let us recall that $F_{m-1}^*(x_i) = \widehat{y}_i^{(m-1)}$. Therefore, Eq. (7.7) can be rewritten as follows:

$$\beta_m^* = \arg\min_\beta \frac{1}{n} \sum_{i=1}^n \left(\mathcal{L}\left(y_i, \widehat{y}_i^{(m-1)} + \beta\, f_m^*(x_i), v_i \right) \right) ,$$

$$= \arg\min_\beta \frac{1}{n} \sum_{i=1}^n v_i \left(y_i - \widehat{y}_i^{(m-1)} - \beta\, f_m^*(x_i) \right)^2 .$$

Cancelling the first order derivative of this last objective function with respect to β leads to the equality

$$0 = -\frac{2}{n} \sum_{i=1}^n v_i\, f_m^*(x_i) \left(y_i - \widehat{y}_i^{(m-1)} - \beta\, f_m^*(x_i) \right) ,$$

which admits (7.11) as solution.   □

7.1.2   The Poisson Case, Standard GBM

Let us assume that Y_i is the ratio of N_i, a Poisson random variable on an exposure v_i. The domain of Y_i is in this case equal to $\mathbb{R}^+$ but the optimal estimator $F_m^*(.)$ sums up the outputs of m neural networks taking values in $\mathbb{R}$. As done in Sect. 1.7, we match the domains of Y_i and of $F_m^*(.)$ by considering an exponential link function, $h(x) = \exp(x)$, between the estimate and the sum of base learners outputs:

$$\widehat{y}_i^{(m)} = F_m^*(x_i) = \exp\left(S_m^*(x_i) \right) = \exp\left(\sum_{j=1}^m \beta_j^* f_j^*(x_i) \right) \qquad i = 1, \ldots n . \qquad (7.12)$$

This assumption guarantees the positivity of predictions. The best criterion for adjusting a prediction model to realizations of a Poisson random variable is the unscaled deviance:

$$\mathcal{L}(y_i, \widehat{y}_i, v_i) = \begin{cases} 2v_i \left(y_i \ln y_i - y_i \ln \widehat{y}_i - y_i + \widehat{y}_i \right) & y_i > 0 \\ 2v_i \widehat{y}_i & y_i = 0 \end{cases} . \qquad (7.13)$$

For this choice of loss function, a direct calculation leads to the following expression for the negative gradient in the mth iteration of the gradient boosting machine:

$$g_i = -\frac{1}{n} \frac{\partial \mathcal{L}(y_i, F_m(\boldsymbol{x}_i), 1)}{\partial F_m(\boldsymbol{x}_i)} \frac{\partial h(S_m(\boldsymbol{x}_i))}{\partial S_m(\boldsymbol{x}_i)} \Bigg|_{F_m(\boldsymbol{x}_i)=F_m^*(\boldsymbol{x}_i)}$$

$$= -\frac{2}{n} \left(\frac{\widehat{y}_i^{(m)} - y_i}{\widehat{y}_i^{(m)}} \right) \widehat{y}_i^{(m)}$$

$$= -\frac{2}{n} \left(\widehat{y}_i^{(m)} - y_i \right) i = 1, \ldots, n . \tag{7.14}$$

In the standard GBM algorithm 7.1, the base learner $f_m^*(\boldsymbol{x})$ is adjusted to $g = (g_i)_{i=1,\ldots,n}$. The optimal neural weights Ω_m^* of $f_m^*(\boldsymbol{x})$ are found by minimizing the weighted sum of squared error (7.9) between negative gradients and outputs of $f_m(.)$. We have mentioned in the previous subsection that choosing this error is equivalent to assume that g_i are realizations of a Normal random variable. Here, this assumption is not satisfied since the negative gradient (7.14) is proportional to N_i which has a Poisson distribution. When the frequency of N_i is high, the Poisson distribution can nevertheless be approached by a normal law according to the central limit theorem. As illustrated in Fig. 7.1, this approximation does not hold anymore when the frequency of the number of events is low, as for insurance claims. For analyzing this type of data, we use the alternative gradient boosting algorithm presented in the next section.

Once that the base learner is fitted to negative gradients, we have to calculate the series of β_m^*. The predictor $F_m(x)$ is initialized as follows:

Proposition 7.12 *The parameter β_0^* in Eq. (7.6) used for initializing the GBM procedure is equal to*

$$\beta_0^* = \ln \left(\frac{\sum_{i=1}^n v_i y_i}{\sum_{i=1}^n v_i} \right) . \tag{7.15}$$

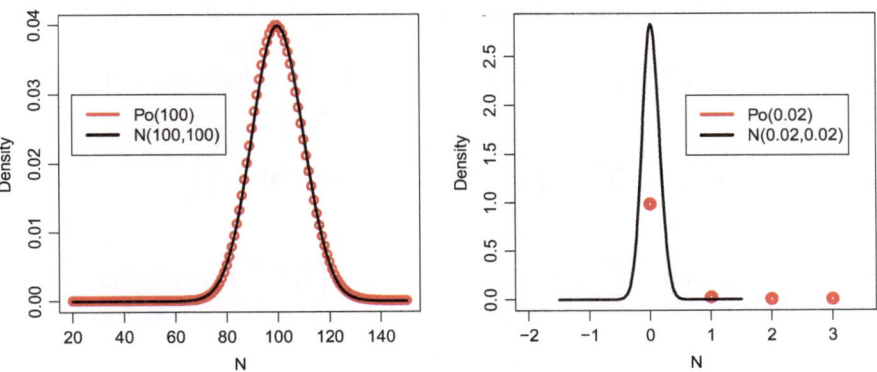

Fig. 7.1 Comparison of Poisson and normal densities with same means and variances

Proof The parameter β_0^* in Eq. (7.6) is in this case solution of the following optimization problem:

$$\beta_0^* = \arg\min_\beta \frac{2}{n} \sum_{i=1}^n v_i \left(1_{y_i > 0} y_i \ln y_i - y_i \beta - y_i + e^\beta \right) .$$

If we derive this objective with respect to β and cancel the derivative, we conclude that β_0^* is solution of $0 = \frac{2}{n} \sum_{i=1}^n v_i \left(e^\beta - y_i \right)$.   □

The introduction of an exponential link between the forecast and the output of $F_m(x)$ prevents us to find an analytical solution for the optimal β_m^*. The next proposition details the numerical procedure for computing β_m^*:

Proposition 7.13 *The parameter β_m^* solution of Eq. (7.7) is computed iteratively. For a level of accuracy ϵ, we construct a series of $\beta_m^{(k)}$ initialized by*

$$\beta_m^{(0)} = \frac{\sum_{i=1}^n v_i \, f_m^*(x_i) \left(y_i - \widehat{y}_i^{(m-1)} \right)}{\sum_{i=1}^n v_i \widehat{y}_i^{(m-1)} f_m^*(x_i)^2} , \tag{7.16}$$

satisfying the relation

$$\beta_m^{(k+1)} = \beta_m^{(k)} - \frac{v(\beta_m^{(k)})}{v'(\beta_m^{(k)})} , \tag{7.17}$$

for $k = 1, \ldots, l$. The function $v(.)$ and its first order derivatives are defined by:

$$v(\beta) = \sum_{i=1}^n v_i \, f_m^*(x_i) \left(\widehat{y}_i^{(m-1)} \exp \left(\beta \, f_m^*(x_i) \right) - y_i \right) , \tag{7.18}$$

$$v'(\beta) = \sum_{i=1}^n v_i \, f_m^*(x_i)^2 \, \widehat{y}_i^{(m-1)} \exp \left(\beta \, f_m^*(x_i) \right) .$$

The series is stopped after l iterations, when $v\left(\beta_m^{(l)} \right) \le \epsilon$ for the first time.

Proof Since $\exp \left(\sum_{j=1}^{m-1} \beta_j^* f_j^*(x_i) \right) = \widehat{y}_i^{(m-1)}$, Eq. (7.7) becomes in the Poisson case:

$$\beta_m^* = \arg\min_\beta \frac{1}{n} \sum_{i=1}^n \left(\mathcal{L} \left(y_i, \widehat{y}_i^{(m-1)} \exp \left(\beta \, f_m^*(x_i) \right) , v_i \right) \right) ,$$

$$= \arg\min_\beta \frac{2}{n} \sum_{i=1}^n v_i \left(1_{y_i > 0} y_i \ln y_i - y_i \ln \left(\widehat{y}_i^{(m-1)} \right) - y_i \beta \, f_m^*(x_i) \right.$$

$$\left. - y_i + \widehat{y}_i^{(m-1)} \exp \left(\beta \, f_m^*(x_i) \right) \right) .$$

Cancelling the derivative of this objective function leads to the equality

$$
0 = \sum_{i=1}^{n} v_i \underbrace{\left(-y_i \, f_m^*(\boldsymbol{x}_i) + \widehat{y}_i^{(m-1)} f_m^*(\boldsymbol{x}_i) \exp\left(\beta \, f_m^*(\boldsymbol{x}_i) \right) \right)}_{v(\beta)},
$$

whose β_m^* is solution. However, this last equation does not admit an analytical solution and we instead solve it with the Newton-Raphson algorithm. In order to limit the computation time, the algorithm is initialized with a starting point obtained after a first-order Taylor's expansion of the exponential function $\exp\left(\beta \, f_m^*(\boldsymbol{x}_i) \right) \approx 1 + \beta \, f_m^*(\boldsymbol{x}_i)$. The equality $v(\beta)=0$ is therefore approached by

$$
\sum_{i=1}^{n} v_i y_i \, f_m^*(\boldsymbol{x}_i) \approx \sum_{i=1}^{n} v_i \widehat{y}_i^{(m-1)} f_m^*(\boldsymbol{x}_i) + \sum_{i=1}^{n} v_i \widehat{y}_i^{(m-1)} f_m^*(\boldsymbol{x}_i)^2 \beta,
$$

which admits (7.16) as solution. Equations (7.17) and (7.18) are the standard iterations of the Newton-Raphson algorithm.   □

7.1.2.1 The Poisson Boosting Method (PBM) for Low Frequencies

In the standard GBM algorithm 7.1, the base learner $f_m^*(\boldsymbol{x})$ is fitted to $g = (g_i)_{i=1,\ldots,n}$ by minimization of a weighted sum of squared errors. This approach is acceptable if N_i can be approached reasonably well by a Gaussian random variable, i.e. when the frequency of N_i is high. In actuarial sciences, this is however not the case since N_i represents a number of rare claims caused by the ith policyholder. In this situation, a powerful alternative method to the classical gradient boosting machine exists. We assume that the estimate of Y_i is a function of the sum of base learners with weights $\beta_j^* = 1$ for $j = 1, \ldots, n$:

$$
\widehat{y}_i^{(m)} = F_m^*(\boldsymbol{x}_i) = \exp\left(\sum_{j=1}^{m} f_j^*(\boldsymbol{x}_i) \right) \quad i = 1, \ldots n. \tag{7.19}
$$

Since $Y_i = \frac{N_i}{v_i}$, the unscaled deviance (7.13) becomes:

$$
\mathcal{L}\left(n_i, \widehat{y}_i, v_i \right) = 2 \left(\widehat{y}_i v_i - 1_{n_i>0} n_i \ln\left(\frac{\widehat{y}_i v_i}{n_i} \right) - n_i \right) \tag{7.20}
$$

$$
= 2 n_i \left(\frac{\widehat{y}_i v_i}{n_i} - 1_{n_i>0} \ln\left(\frac{\widehat{y}_i v_i}{n_i} \right) - 1 \right),
$$

where n_i is here the realization of N_i (and should not be confused with the number of neurons in base learners). The optimal base learner at iteration m is therefore solution of:

$$f_m^*(x) = \arg \min_{f_m(.)} \frac{1}{n} \sum_{i=1}^{n} \mathcal{L}\left(n_i, \widehat{y}_i^{(m-1)} \exp\left(f_m(x_i)\right), v_i\right).$$

The sum of unscaled deviances in the objective function is rewritten as follows:

$$\frac{1}{n} \sum_{i=1}^{n} \mathcal{L}\left(n_i, \widehat{y}_i^{(m-1)} \exp\left(f_m(x_i)\right), v_i\right)$$

$$= \frac{2}{n} \sum_{i=1}^{n} n_i \left(\frac{\widehat{y}_i^{(m-1)} v_i e^{f_m(x_i)}}{n_i} - 1_{n_i>0} \ln\left(\frac{\widehat{y}_i^{(m-1)} v_i e^{f_m(x_i)}}{n_i}\right) - 1\right)$$

$$= \frac{2}{n} \sum_{i=1}^{n} n_i \left(\frac{\tilde{v}_i^{(m)} e^{f_m(x_i)}}{n_i} - 1_{n_i>0} \ln\left(\frac{\tilde{v}_i^{(m)} e^{f_m(x_i)}}{n_i}\right) - 1\right)$$

$$= \frac{1}{n} \sum_{i=1}^{n} \mathcal{L}\left(n_i, \exp\left(f_m(x_i)\right), \tilde{v}_i^{(m)}\right)$$

where $\tilde{v}_i^{(m)} = \widehat{y}_i^{(m-1)} v_i$ for $i = 1, \ldots, n$ are modified exposures. Here, $\exp\left(f_m(x_i)\right)$ is an estimate of the frequency of N_i but this time computed with different exposures. This suggests the GBM algorithm 7.2.

7.1.3   The Gamma Case

Let us assume that $Y_i \in \mathbb{R}^+$ has a Gamma distribution. As in the Poisson case, we match the domains of Y_i and of neural networks outputs with an exponential transformation:

$$\widehat{y}_i^{(m)} = F_m^*(x_i) = \exp\left(\sum_{j=1}^{m} \beta_j^* f_j^*(x_i)\right) \quad i = 1, \ldots, n. \tag{7.22}$$

The loss function in this case is the Gamma unscaled deviance equal to:

$$\mathcal{L}(y_i, \widehat{y}_i, v_i) = \begin{cases} 2v_i\left(\frac{y_i}{\widehat{y}_i} - 1 - \ln\left(\frac{y_i}{\widehat{y}_i}\right)\right) & y_i > 0 \\ 0 & y_i = 0 \end{cases}. \tag{7.23}$$

Algorithm 7.2 Gradient boosting machine (GBM) for low frequency Poisson observations

---

Initialization:

Choose an architecture for neural base learners.
Initialize the estimator $F_0^*(x) = \exp\left(f_0^*(x)\right)$ where

$$f_0^*(x) = \beta_0^* = \ln\left(\frac{\sum_{i=1}^n n_i}{\sum_{i=1}^n v_i}\right). \tag{7.21}$$

Main procedure:

For $m = 1$ to maximum epoch, M

1. Calculate the modified exposure $\tilde{v}_i^{(m)}$ for $i = 1, \ldots, n$

$$\tilde{v}_i^{(m)} = \widehat{y}_i^{(m-1)} v_i = F_{m-1}^*(x_i) v_i.$$

2. Estimate the weights Ω_m of a neural network $f_m(x)$:

$$f_m^*(x_i) = \arg\min_{f_m(.)} \frac{1}{n} \sum_{i=1}^n \mathcal{L}\left(n_i, \exp\left(f_m(x_i)\right), \tilde{v}_i^{(m)}\right).$$

3. Update the estimator $F_m^*(x)$ as follows:

$$F_m^*(x) = F_{m-1}^*(x) \exp\left(f_m(x)\right)$$

End loop on epochs

For this loss function, the negative gradient in the mth iteration of the gradient boosting machine is given by:

$$g_i = -\frac{1}{n} \frac{\partial \mathcal{L}\left(y_i, F_m(x_i), 1\right)}{\partial F_m(x_i)} \frac{\partial h\left(S_m(x_i)\right)}{\partial S_m(x_i)}\Bigg|_{F_m(x_i)=F_m^*(x_i)}$$

$$= -\frac{2\widehat{y}_i^{(m)}}{n} 1_{y_i>0} \frac{\partial}{\partial F_m(x_i)} \left(\frac{y_i}{F_m(x_i)} - 1 - \ln(y_i) + \ln F_m(x_i)\right)\Bigg|_{F_m(x_i)=F_m^*(x_i)}$$

$$= -\frac{2}{n} \frac{\left(\widehat{y}_i^{(m)} - y_i\right)}{\widehat{y}_i^{(m)}} i = 1, \ldots, n. \tag{7.24}$$

This time, the negative gradient is proportional to the relative error of forecast with respect to the predictor, weighted by the exposure. The base learner $f_m^*(x)$ in Algorithm 7.1 is next fitted to negative gradients $g = (g_i)_{i=1,\ldots,n}$ by minimization of weighted quadratic errors:

$$f_m^*(x) = \arg\min_{f_m(.)} \frac{1}{n} \sum_{i=1}^n v_i \left(g_i - f_m(x_i)\right)^2.$$

Choosing this optimization criterion is equivalent to assume that g_i are realizations of a Normal random variable. Since the negative gradient (7.14) is proportional to a Gamma random variable, this assumption does not hold. However, a Gamma random variable with an integer shape parameter $\alpha \in \mathbb{N}$ is the sum of α exponential random variables. According to the central limit theorem, if the shape parameter is sufficiently high, the Normal approximation is satisfactory.

The initialization of Algorithm 7.1 is similar to the Poisson case:

Proposition 7.14 *The parameter β_0^* in Eq. (7.6) used for initializing the GBM procedure is equal to*

$$\beta_0^* = \ln\left(\frac{\sum_{i=1}^n 1_{y_i>0} v_i y_i}{\sum_{i=1}^n 1_{y_i>0} v_i} \right). \tag{7.25}$$

Proof Since $\widehat{y}_i^{(0)} = e^{\beta_0}$, β_0^* is solution of the following optimization problem:

$$\beta_0^* = \arg\min_\beta \frac{2}{n} \sum_{i=1}^n v_i 1_{y_i>0} \left(y_i e^{-\beta} - 1 - \ln(y_i) + \beta \right).$$

Cancelling the derivative leads to the equation $\sum_{i=1}^n 1_{y_i>0} v_i \left(1 - y_i e^{-\beta}\right) = 0$ whose β_0^* is solution.   □

The series of optimal β_m^* is found numerically with the following iterative procedure:

Proposition 7.15 *The parameter β_m solution of Eq. (7.7) is computed iteratively. For a level of accuracy ϵ, we construct a series of $\beta_m^{(k)}$ initialized by*

$$\beta_m^{(0)} = \frac{\sum_{i=1}^n 1_{y_i>0} v_i f_m^*(x_i) \left(\frac{y_i}{\widehat{y}_i^{(m-1)}} - 1 \right)}{\sum_{i=1}^n 1_{y_i>0} v_i f_m^*(x_i)^2 \frac{y_i}{\widehat{y}_i^{(m-1)}}}, \tag{7.26}$$

satisfying the relation

$$\beta_m^{(k+1)} = \beta_m^{(k)} - \frac{v(\beta_m^{(k)})}{v'(\beta_m^{(k)})}, \tag{7.27}$$

for $k = 1, \ldots, l$. The function $v(.)$ and its first order derivatives are defined by:

$$v(\beta) = \sum_{i=1}^n 1_{y_i>0} v_i f_m^*(x_i) \left(1 - \frac{y_i}{\widehat{y}_i^{(m-1)}} \exp\left(-\beta f_m^*(x_i)\right) \right), \tag{7.28}$$

$$v'(\beta) = \sum_{i=1}^n 1_{y_i>0} v_i \frac{y_i}{\widehat{y}_i^{(m-1)}} f_m^*(x_i)^2 \exp\left(-\beta f_m^*(x_i)\right).$$

The series is stopped after l iterations, when $v\left(\beta_m^{(l)}\right) \le \epsilon$ for the first time.

Proof Since $\exp\left(\sum_{j=1}^{m-1}\beta_j^* f_j^*(x_i)\right) = \widehat{y}_i^{(m-1)}$, Eq. (7.7) becomes in the Gamma case:

$$\beta_m^* = \arg\min_\beta \frac{1}{n}\sum_{i=1}^{n}\left(\mathcal{L}\left(y_i, \widehat{y}_i^{(m-1)}\exp\left(\beta\, f_m^*(x_i)\right), \nu_i\right)\right).$$

$$= \arg\min_\beta \frac{2}{n}\sum_{i=1}^{n} 1_{y_i>0}\nu_i\left(\frac{y_i}{\widehat{y}_i^{(m-1)}}\exp\left(-\beta\, f_m^*(x_i)\right)-1\right.$$

$$\left. -\ln\left(y_i\right)+\ln\left(\widehat{y}_i^{(m-1)}\right)+\beta\, f_m^*(x_i)\right).$$

Therefore β_m^* is solution of the following equality:

$$0 = \underbrace{\sum_{i=1}^{n} 1_{y_i>0}\nu_i f_m^*(x_i)\left(1-\frac{y_i}{\widehat{y}_i^{(m-1)}}\exp\left(-\beta\, f_m^*(x_i)\right)\right)}_{v(\beta)}.$$

As in the Poisson case, β_m^* has no analytical expression. Instead, we compute it numerically with the Newton-Raphson algorithm. The algorithm is initialized with a starting point obtained after a first order Taylor's development of the exponential function $\exp\left(-\beta\, f_m^*(x_i)\right) \approx 1 - \beta\, f_m^*(x_i)$. The equality $v(\beta) = 0$ becomes

$$\sum_{i=1}^{n} 1_{y_i>0}\nu_i f_m^*(x_i) \approx \sum_{i=1}^{n} 1_{y_i>0}\nu_i f_m^*(x_i)\frac{y_i}{\widehat{y}_i^{(m-1)}}\left(1-\beta\, f_m^*(x_i)\right)$$

$$= \sum_{i=1}^{n} 1_{y_i>0}\nu_i f_m^*(x_i)\frac{y_i}{\widehat{y}_i^{(m-1)}} - \beta\sum_{i=1}^{n} 1_{y_i>0}\nu_i f_m^*(x_i)^2\frac{y_i}{\widehat{y}_i^{(m-1)}}$$

which admits Eq. (7.26) as solution. Equations (7.27) and (7.28) are the standard iterations of the Newton-Raphson algorithm. $\qquad\square$

7.1.4 The Binomial Case

In this last case, we assume that $\nu_i \times Y_i$ is a Binomial random variable $Bin(\nu_i, p_i)$. Here $Y_i \in [0, 1]$ and is such that $\mathbb{E}(Y_i) = p_i$. In order to match the domains of Y_i and of base learners, we consider a logistic link function, denoted by $\mathrm{lgc}(.)$:

$$\widehat{y}_i^{(m)} = F_m^*(x_i) = \mathrm{lgc}\left(S_m^*(x_i)\right)$$

$$= \frac{1}{1+\exp\left(-S_m^*(x_i)\right)}.$$

By construction, $S_m^*(x_i) = \sum_{j=1}^m \beta_j^* f_j^*(x_i)$ is the logit of $\widehat{y}_i^{(m)}$:

$$S_m^*(x_i) = \ln\left(\frac{\widehat{y}_i^{(m)}}{1 - \widehat{y}_i^{(m)}}\right) = \text{logit}\left(\widehat{y}_i^{(m)}\right).$$

We have also two interesting relations used later:

$$1 - \widehat{y}_i^{(m)} = \frac{\exp\left(-S_m^*(x_i)\right)}{1 + \exp\left(-S_m^*(x_i)\right)},$$

$$\frac{\partial \widehat{y}_i^{(m)}}{\partial S_m(x_i)} = \left(1 - \widehat{y}_i^{(m)}\right) \widehat{y}_i^{(m)}. \tag{7.29}$$

The loss function is the unscaled deviance which is in the binomial case equal to:

$$\mathcal{L}\left(y_i, \widehat{y}_i, v_i\right) = \tag{7.30}$$

$$2v_i\left(1_{y_i>0} y_i \ln\left(\frac{y_i}{\widehat{y}_i}\right) + 1_{y_i<1}(1 - y_i) \ln\left(\frac{1 - y_i}{1 - \widehat{y}_i}\right)\right).$$

the negative gradient in the mth iteration of the gradient boosting machine is given by:

$$g_i = -\frac{1}{n} \frac{\partial}{\partial S_m(x_i)} \mathcal{L}\left(y_i, \text{lgc}\left(S_m(x_i)\right), 1\right)\Big|_{S_m(x_i)=S_m^*(x_i)}$$

$$= -\frac{2}{n} \frac{\partial}{\partial S_m(x_i)} \left(1_{y_i>0} y_i \ln(y_i) - y_i \ln\left(\widehat{y}_i^{(m)}\right) + \right.$$

$$\left. 1_{y_i<1}(1 - y_i) \ln(1 - y_i) - (1 - y_i) \ln\left(1 - \widehat{y}_i^{(m)}\right)\right)\Big|_{S_m(x_i)=S_m^*(x_i)}.$$

From Eq. (7.29), we obtain after simplification that:

$$g_i = -\frac{2}{n}\left(-\frac{y_i}{\widehat{y}_i^{(m)}}\left(1 - \widehat{y}_i^{(m)}\right) \widehat{y}_i^{(m)} + \frac{1 - y_i}{1 - \widehat{y}_i^{(m)}}\left(1 - \widehat{y}_i^{(m)}\right) \widehat{y}_i^{(m)}\right)$$

$$= -\frac{2}{n}\left(\widehat{y}_i^{(m)} - y_i\right) \text{ for } y_i \in [0, 1], \ i = 1, \dots, n.$$

As for Poisson and Normal distributions, the GBM builds a new weak base-learner trained on residual errors of prediction by minimization of weighted quadratic errors:

$$f_m^*(x) = \arg\min_{f_m(.)} \frac{1}{n} \sum_{i=1}^n v_i \left(g_i - f_m(x_i)\right)^2.$$

Again, choosing this criterion introduces a bias since the negative gradient is proportional to a binomial random variable. However, if the exposure v_i is sufficiently high, we can consider a Normal approximation for a Binomial law.

Algorithm 7.1 is initialized as follows:

Proposition 7.16 *The parameter β_0 in Eq. (7.6) used for initializing the GBM procedure is equal to*

$$\beta_0 = logit\left(\frac{\sum_{i=1}^{n} v_i y_i}{\sum_{i=1}^{n} v_i}\right). \tag{7.31}$$

Proof Since $\widehat{y}_i^{(0)} = \lg c\left(S_m^*(x_i)\right)$, β_0 is solution of the following optimization problem:

$$\beta_0 = \arg\min_{\beta} \frac{2}{n} \sum_{i=1}^{n} v_i \left(1_{y_i>0} y_i \ln(y_i) - y_i \ln\left(\frac{1}{1+e^{-\beta}}\right)\right.$$

$$+ 1_{y_i<1} (1 - y_i) \ln(1 - y_i) - (1 - y_i) \ln\left(\frac{e^{-\beta}}{1+e^{-\beta}}\right)\bigg)$$

$$= \arg\min_{\beta} \frac{2}{n} \sum_{i=1}^{n} v_i \left(1_{y_i>0} y_i \ln(y_i) + 1_{y_i<1} (1 - y_i) \ln(1 - y_i)\right.$$

$$+ (1 - y_i) \beta + \ln\left(1 + e^{-\beta}\right)\bigg)$$

Cancelling the derivative of this last objective function with respect to β leads to the equality:

$$0 = \frac{2}{n} \sum_{i=1}^{n} v_i \left((1 - y_i) - \frac{e^{-\beta}}{1+e^{-\beta}}\right).$$

As this last equation may be rewritten as:

$$\frac{\sum_{i=1}^{n} v_i (1 - y_i)}{\sum_{i=1}^{n} v_i} = 1 - \frac{1}{1+e^{-\beta}},$$

we infer that

$$\beta_0 = logit\left(1 - \frac{\sum_{i=1}^{n} v_i (1 - y_i)}{\sum_{i=1}^{n} v_i}\right),$$

and can conclude. $\square$

The optimal β_m that minimizes the distance between observations and the logistic function of predictors is found numerically with the following iterative method:

Proposition 7.17 *The parameter β_m solution of Eq. (7.7) is computed iteratively. For a level of accuracy ϵ, we construct a series of $\beta_m^{(k)}$ initialized by*

$$\beta_m^{(0)} = \frac{\sum_{i=1}^{n} v_i \left(4 y_i - 2 - F_{m-1}^*(x_i)\right) f_m^*(x_i)}{\sum_{i=1}^{n} v_i f_m^*(x_i)^2}, \tag{7.32}$$

satisfying the relation

$$\beta_m^{(k+1)} = \beta_m^{(k)} - \frac{v(\beta_m^{(k)})}{v'(\beta_m^{(k)})}, \tag{7.33}$$

for $k = 1, \ldots, l$. If we denote $\widehat{y}_i^{(m)}(\beta) = lgc\left(S_{m-1}^*(x_i) + \beta \, f_m^*(x_i)\right)$, the function $v(.)$ and its first order derivatives are defined by:

$$v(\beta) = \sum_{i=1}^{n} v_i \left(\widehat{y}_i^{(m)}(\beta) - y_i\right) f_m^*(x_i) \tag{7.34}$$

$$v'(\beta) = \sum_{i=1}^{n} v_i f_m^*(x_i)^2 \left(1 - \widehat{y}_i^{(m)}(\beta)\right) \widehat{y}_i^{(m)}(\beta).$$

The series is stopped after l iterations, when $v\left(\beta_m^{(l)}\right) \le \epsilon$ for the first time.

Proof From the definition (7.30), we infer that β_m^* solves:

$$\beta_m^* = \arg\min_{\beta} \frac{1}{n} \sum_{i=1}^{n} \left(\mathcal{L}\left(y_i, \widehat{y}_i^{(m)}, v_i\right)\right).$$

$$= \arg\min_{\beta} \frac{2}{n} \sum_{i=1}^{n} v_i \left(1_{y_i > 0} y_i \ln(y_i) - y_i \ln\left(\widehat{y}_i^{(m)}(\beta)\right)\right.$$

$$\left. + 1_{y_i < 1} (1 - y_i) \ln(1 - y_i) - (1 - y_i) \ln\left(1 - \widehat{y}_i^{(m)}(\beta)\right)\right).$$

If we cancel the derivative of this last equation with respect to β, we infer that β_m^* is solution of

$$0 = \sum_{i=1}^{n} v_i \left(\frac{1 - y_i}{1 - \widehat{y}_i^{(m)}(\beta)} - \frac{y_i}{\widehat{y}_i^{(m)}(\beta)}\right) \frac{\partial \widehat{y}_i^{(m)}(\beta)}{\partial \beta} \tag{7.35}$$

$$= \underbrace{\sum_{i=1}^{n} v_i \left(\frac{\widehat{y}_i^{(m)}(\beta) - y_i}{\left(1 - \widehat{y}_i^{(m)}(\beta)\right) \widehat{y}_i^{(m)}(\beta)}\right) \frac{\partial \widehat{y}_i^{(m)}(\beta)}{\partial \beta}}_{v(\beta)}.$$

On the other hand, using the relation (7.29) allows us to develop the derivative of $\widehat{y}_i^{(m)}(\beta)$ with respect to β as follows:

$$\frac{\partial \widehat{y}_i^{(m)}(\beta)}{\partial \beta} = \frac{f_m^*(x_i) \exp\left(-S_m^*(x_i) - \beta\, f_m^*(x_i)\right)}{\left(1 + \exp\left(-S_{m-1}^*(x_i) - \beta\, f_m^*(x_i)\right)\right)^2} \tag{7.36}$$

$$= f_m^*(x_i)\left(1 - \widehat{y}_i^{(m)}(\beta)\right)\widehat{y}_i^{(m)}(\beta).$$

I we inject this last expression in Eq. (7.35), the function $v(\beta)$ becomes:

$$v(\beta) = \sum_{i=1}^n v_i \left(\widehat{y}_i^{(m)}(\beta) - y_i\right) f_m^*(x_i).$$

Unfortunately, the equation $v(\beta) = 0$ must be solved numerically with the Newton Raphson algorithm. In order to speed the convergence of this method, we find a good starting point with the first order Taylor's development of the logistic function: $\lg c(x) = \frac{1}{2} + \frac{1}{4}x + \mathcal{O}(x^2)$. The equality $v(\beta) = 0$ is next approached by

$$0 = \sum_{i=1}^n v_i \left(\frac{1}{2} + \frac{1}{4}S_{m-1}^*(x_i) + \frac{1}{4}\beta\, f_m^*(x_i) - y_i\right) f_m^*(x_i)$$

that admits Eq. (7.32) as solution. Equations (7.33) and (7.34) are the standard iterations of the Newton-Raphson algorithm where

$$v'(\beta) = \frac{\partial h(\beta)}{\partial \beta} = \sum_{i=1}^n v_i \frac{\partial \widehat{y}_i^{(m)}(\beta)}{\partial \beta} f_m^*(x_i).$$

$\square$

7.2 Analysis of Claims Frequencies with GBM

We test the gradient boosting machines of Sects. 7.1.2 and 7.1.2.1 for predicting the claims frequency of motorcycle drivers. We use again the dataset from the company *Wasa* (see Sect. 1.11 of Chap. 1 for details). The (scaled) ages of the vehicle and of the owner are quantitative variables used as input of the network. The other covariates are categorical variables converted in binary modalities: gender, geographic area and vehicle class. An insurance contract is therefore described by 2 quantitative variables and 13 binary modalities. We use as predictor a shallow neural network illustrated in Fig. 7.2 with a hidden layers counting 5 neurons, NN(5,1) and 86 neural weights. The activation functions of the hidden and output layers are respectively sigmoidal and linear. The estimated claims frequency is the exponential of the network output signal to ensure the positivity of the network prediction.

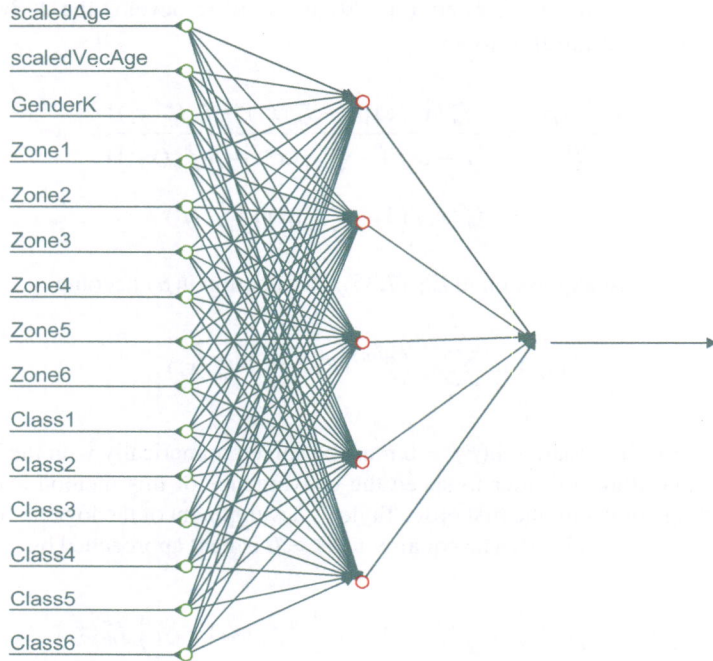

Fig. 7.2 Architecture of the neural network, NN(5,1) used in the gradient boosting machine algorithm

7.2.1 PBM for Claims Frequencies

We run 10 iterations the Poisson Boosting Machine. At each epoch, we use KERAS for fitting a NN(5,1) network with 2000 steps of the RMSprop algorithm and use the randomization techniques with shuffled batches of 5000 contracts. Table 7.1 reports the statistics of goodness of fit after 1, 5 and 10 PBM epochs and for a generalized linear model (GLM). The deviance computed with a single neural network is 14% lower than the one obtained with a GLM. As we could expect, the AIC and BIC are slightly less good than those computed with a GLM due to the high number of parameters. After 5 epochs, the deviance falls to 5672.42. This correspond to a relative improvement of only 0.5% compared to the NN(5,1) network. Running more than 5 iterations of the PBM does not improve the deviance and only worsens the AIC and BIC. Based on this observation, we may think that the PBM has little added value compared to a single neural network. This statement must be nuanced on account of the characteristics of our database. Here, a contract is specified by a limited number of explanatory variables and a simple shallow network is sufficient to challenge a GLM. For datasets with a larger number of covariates, the PBM is clearly an alternative to consider. On the other hand, we will see that for other loss

Table 7.1 Statistics of calibration for the GLM and PBM

	GLM	Number of epochs, PBM			
		0	1	5	10
Deviance	5781.36	6647.56	5700.55	5672.42	5669.17
Log. Lik.	−3565.11	−3998.06	−3524.56	−3510.50	−3508.87
AIC	7162.23	7998.13	7223.12	7882.99	8739.74
BIC	7306.90	8007.17	8009.76	11,780.05	16,524.81

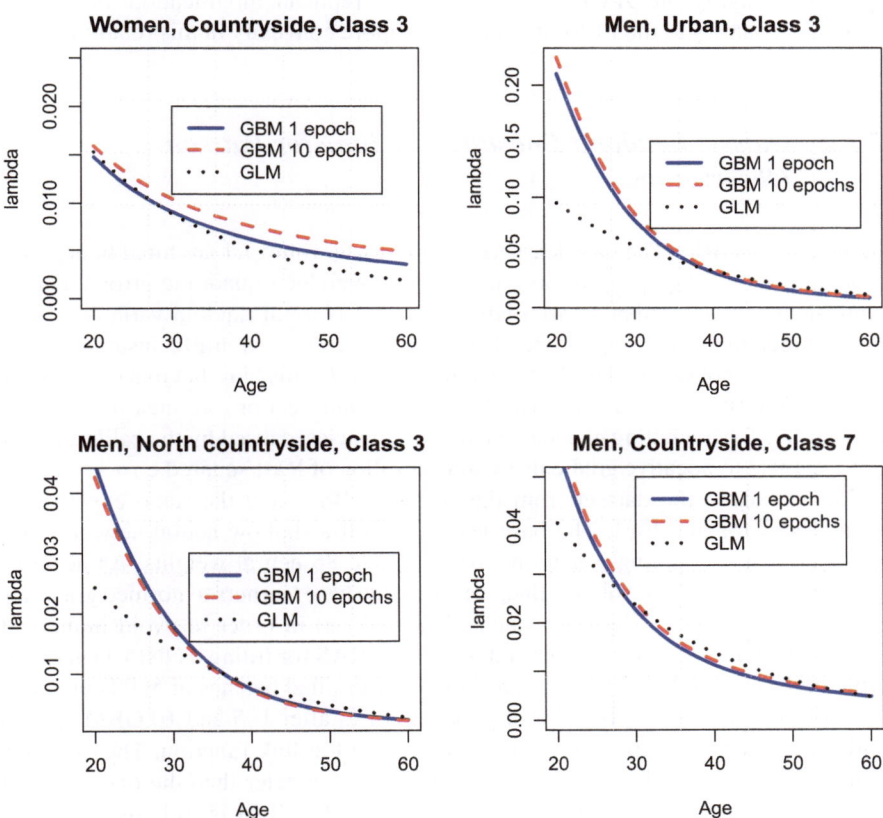

Fig. 7.3 Comparison of claims frequencies forecast with the GLM and PBM after 1 and 10 epochs

functions than the Poisson deviance, several GBM iterations are needed to achieve an acceptable accuracy.

Figure 7.3 compares forecast claims frequencies computed with a GLM ("log" link function) and PBM after one and ten epochs. Four categories of policyholders are considered: (1) women living in the countryside and driving a medium powerful vehicle (class 3) (2) men driving a class 3 motorcycle in an urban environment

(3) men of northern countryside owning a class 3 vehicle and (4) men driving the most powerful motorcycle (class 7) in the countryside. Whatever the model, the frequency of claims falls with the owner's age. Expected claims frequencies for profiles (1) and (4) and computed with the GLM or PBM are quite close. For policyholders of type (2) and (3) less than 35 years old, claims frequencies forecast with the PBM are significantly higher than those obtained with a GLM and nearly twice as big for the youngest drivers. In a GLM, the relation between drivers' age and expected frequencies is by construction an exponential decreasing curve with a constant decaying rate. A GLM fails therefore to replicate modifications of this rate with age, contrary to the PBM that allows for more convexity in this relation.

7.2.2   Classic Gradient Boosting Machine for Poisson Observations

In the classic GBM, The base learners $f_m^*(x)$ in Algorithm 7.1 are fitted to negative gradients $g = (g_i)_{i=1,...,n}$ by minimization of weighted quadratic errors. This is equivalent to assume that g_i are realizations of a Normal random variable. When the number of events, N_i, is small as it is the case in non-life insurance, this assumption is not satisfied as discussed in Sect. 7.1.2. This bias that makes in theory the classic GBM less efficient than the PBM. In this section, we measure the loss of accuracy for predicting expected claims frequencies, caused by the calibration of base learners to negative gradients by minimization of least squared errors.

We use again the dataset from the company *Wasa* and the same explanatory variables as for the PBM. The base learner is still a shallow neural network with a hidden layers counting 5 neurons, NN(5,1) and 86 neural weights. As negative gradients are small in absolute value, we adjust the base learners to normed gradients (we divide $(g_i)_{i=1,...,n}$ by their standard deviations) in order to avoid numerical issues. The GBM is run ten times and we use KERAS for fitting a NN(5,1) network with 2000 steps of the RMSprop algorithm with shuffled batches of 5000 contracts.

Table 7.2 reports the statistics of goodness of fit after 1, 5 and 10 GBM epochs and for a generalized linear model (GLM) with a log link function. The deviance computed with a single epoch of the GBM is 4.8% greater than the one obtained with a GLM. After five iterations, the deviance of the GBM is still slightly above

Table 7.2 Statistics of calibration for the GLM and GBM

| | GLM | Number of epochs | | | |
		0	1	5	10
Deviance	5781.36	6647.56	6058.00	5801.64	5699.97
Log. Lik.	−3565.11	−3998.06	−3703.29	−3575.10	−3524.27
AIC	7162.23	7998.13	7580.57	8012.21	8770.54
BIC	7306.90	8007.17	8367.21	11,909.27	16,555.61

Fig. 7.4 Evolution of the
deviance with respect to the
number of GBM epochs

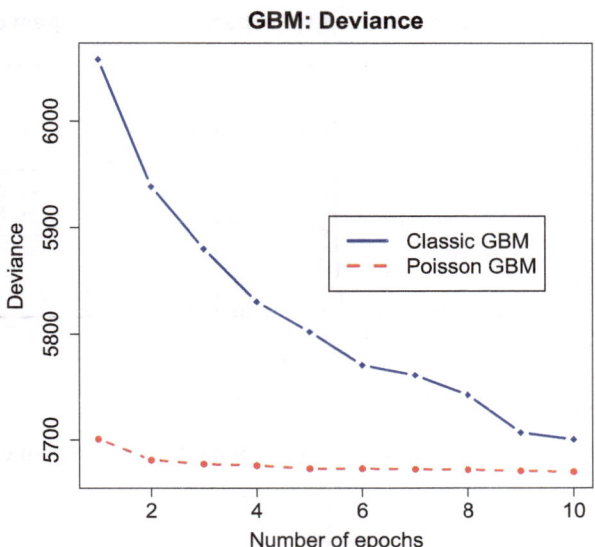

the one of the GLM. If we only use the deviance as criterion of accuracy, after six
epochs the GBM achieves the same performance as the GLM (deviance of 5770)
and clearly outperforms it after ten iterations.

Figure 7.4 compares deviances obtained with the PBM and the GBM for one to
ten epochs. Even if each iteration of the GBM significantly reduces the deviance, the
PBM converges much quicker to a minimum. In order to understand the reasons of
this slow convergence, we plot in Fig. 7.5 the empirical densities of (non-normed)
negative gradients $(g_i)_{i=1,\dots,n}$ and values $(f_m(x_i))_{i=1,\dots,n}$ of the base learner for
the first, second, third and tenth GBM iterations. The distance between these
distributions informs us about the error of calibration. We see that even after 10
epochs, the GBM fails to find a base learner replicating the distribution of negative
gradients. This is partly due to the chosen architecture of the base neural network
that is probably too basic for allowing a better fit. But this is also the consequence of
the chosen criterion for adjusting the base learners. Minimizing weighted quadratic
errors is indeed equivalent to assume that $(g_i)_{i=1,\dots,n}$ are realizations of Normal
random variables whereas in our numerical illustration, this is not the case.

7.2.3   GBM Applied to Claims Costs Prediction

In this section we apply the GBM for regressing the average claims cost on policy's
characteristics. The database from the Swedish the company *Wasa* counts 666
policies that have submitted a strictly positive claim. We consider as explanatory
variables: the scaled vehicle and policyholder ages, the gender, the geographic area

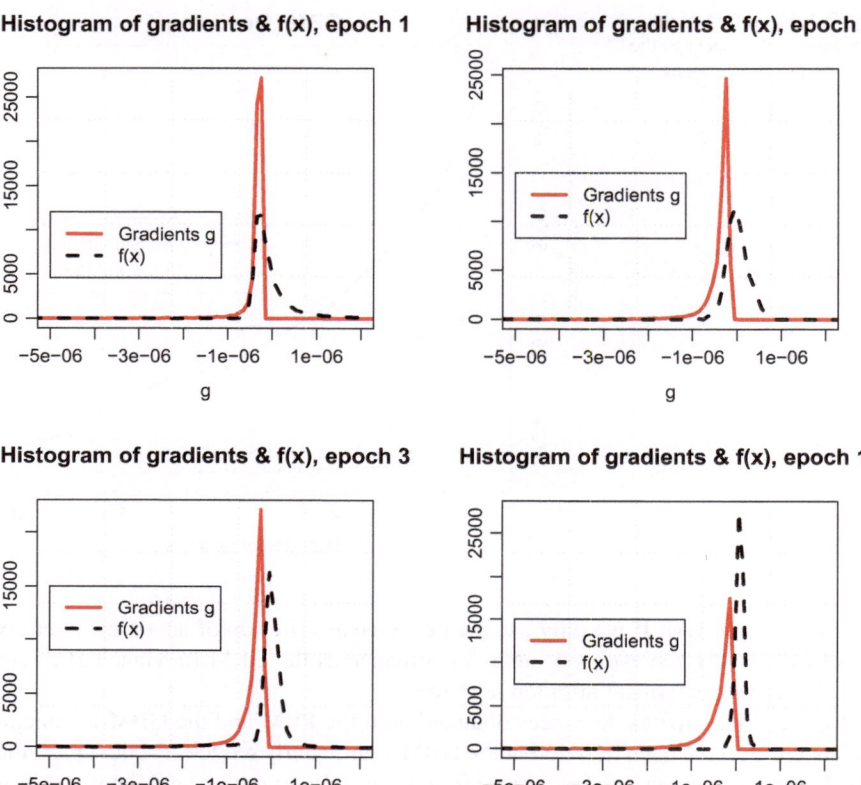

Fig. 7.5 Comparison of empirical densities of negative gradients $(g_i)_{i=1,\dots,n}$ and base learners $(f_m(\boldsymbol{x}_i))_{i=1,\dots,n}$ for epochs 1, 2, 3 and 10

and the power class of the motorcycle. Totally, we have two quantitative variables and 13 modalities per policy.

We use as predictor a shallow neural network with a hidden layers counting 5 neurons, NN(5,1) and 86 neural weights. The activation functions of the hidden and output layers are respectively the hyperbolic tangent and linear functions.

We consider that claims have a Gamma distribution. The expected claim cost is related to the output of the neural base learner by an exponential link. The GBM is run five times and the NN(5,1) network is fitted in KERAS with 4000 iterations of the RMSprop algorithm with shuffled batches of 5000 contracts.

Table 7.3 reports the statistics about the goodness of fit obtained with a generalized linear model and the gradient boosting machine up to 5 iterations. After only one epochs, the GBM outperforms the GLM in terms of deviance and AIC. Increasing the number of GBM iterations improves the deviance but after 4 iterations, this improvement becomes marginal (less than 0.2% in relative value).

Table 7.3 Statistics of calibration, Gamma GBM

	GLM	GBM, number of epochs				
		1	2	3	4	5
Deviance	1250.62	1165.74	1121.58	1086.33	1073.37	1071.40
Log. Lik.	−7384.75	−7263.63	−7233.32	−7306.41	−7303.78	−7368.33
AIC	14,801.51	14,701.26	14,812.64	15,130.82	15,297.56	15,598.66
BIC	14,873.53	15,092.86	15,591.37	16,296.66	16,850.51	17,538.72

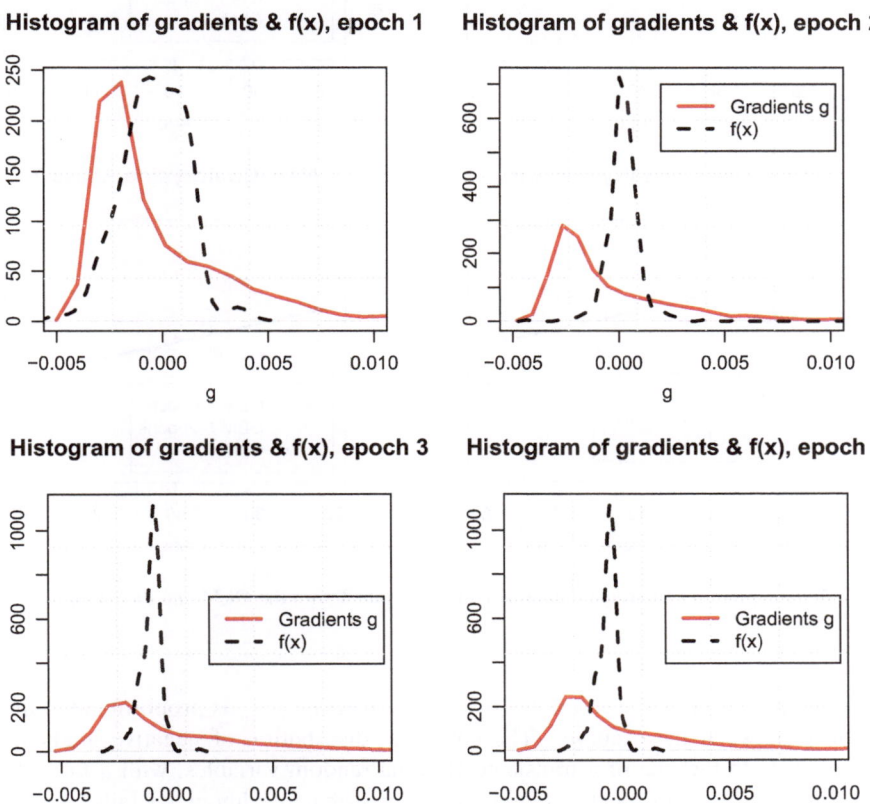

Fig. 7.6 Comparison of empirical densities of negative gradients $(g_i)_{i=1,\ldots,n}$ and base learners $(f_m(x_i))_{i=1,\ldots,n}$ for epochs 1, 2, 3 and 4

By construction, negative gradients $(g_i)_{i=1,\ldots,n}$ do not have a Gaussian distribution, contrary to what is implicitly assumed by the least squared error criterion chosen for estimating the base learner. To evaluate the impact of this bias, Fig. 7.6 shows the empirical densities of (non-normed) negative gradients $(g_i)_{i=1,\ldots,n}$ and values $(f_m(x_i))_{i=1,\ldots,n}$ of the base learner for the first four GBM iterations. The

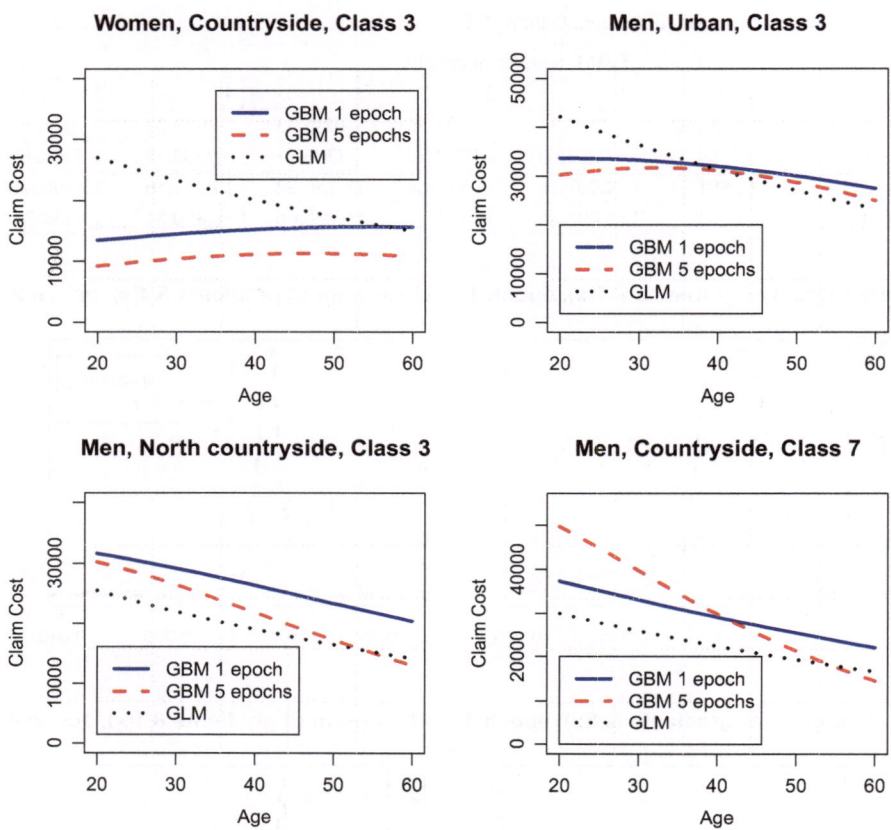

Fig. 7.7 Comparison of expected claim amounts computed with the GLM and the BGM with 1 and 5 epochs

error of calibration between $(f_m(\boldsymbol{x}_i))_{i=1,\dots,n}$ and $(g_i)_{i=1,\dots,n}$ is proportional to the distance between these curves. The empirical distribution of negative gradients seems close to the one of a translated Gamma random variables, with a fat right tail. Base learners are nearly symmetrically distributed without fat tails. As for the Poisson GBM, the spread between these empirical distributions can partly be explained by the implicit assumption that $(g_i)_{i=1,\dots,n}$ are realizations of normal random variables while they are not in practice.

Figure 7.7 compares expected claims costs computed with a GLM ("log" link function) and a Gamma GBM after one and five epochs. We consider the same four categories of policyholders as in Sect. 7.2.1. For women living in the countryside and driving a medium powerful vehicle (class 3), GLM and GBM yield very different predictions. Within the GLM approach, the expected claim cost decreases with age whereas the GBM forecasts lower and nearly constant claims costs (around 10,000 SEK). For men driving a class 3 vehicle in an urban environment, both models predict similar claim amounts, at least above 35 years old. For younger drivers, GBM claims costs are 25% lower than GLM ones. For men of northern countryside owning a class 3 vehicle, the GBM with 5 epochs computes higher claims costs than the GLM but predictions of both models converge for ages above 50 years old. The same observation holds for men driving class 7 motorcycles in the countryside. Nevertheless, the discrepancy between GLM and GBM expected claim amounts is much more important for the youngest drivers. For a 20 years driver, the GLM concludes that the expected cost of one claims is around 29,962 SEK. For the same driver, the GBM predicts that a claim costs on average 61,285 SEK, twice as high than the GLM.

7.3 Further Readings

Boosting machines have initially been developped for regression and classification trees by Freund and Schapire (1996), Friedman et al. (1998) or Friedman (1999). The adaBoost is a machine learning meta-algorithm developed by Freund and Schapire (1997) that may be used in conjunction with many other types of base learners. Mason et al. (1999) provide a more general functional gradient boosting perspective. They view boosting algorithms as iterative functional gradient descent procedures. That is, algorithms that optimize a cost function over function space by iteratively choosing a function that points in the negative gradient direction. GBMs have shown considerable success in practical applications and in various machine-learning and data-mining challenges. Bissacco et al. (2007) use a GBM for recognizing human pose in video sequences. Hutchinson et al. (2011) incorporate boosted regression trees into ecological latent variable models. The approach is presented in the context of occupancy-detection modeling, where the goal is to model the distribution of a species from imperfect detections. Pittman and Brown (2011) use boosted regression trees for predicting fish species distributions across coral reef seascapes. XGBoost is a research project started by Chen and Guestrin (2016) on scalable tree boosting methods. After winning the Higgs Machine Learning Challenge, it became well known in the ML competition circles. Soon after, the Python and R packages were built.

References

Bissacco A, Yang MH, Soatto S (2007) Fast human pose estimation using appearance and motion via multi-dimensional boosting regression. In: IEEE conference on computer vision and pattern recognition, CVPR'07
Chen T, Guestrin C (2016) XGBoost: a scalable tree boosting system. In: Proceedings of the 22nd ACM SIGKDD international conference on knowledge discovery and data mining, San Francisco, August 13–17, 2016. ACM, New York, pp 785–794

Freund Y, Schapire R (1996) Experiments with a new boosting algorithm. In: Machine learning: proceedings of the thirteenth international conference, pp 148–156

Freund Y, Schapire R (1997) A decision-theoretic generalization of on-line learning and an application to boosting. J Comput Syst Sci 55:119–139

Friedman JH (1999) Greedy function approximation: a gradient boosting machine. Technical report, Dept. of Statistics, Stanford University

Friedman JH, Hastie T, Tibshirani R (1998) Additive logistic regression: a statistical view of boosting. Technical report, Dept. of Statistics, Stanford University

Hutchinson RA, Liu LP, Dietterich TG (2011) Incorporating boosted regression trees into ecological latent variable models. In: Twenty-fifth conference on artificial intelligence, AAAI'11, San Francisco, pp 1343–1348

Mason L, Baxter J, Bartlett PL, Frean M (1999) Boosting algorithms as gradient descent. In: Solla SA, Leen TK, Muller K (eds) Advances in neural information processing system, vol 12. MIT Press, Cambridge, pp 512–518

Pittman SJ, Brown KA (2011) Multi-scale approach for predicting fish species distributions across coral reef seascapes. PLoS ONE 6(5):e20583. https://doi.org/10.1371/journal.pone.0020583

Chapter 8
Time Series Modelling with Neural Networks

The main objective of time series analysis is to provide mathematical models that offer a plausible description for a sample of data indexed by time. Time series modelling may be applied in many different fields. In finance, it is used for explaining the evolution of asset returns. In actuarial sciences, it may be used for forecasting the number of claims caused by natural phenomenons or for claims reserving.

8.1 Time Series Analysis in a Nutshell

An univariate time series is a vector of real numbers indexed by time: $x = (x_t)_{t=1,2,...}$. These observations are realizations of a sequence of random variables indexed by time and denoted by $(X_t)_{t=1,2,...}$. This set of random variables is called a discrete stochastic process. If the probability density function of X_t is noted $f_t(x)$, the mean function μ_t is calculated as:

$$\mu_t = \mathbb{E}(X_t) = \int_{-\infty}^{+\infty} x \, f_t(x) \, dx \,,$$

provided it exists. The linear dependence between observations of the stochastic process at time t and s is measured by the autocovariance function.

Definition 8.19 The autocovariance function, $\gamma(t, s)$, is defined as the covariance between X_t and X_s:

$$\gamma(t, s) = \mathbb{E}[(X_t - \mu_t)(X_s - \mu_s)] \; \forall t, s \in \mathbb{N}.$$

This function provides us information about the smoothness of the time series. Smooth time series show a high autocovariances even when t and s are far apart.

© Springer Nature Switzerland AG 2019
M. Denuit et al., *Effective Statistical Learning Methods for Actuaries III*,
Springer Actuarial, https://doi.org/10.1007/978-3-030-25827-6_8

On the contrary, highly volatile series have an autocovariance function quickly converging to zero when t moves away from s. The variance of X_t is equal to $\gamma(t, t)$. As it is more convenient to work with a measure of association in the interval $[-1, 1]$, we define the autocorrelation of a stochastic process.

Definition 8.20 The autocorrelation function, $\rho(t, s)$, is defined as the ratio:

$$\rho(t, s) = \frac{\gamma(t, s)}{\sqrt{\gamma(t, t)\gamma(s, s)}} \; \forall t, s \in \mathbb{N}.$$

The statistical analysis of a stochastic process requires a constraint of regularity using a concept called stationarity. If this condition is not satisfied, the process is too unstable for finding a plausible mathematical model.

Definition 8.21 A strictly stationary stochastic process is such that the probability law of a sample $\{X_{t_1}, X_{t_2}, \ldots, X_{t_k}\}$ is the same as the one of the time shifted set $\{X_{t_1+h}, X_{t_2+h}, \ldots, X_{t_k+h}\}$. That is

$$P(X_{t_1} \leq c_1, \ldots, X_{t_k} \leq c_k) = P(X_{t_1+h} \leq c_1, \ldots, X_{t_k+h} \leq c_k)$$

for all $k = 1, 2, \ldots$, for all times $t_1, t_2, \ldots, t_k \in \mathbb{N}$, for $h \in \mathbb{N}$ and for $c_1, \ldots, c_k \in \mathbb{R}$.

However, this definition of stationarity is often too constraining for most of applications. Furthermore, assessing the strictly stationarity of a time series is a challenging task. In practice, we adopt a milder version of this definition that imposes conditions on the first two moments of the time series.

Definition 8.22 The process $(X_t)_{t\in\mathbb{N}}$ is a weakly stationary stochastic process if its variance is finite and such that

1. The mean value function μ_t is constant and therefore does not depends upon time,
2. The autocovariance function $\gamma(t, s)$ depends only on the difference $|s - t|$.

The stationarity of a time series may be checked with the Dickey–Fuller test. We will come back later on this point. By construction, the strictly stationarity implies the weak stationarity but the reverse is not true. To conclude this section, we define a very popular process for practical applications.

Definition 8.23 The process $(X_t)_{t\in\mathbb{N}}$ is a Gaussian process if the n-dimensional vector $(X_{t_1}, X_{t_2}, \ldots, X_{t_n})$ for every $t_1, \ldots, t_n$ has a multivariate normal distribution.

For this particular process, the weakly stationarity entails the strictly stationarity because the multivariate Normal distribution is fully defined by its mean and covariance matrix.

8.1.1 Seasonality and Time Trend

In many different applications, time series exhibit a trend or seasonality due to seasonal effects. This dependence upon calendar time makes the stochastic process non-stationary. It is nevertheless possible to convert the time series to a stationary one either by subtracting a function of time from data or either by studying the variations of the process.

Let us consider a basic stochastic process $(X_t)_{t \in \mathbb{N}}$ that is the sum of a time dependent function $g(t)$ and of a stationary process $(Y_t)_{t \in \mathbb{N}}$

$$X_t = g(t) + Y_t .$$

Removing $g(t)$ from X_t will therefore make the process stationary. Since the function $g(t)$ is in practice unknown, an obvious solution consists to approximate $g(t)$ by a polynomial function $\tilde{g}(t)$ of order p:

$$\tilde{g}(t) = \beta_0 + \beta_1 t + \beta_2 t^2 + \ldots + \beta_p t^p . \tag{8.1}$$

Under the assumption that $(Y_t)_{t \in \mathbb{N}}$ is a Gaussian process with a null mean, the coefficients $\boldsymbol{\beta} = (\beta_0, \ldots, \beta_p)^\top$ are adjusted by linear regression. Let us assume that the stochastic process $(X_t)_{t \in \mathbb{N}}$ is observed at times $t_k = 0, \ldots, T$. The time series vector is denoted by $\boldsymbol{x} = (x_0, ..x_{t_k}, \ldots, x_T)^\top$. If $z_k = (1, t_k, t_k^2, \ldots, t_k^p)$ and $\boldsymbol{Z}$ is the $(T+1) \times p$-matrix of $(z_k)_{k=1,\ldots T}$, the regression equation (8.1) becomes $\boldsymbol{x} = \boldsymbol{Z}\boldsymbol{\beta}$. The coefficients $\boldsymbol{\beta}$ obtained by log-likelihood maximization (or by least squares minimization) are given by:

$$\boldsymbol{\beta} = \left(\boldsymbol{Z}^\top \boldsymbol{Z} \right)^{-1} \boldsymbol{Z}^\top \boldsymbol{x}, \tag{8.2}$$

if the matrix $\boldsymbol{Z}^\top \boldsymbol{Z}$ is non-singular.

In case of seasonality, we can employ a periodic regression function of the form:

$$\tilde{g}(t) = \beta_0 + \beta_1 \cos(2\pi \omega_1 t) + \beta_2 \sin(2\pi \omega_1 t) \tag{8.3}$$
$$+ \ldots + \beta_{p-1} \cos\left(2\pi \omega_{\frac{p}{2}} t \right) + \beta_p \sin\left(2\pi \omega_{\frac{p}{2}} t \right)$$

where p is even and $(\omega_1, \ldots, \omega_p)$ are pre-specified frequencies. The parameters $\boldsymbol{\beta}$ are obtained with Eq. (8.2) in which $\boldsymbol{Z}$ is the matrix of

$$z_k = \left(1, \cos(2\pi \omega_1 t_k), \sin(2\pi \omega_1 t_k), \ldots, \cos\left(2\pi \omega_1 t_{\frac{p}{2}} \right), \sin\left(2\pi \omega_1 t_{\frac{p}{2}} \right) \right) .$$

We test this last approach on the time series of minimum daily temperatures in Brussels (Uccle, Belgium) from the 1/8/2009 to the 31/12/2017. In the first graph of Fig. 8.1, we clearly observe the seasonality in the evolution of temperatures. In order

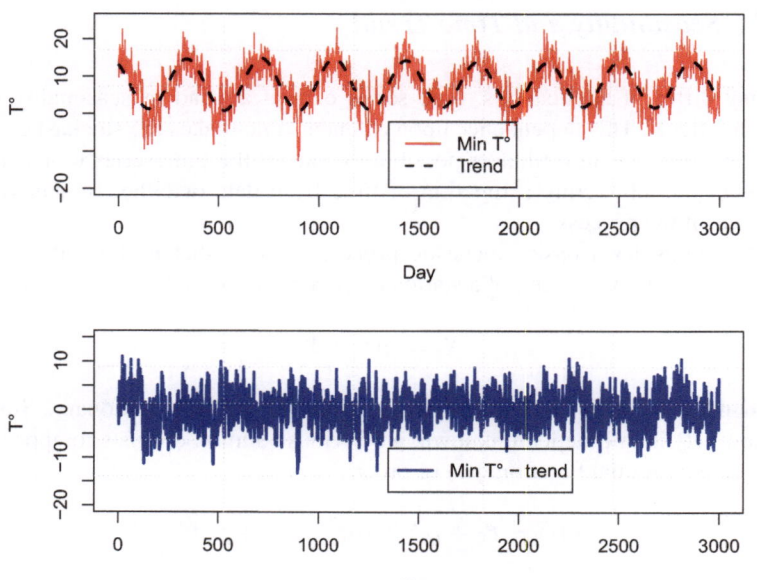

Fig. 8.1 First graph: daily minimum temperatures in Brussels (Uccle, Belgium) from the 1/8/2009 to the 31/12/2017 and its trend. Second graph: detrended time series

Table 8.1 Linear regression of daily minimal temperatures in Brussels, on sinus and cosinus functions

		Estimate	Standard error	T-test statistics	p-Value
	Intercept	7.89	0.067	118.08	$<10{-}12$
β_1	$\cos(2\pi\,0.9\,t)$	0.906	0.1	9.017	$<10{-}12$
β_3	$\cos(2\pi\,1.0\,t)$	-0.973	0.098	-9.959	$<10{-}12$
β_5	$\cos(2\pi\,1.1\,t)$	4.978	0.099	50.136	$<10{-}12$
β_2	$\sin(2\pi\,0.9\,t)$	-3.629	0.1	-36.283	$<10{-}12$
β_4	$\sin(2\pi\,1.0\,t)$	-0.698	0.098	-7.141	$<10{-}12$
β_6	$\sin(2\pi\,1.1\,t)$	0.824	0.1	8.229	$<10{-}12$

to detrend the time series, we linearly regress observations on 3 cosinus and sinus functions of time (expressed on a yearly basis), with frequencies 0.9, 1.0 and 1.1. The result of this regression are reported in Table 8.1. All coefficients of regression are highly significant. The average trend and detrended time series are respectively plotted in the first and second graphs of Fig. 8.1. We observe that the mixture of these cosinus-sinus functions captures most of the seasonality of temperatures.

There exists many other regression techniques, e.g. like kernel smoothing, that allows for detrending time series. We refer the interested reader to Schumway and Stoffer (2011, Chapter 1) for a detailed presentation of alternatives.

8.1.2 Autoregressive Average Models

In classical regression models, a dependent variable is explained by a set of current independent variables. In time series analysis, it is often necessary to allow the dependent variable to be influenced by its past recent values. An autoregressive model is based on the principle that the current value of a series, x_t is explainable by a function of p past values $(x_{t-1}, x_{t-2}, \ldots, x_{t-p})$. To ease the reading, we do not differentiate anymore the stochastic process $(X_t)_{t \in \mathbb{N}}$ from its time series $(x_t)_{t \in \mathbb{N}}$ and we denote by $\boldsymbol{x}_t = (x_t, x_{t-1}, \ldots, x_{t-p+1})$ the history of the process from time $t - p + 1$ up to time t.

Definition 8.24 An autoregressive model of order p, denoted AR(p) is of the form

$$x_t = \eta_1 x_{t-1} + \ldots + \eta_p x_{t-p} + w_t , \qquad (8.4)$$

where $(\eta_1, \ldots, \eta_p) \in \mathbb{R}^p$ and $(w_t)_{t \in \mathbb{N}}$ is a stochastic process with a null mean, $\mathbb{E}(w_t) = 0$.

In practice, we often assume that w_t is a Gaussian white noise.

Definition 8.25 A process $(w_t)_{t \in \mathbb{N}}$ is a Gaussian white noise if w_t are independent realizations of a Normal random distribution with null mean and variance equal to σ_w^2.

But we can consider for w_t any other distribution with null mean and finite variances. By construction, the expectation of x_t conditionally to $\boldsymbol{x}_{t-1}$ is equal to

$$\widehat{x_t} = \mathbb{E}(x_t | \boldsymbol{x}_{t-1}) = \eta_1 x_{t-1} + \ldots + \eta_p x_{t-p} ,$$

and its variance is $\mathbb{V}(x_t | \boldsymbol{x}_{t-1}) = \sigma_w^2$. In view of Eq. (8.4), the AR(p) is a regression model of x_t on its past values. Here, $\hat{x}_t$ estimates therefore x_t, conditionally to the sample path of the process up to time $t - 1$. Another useful representation of an AR(p) model is based on the backshift operator. This backshift operator is defined as follows:

Definition 8.26 Let us consider a stochastic process $(x_t)_{t \in \mathbb{N}}$. The backshift operator is such that:

$$Bx_t = x_{t-1} ,$$

and extend it to powers

$$B^k x_t = B^{k-1}(B(x_t))$$
$$= B^{k-1}(x_{t-1})$$
$$= \ldots = x_{t-k} .$$

This backshift operator is also closely related to another important operation in time series analysis: the differencing. Differencing the data often allows converting a non-stationary process in a stationary series. The first difference operator is noted

$$\nabla x_t = x_t - x_{t-1} ,$$

and eliminates any linear trend. Whereas the d-difference operator is defined by

$$\nabla^d x_t = \nabla^{d-1} x_t - \nabla^{d-1} x_{t-1} .$$

Therefore, for $d = 2$, we have that $\nabla^2 x_t = x_t - 2x_{t-1} + x_{t-2}$. The differencing operators are rewritten in term of the backshift operator as follows:

$$\nabla^d = (1 - B)^d x_t .$$

The autoregressive model may then be rewritten in term of the autoregressive operator:

Definition 8.27 The autoregressive operator is defined to be:

$$\eta(B) = 1 - \eta_1 B - \eta_2 B^2 - \ldots - \eta_p B^p ,$$

and the AR(p) model admits the concise representation:

$$\eta(B)x_t = w_t . \tag{8.5}$$

In order to understand the conditions that ensure the stationarity of an autoregressive process, we focus on the AR(1) model: $x_t = \eta x_{t-1} + w_t$. If we iterate backwards k times, x_t becomes:

$$x_t = \eta \left(\eta x_{t-2} + w_{t-1} \right) + w_t$$

$$= \eta^k x_{t-k} + \sum_{j=0}^{k-1} \eta^j w_{t-j} . \tag{8.6}$$

This last equation suggests that if $|\eta| < 1$, the process x_t is an infinite sum

$$x_t = \sum_{j=0}^{\infty} \eta^j w_{t-j} , \tag{8.7}$$

and therefore x_t is a stationary process as $\eta^j \to 0$ when $j \to \infty$. If $|\eta| \geq 1$, from Eq. (8.6), we see that x_t has an infinite amplitude when $t \to \infty$. Hence, the process cannot be stationary. From Eq. (8.7), we also deduce that the expectation of x_t is equal to

$$\mathbb{E}(x_t) = \sum_{j=0}^{\infty} \eta^j \mathbb{E}(w_{t-j}) = 0.$$

Since $\mathbb{C}(w_t, w_s) = 0$ if $s \neq t$ and $\mathbb{C}(w_t, w_s) = \sigma_w^2$, the autocovariance function is given by

$$\gamma(h) = \mathbb{C}(x_{t+h}, x_t)$$

$$= \mathbb{E}\left(\sum_{j=0}^{\infty} \eta^j w_{t+h-j} \times \sum_{k=0}^{\infty} \eta^k w_{t-k}\right)$$

$$= \sigma_w^2 \sum_{j=0}^{\infty} \eta^{h+j} \eta^j = \sigma_w^2 \eta^h \sum_{j=0}^{\infty} \eta^{2j}$$

$$= \frac{\sigma_w^2 \eta^h}{1 - \eta^2} \quad h \in \mathbb{N}.$$

We immediately infer that the autocorrelation function of an AR(1) process is a power of h:

$$\rho(h) = \frac{\gamma(h)}{\gamma(0)} = \eta^h \quad h \in \mathbb{N}.$$

If the process is stationary ($|\eta| < 1$), the autocorrelation vanishes when $h \to \infty$. This characteristic is common to all autocorrelation functions of AR(p) process, whatever the order p. To illustrate this, we have simulated three autoregressive processes of order 1, 3 and 5 such as described by the following equation:

$$AR(p): \quad x_t = \sum_{j=1}^{p} 0.15 x_{t-j} + w_t \quad p = 1, 3, 5,$$

with a standard deviation of $\sigma_w = 0.05$ for w_t. Ten thousand sample paths are computed. Figure 8.2 shows the autocorrelation functions of these three processes for 1 up to 20 lags of time. We clearly observe that the speed at which the autocorrelation decays is inversely proportional to the autoregression order. Before studying the conditions that ensure the stationarity of an AR process, we develop the polynomials property of an AR(1) model. Let us consider the AR(1) model written in its operator form: $\eta(B)x_t = w_t$ where $\eta(B) = 1 - B$ and $|\eta| < 1$. According to Eq. (8.7), x_t can also be written as the sum

$$x_t = \sum_{j=0}^{\infty} \psi_j w_{t-j} = \psi(B)w_t, \tag{8.8}$$

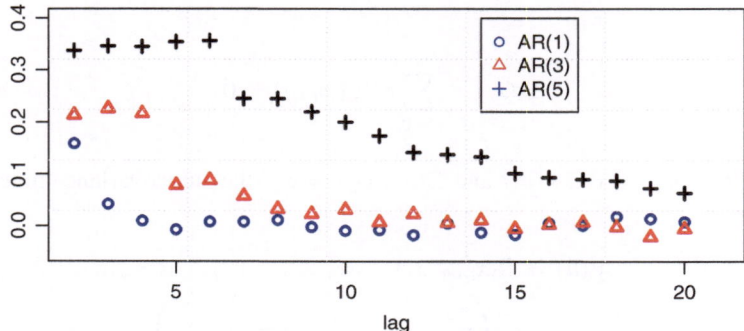

Fig. 8.2 Autocorrelation functions of 10,000 simulated paths of AR(1), AR(3) and AR(5) processes

where the operator $\psi(B) = \sum_{j=0}^{\infty} \psi_j B^j$ and $\psi_j = \eta^j$. A direct consequence of Eq. (8.8) is that

$$\eta^{-1}(B) = \psi(B) = 1 + \eta B + \eta^2 B^2 + \ldots + \eta^j B^j + \ldots$$

Therefore, we have that

$$\eta^{-1}(B)\eta(B)x_t = \eta^{-1}(B)w_t$$

$$x_t = \psi(B)w_t \,.$$

Working with operators is like working with polynomials. Indeed, if we consider the polynomial $\eta(z) = 1 - \eta z$ where z is a complex number and $|\eta| < 1$ then

$$\eta^{-1}(z) = 1 + \eta z + \eta^2 z^2 + \ldots + \eta^j z^j + \ldots \quad |z| \le 1$$

$$= \frac{1}{1 - \eta z} \,.$$

This means that the backshift operator may be considered as a complex number of module $|z| \le 1$, in the AR(1) autoregressive operator. This result may be extended to any stationary $AR(p)$ model and will serve us later to study the stationarity of autoregressive models in Sect. 8.1.4.

8.1.3 Moving Average Models

Moving average models offer an interesting alternative to autoregressive processes. In this category of models, the observation at time t is a linear combination of past white noises:

Definition 8.28 The moving average model of order q, denoted by MA(q), is defined as

$$x_t = w_t + \theta_1 w_{t-1} + \theta_2 w_{t-2} + \ldots + \theta_q w_{t-q}, \tag{8.9}$$

where $(\theta_1, \ldots, \theta_q) \in \mathbb{R}^q$ and $(w_t)_{t \in \mathbb{N}}$ is a sequence of independent random variables with variance $\sigma_w^2 < \infty$ and null mean, $\mathbb{E}(x_t) = 0$.

In most of situations, we assume that $(w_t)_{t \in \mathbb{N}}$ is a Gaussian white noise, i.e. $w_t \sim N(0, \sigma_w^2)$. The MA(q) process may also be rewritten in a more concise form with an operator:

Definition 8.29 The moving average operator is defined by:

$$\theta(B) = 1 + \theta_1 B + \theta_2 B^2 + \ldots + \theta_p B^p,$$

and the AR(p) model admits the concise representation:

$$x_t = \theta(B) w_t.$$

If we consider a moving average model of order 1, $x_t = w_t + \theta w_{t-1}$, it is easy to check that its expectation is null. Since white noises are independent, the autocovariance function is equal to

$$\gamma(h) = \mathbb{C}(x_{t+h}, x_t) = \begin{cases} \left(1 + \theta^2\right) \sigma_w^2 & h = 0, \\ \theta \sigma_w^2 & h = 1, \\ 0 & h > 1, \end{cases}$$

and the autocorrelation function is in this case given by

$$\rho(h) = \begin{cases} \frac{\theta}{1+\theta^2} & h = 1, \\ 0 & h > 1. \end{cases}$$

This means that x_t is correlated with x_{t-1} but not with previous occurrences, $x_{t-2}, x_{t-3}, \ldots$. Using a similar approach, we infer the autocovariance function of a MA(q) process:

$$\gamma(h) = \mathbb{C}(x_{t+h}, x_t) = \begin{cases} \sigma_w^2 \sum_{j=0}^{q-h} \theta_j \theta_{j+h} & 0 \leq h \leq q, \\ 0 & h > q, \end{cases}$$

whereas the autocorrelation function is in this case given by

$$\rho(h) = \begin{cases} \frac{\sum_{j=0}^{q-h} \theta_j \theta_{j+h}}{1 + \theta_1^2 + \ldots + \theta_q^2} & 1 \leq h \leq q, \\ 0 & h > q. \end{cases}$$

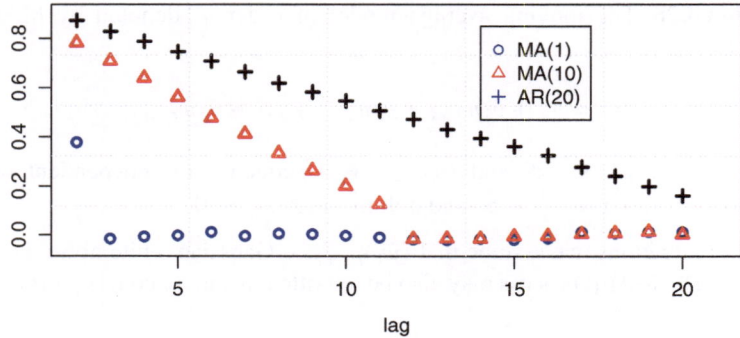

Fig. 8.3 Autocorrelation functions of 10,000 simulated paths of MA(1), MA(10) and MA(20) processes

Contrary to AR(p) processes that have long decaying autocorrelations, MA(q) models display an autocorrelation function that is cut off after q lags. To illustrate this point, we have simulated three moving average processes of order 1, 10 and 20 such as described by the following equation:

$$MA(q): \quad x_t = w_t + \sum_{j=1}^{q} 0.5 w_t \quad q = 1, 10, 20,$$

with a standard deviation $\sigma_w = 0.05$ for w_t. We simulate 10,000 sample paths. Figure 8.3 shows the empirical autocorrelations of these processes. We clearly observe that the correlation vanishes after $q = 1, 10, 20$ lags depending upon the configuration of the process. A moving average model is called invertible, if it can be represented as an AR(1) model. For example, let us consider the MA(1) model, $x_t = w_t + \theta w_{t-1}$. We can rewrite it as $w_t = -\theta w_{t-1} + x_t$. If $|\theta| < 1$, using the same reasoning as for the AR(1) allows to rewrite w_t as an infinite sum of x_{t-j}:

$$w_t = \sum_{j=0}^{\infty} (-\theta)^j x_{t-j} \tag{8.10}$$

which is the infinite representation of an AR(1) model.

On the other hand, the polynomial $\theta(z)$ corresponding to the moving average operator $\theta(B)$ is used later for studying conditions of stationarity of MA and AR processes. If we remember that the MA(1) process is equal to $x_t = \theta(B)w_t$ where $\theta(B) = 1 + \theta B$. From Eq. (8.10), if $|\theta| < 1$, the inverse of $\theta(B)$ is denoted by $\chi(B) = \theta^{-1}(B)$ and is such that:

$$\theta^{-1}(B)\theta(B)x_t = \chi(B)w_t.$$

If we consider the polynomial $\theta(z) = 1 + \theta z$, we have for $|z| \leq 1$ that

$$\chi(z) = \theta^{-1}(z) = \sum_{j=0}^{\infty} (-\theta)^j z^j$$

$$= \frac{1}{1 + \theta z}.$$

As for the AR(1), the backshift operator may be considered as a complex number z with $|z| \leq 1$, in the moving average operator.

8.1.4 Autoregressive Moving Average Models

Autoregressive average models combine AR and MA processes as follows:

Definition 8.30 A time series $(x_t)_{t \in \mathbb{Z}}$ is ARMA(p,q) if it is stationary and is driven by the following dynamics

$$x_t = \eta_1 x_{t-1} + \ldots + \eta_p x_{t-p} + w_t + \theta_1 w_{t-1} + \ldots + \theta_q w_{t-q} \qquad (8.11)$$

where $\eta_p \neq 0$, $\theta_q \neq 0$ and w_t is a Gaussian white noise (or any other random variable) of null mean and variance $\sigma_w^2 > 0$. Here, p and q are respectively called the autoregressive and moving average orders.

An ARMA(p,q) model can be rewritten in a concise form using the autoregressive and moving average operators as follows:

$$\eta(B)x_t = \theta(B)w_t . \qquad (8.12)$$

In order to study the conditions guaranteeing the stationarity of x_t, we define the AR and MA polynomials as follows

$$\eta(z) = 1 - \eta_1 z - \ldots - \eta_p z^p , \quad \eta_p \neq 0,$$
$$\theta(z) = 1 + \theta_1 z + \ldots + \theta_q z^q , \quad \theta_q \neq 0,$$

where z is a complex number. These polynomials are used for defining the concept of causality:

Definition 8.31 An ARMA(p,q) model is causal if the time series $(x_t)_{t \in \mathbb{Z}}$ driven by Eq. (8.11) can be rewritten as a one-sided linear process:

$$x_t = \sum_{j=0}^{\infty} \psi_j w_{t-j} = \psi(B)w_t \qquad (8.13)$$

where $\psi(B) = \sum_{j=0}^{\infty} \psi_j B^j$ with $\sum_{j=0}^{\infty} |\psi_j| < \infty$ and $\psi_0 = 1$.

If we remember the AR(1) example of Sect. 8.1.2, the process $x_t = \eta x_{t-1} + w_t$ is causal if and only if $|\eta| < 1$. Equivalently, the process is causal if and only if the root of the AR(1) polynomial, $\eta(z) = 1 - \eta z$, is greater than one in absolute value. By construction, a causal process is weakly stationary: its mean is constant and its autocovariance is equal to

$$\gamma(h) = \mathbb{C}(x_{t+h}, x_t) = \sigma_w^2 \sum_{j=0}^{\infty} \psi_j \psi_{j+h}$$

depends solely of h. Since the backshift operator can be treated as a complex number in autoregressive and moving average operators, the ARMA(p,q) model expressed in its concise form (8.12) is equivalent to

$$x_t = \frac{\theta(B)}{\eta(B)} w_t,$$

and a comparison with condition (8.13) fulfilled by a causal ARMA process leads to the following proposition:

Proposition 8.18 *An ARMA(p,q) process is causal if and only of $\eta(z) \neq 0$ for $|z| \leq 1$. The coefficients $(\psi_j)_{j \in \mathbb{N}}$ of $\psi(B)$ are solution of the following equality:*

$$\psi(z) = \sum_{j=0}^{\infty} \psi_j z^j = \frac{\theta(z)}{\eta(z)} \quad |z| \leq 1. \tag{8.14}$$

The condition $\eta(z) \neq 0$ for $|z| \leq 1$ means that the roots of $\eta(z)$ lie outside the unit circle: $\eta(z) = 0$ when $|z| > 1$. Otherwise, we say that $\eta(z)$ admits a unit root.

For a ARMA(p,q), the coefficients ψ_j are found with a recursive difference equation. By construction, the ψ_j are such that $\psi(z)\eta(z) = \theta(z)$ and developing this equality give us:

$$\left(1 - \eta_1 z - \ldots - \eta_p z^p\right)\left(\psi_0 + \psi_1 z + \psi_2 z^2 + ..\right) = (1 + \theta_1 z + \ldots + \theta_q z^q).$$

Matching coefficients leads to

$$\psi_0 = 1,$$
$$\psi_1 - \eta_1 \psi_0 = \theta_1,$$
$$\psi_2 - \eta_1 \psi_1 - \eta_2 \psi_0 = \theta_2,$$
$$\psi_3 - \eta_1 \psi_2 - \eta_2 \psi_1 - \eta_3 \psi_0 = \theta_3.$$

Therefore, we deduce that $\left(\psi_j\right)_{j\in\mathbb{N}}$ are solutions of:

$$\psi_j - \sum_{k=1}^{p} \eta_k \psi_{j-k} = 0, \quad j \geq \max(p, q+1),$$

$$\psi_j - \sum_{k=1}^{j} \eta_k \psi_{j-k} = \theta_j, \quad 0 \leq j < \max(p, q+1).$$

8.1.5 Estimation of ARMA Models

The calibration of Gaussian ARMA models is done by log-likelihood maximization. The likelihood is computed iteratively and requires to evaluate the expectation and variance of x_{t+m}, conditionally to the information about the time series up to time t. This information is here denoted by $x_t = \{x_t, x_{t-1}, \ldots x_1\}$.

We assume that $\mathbb{E}(x_t) = 0$. Otherwise, when $\mathbb{E}(x_t) = \mu \neq 0$, we consider the translated process $x_t' = x_t - \mu$. In an ARMA model, the expectation $\mathbb{E}(x_{t+m}|x_t)$ does not admit simple analytical expression but we may find a linear estimator, $\widehat{x}_{t+m} = g_m(x_t)$ of the form:

$$g_m(x_t) = \sum_{k=1}^{t} \eta_{tk} x_k,$$

where $\eta_{t1}, \ldots, \eta_{tt}$ are real numbers minimizing the mean square prediction error, $\mathbb{E}\left((x_{t+m} - \widehat{x}_{t+m})^2\right)$. Using the projection theorem, the vector $(x_{t+m} - \widehat{x}_{t+m})$ is orthogonal to all x_k for $k = 1, \ldots, t$ in the set of random variables endowed with a scalar product that is the cross expectation. More precisely, we have the next proposition:

Proposition 8.19 *Given* $x_t = \{x_t, x_{t-1}, \ldots x_1\}$, *the best linear estimator* $\widehat{x}_{t+m} = g_m(x_t)$ *of* $\mathbb{E}(x_{t+m}|x_t)$ *has coefficients,* $\eta_{t1}, \ldots, \eta_{tt}$ *satisfying:*

$$\mathbb{E}\left((x_{t+m} - \widehat{x}_{t+m}) x_k\right) = 0 \quad k = 1, \ldots, t. \tag{8.15}$$

If we focus on the one-step-ahead prediction of x_{t+1} knowing x_t, the coefficients $(\eta_{tk})_{k=1,\ldots,t}$ are solutions of the system of equations (8.15) and may be reformulated in terms of the autocovariance function, $\gamma(.)$:

$$\sum_{j=1}^{t} \eta_{tj} \gamma(k-j) = \gamma(k) \quad k = 1, \ldots, n. \tag{8.16}$$

Let $\boldsymbol{\Gamma}_t = \{\gamma(k - j)\}_{j,k=1}^t$ be a $(t \times t)$- matrix and $\boldsymbol{\eta}_t = (\eta_{t1}, \ldots, \eta_{tt})^\top$, $\boldsymbol{\gamma}_t = (\gamma(1), \ldots, \gamma(t))^\top$ be two vectors of dimension t. Therefore, the system (8.16) becomes

$$\boldsymbol{\Gamma}_t \boldsymbol{\eta}_t = \boldsymbol{\gamma}_t \,,$$

and the linear predictor $\widehat{x}_{t+1}$ is equal to:

$$\widehat{x}_{t+1} = \boldsymbol{\eta}_t \boldsymbol{x}_t \tag{8.17}$$
$$= \boldsymbol{\Gamma}_t^{-1} \boldsymbol{\gamma}_t \boldsymbol{x}_t \,.$$

The mean square one-step-ahead error of $\widehat{x}_{t+1}$ is hence equal to:

$$\mathbb{E}\left((x_{t+1} - \widehat{x}_{t+1})^2\right) = \mathbb{E}\left((x_{t+1} - \boldsymbol{\eta}_t \boldsymbol{x}_t)^2\right) \tag{8.18}$$
$$= \gamma_0 - \boldsymbol{\Gamma}_t^{-1} \boldsymbol{\gamma}_t \,.$$

In practice, we do not need to invert the matrix $\boldsymbol{\Gamma}_t$. Instead, Durbin (1960) have proposed an iterative algorithm for computing both the $\boldsymbol{\eta}_t$ and the mean square error of $\widehat{x}_{t+1}$. This algorithm being beyond the scope of this introduction, we do not present it.

We use the expressions (8.17) and (8.18) for calibrating an ARMA(p,q) model to the time-series $\{x_1, \ldots, x_n\}$ by log-likelihood maximization. Let us denote by $\boldsymbol{\Theta} = (\mu, \eta_1, \ldots, \eta_p, \theta_1, \ldots, \theta_q)$ the vector of parameters, where $\mathbb{E}(x_t) = \mu$. Using the Bayes rule, the log-likelihood $l(\boldsymbol{\Theta}, \sigma_w^2)$ may be rewritten as:

$$l(\boldsymbol{\Theta}, \sigma_w^2) = \sum_{t=1}^n \ln f(x_t | \boldsymbol{x}_{t-1}) \,,$$

where $f(.)$ is the conditional distribution of $x_t | \boldsymbol{x}_{t-1}$. This distribution is Gaussian with a mean and variance respectively approached by (8.17) and (8.18). The parameters $\boldsymbol{\Theta}$ and σ_w are found by numerical minimization of this log-likelihood:

$$\{\boldsymbol{\Theta}, \sigma_w^2\} = \arg\max \sum_{t=1}^n \ln f(x_t | \boldsymbol{x}_{t-1}) \,.$$

As in Chap. 1, the overall quality of the model is judged with Akaike or Bayesian information criterion (AIC and BIC):

$$\text{AIC} = 2m - 2l(\widehat{\boldsymbol{\Theta}}, \widehat{\sigma}_w^2) \,,$$

$$\text{BIC} = \ln(n)m - 2l(\widehat{\boldsymbol{\Theta}}, \widehat{\sigma}_w^2)$$

Table 8.2 Statistics of goodness of fit for an ARMA(p, q) process adjusted to the time series of detrended temperatures in Brussels from the 1/8/2009 to 8/8/2016

		0	1	2	3	4
	q					
ARMA(1,.)	SSE	16,884.58	16,821.21	16,783.64	16,768.87	16,746.43
	log-lik.	−5935.32	−5930.62	−5927.83	−5926.74	−5925.06
	AIC	11,876.64	11,869.25	11,865.67	11,865.47	11,864.13
	BIC	11,886.29	11,884.72	11,886.96	11,892.59	11,897.07
ARMA(2,.)	SSE	16,828.98	16,753.18	16,800.32	16,736.94	16,752.38
	log-lik.	−5931.2	−5925.57	−5929.07	−5924.37	−5925.51
	AIC	11,870.4	**11,861.14**	11,870.15	11,862.74	11,867.02
	BIC	11,885.87	11,882.44	11,897.27	11,895.69	11,905.79
ARMA(3,.)	SSE	16,797.34	16,745.87	16,748.41	16,732.85	16,733.25
	log-lik.	−5928.85	−5925.04	−5925.21	−5924.07	−5924.08
	AIC	11,867.7	11,862.07	11,864.43	11,864.13	11,866.17
	BIC	11,889	11,889.19	11,897.37	11,902.9	11,910.76
ARMA(4,.)	SSE	16,787.44	16,738.48	16,732.46	16,682.29	16,728.49
	log.lik	−5928.12	−5924.49	−5924.04	−5921.8	−5923.73
	AIC	11,868.23	11,862.97	11,864.07	11,861.59	11,867.46
	BIC	11,895.35	11,895.92	11,902.84	11,906.19	11,917.88
ARMA(5,.)	SSE	16,757.03	16,742.41	16,732.62	16,732.75	16,679.35
	log-lik.	−5925.85	−5924.77	−5924.05	−5924.06	−5920.13
	AIC	11,865.71	11,865.53	11,866.1	11,868.12	11,862.26
	BIC	11,898.65	11,904.3	11,910.69	11,918.53	11,918.5

Bold values indicate the best fit

where m is the number of parameters. Given a set of models, the preferred model is the one with the lowest AIC or BIC. The AIC and BIC reward goodness of fit (assessed by the likelihood function), but penalize models with a large number parameters.

Table 8.2 presents the statistics of goodness of fit for an ARMA(p,q) adjusted to detrended temperatures measured in Brussels from the 1/8/2009 to 8/8/2016. The trend is a mixture of sinus and cosinus functions as studied in Sect. 8.1.1. The SSE is the sum of squared errors between forecast and observed temperatures. The model achieving the lowest AIC and BIC is the ARMA(2,1). However, ARMA(3,1) and ARMA(4,1) models are also relevant alternatives for explaining the evolution of temperatures even if their AIC and BIC is slightly penalized by a higher number of parameters than the ARMA(2,1).

8.1.6   The Dickey–Fuller Test

Whatever the method chosen for modelling a time series, it is important to check the stationarity of data. Stationarity entails that the first and second moments are constant through time. Since the statistical inference is based on the assumption of fixed moments, it is essential to check that observations to which we fit a model are stationary. The most common test is the one proposed by Dickey and Fuller (1979). To understand this test, let us consider an AR(1) process

$$x_t = \eta x_{t-1} + w_t$$

If $\eta = 1$, the AR(1) process is a random walk and if $|\eta| < 1$, the process is causal and may be rewritten as an infinite sum of Gaussian white noises. The unit root test check if x_t is a random walk or is causal

$$\begin{cases} H_0: & \eta = 1 \\ H_1: & |\eta| < 1 \end{cases}.$$

If $\widehat{\eta}$ is an estimator of η, one solution consists to study the statistical behaviour of $\widehat{\eta} - 1$ under the null hypothesis. For a time-series $\{x_t\}_{t=1,\ldots,n}$, the mean square estimator of η under H_0 (which is also the maximum log-likelihood estimator) is equal to

$$\widehat{\eta} = \frac{\frac{1}{n}\sum_{t=1}^{n} x_t x_{t-1}}{\frac{1}{n}\sum_{t=1}^{n} x_{t-1}^2} = 1 + \frac{\frac{1}{n\sigma_w^2}\sum_{t=1}^{n} w_t x_{t-1}}{\frac{1}{n\sigma_w^2}\sum_{t=1}^{n} x_{t-1}^2}. \tag{8.19}$$

If we consider the square of $x_t = x_{t-1} + w_t$, we infer that under H_0,

$$w_t x_{t-1} = \frac{1}{2}\left(x_t^2 - x_{t-1}^2 - w_t^2\right).$$

Therefore, the numerator of Eq. (8.19) may be developped as:

$$\frac{1}{n\sigma_w^2}\sum_{t=1}^{n} w_t x_{t-1} = \frac{1}{2}\frac{1}{n\sigma_w^2}\left(x_n^2 - \sum_{t=1}^{n} w_t^2\right).$$

Under H_0, x_n is $N(0, n\sigma_w^2)$ and $\frac{1}{n\sigma_w^2}x_n^2$ is a chi-square random variable with one degree of freedom. As w_t is $N(0, \sigma_w^2)$, the sum $\frac{1}{n}\sum_{t=1}^{n} w_t^2$ converges to the variance σ_w^2 if $n \to \infty$. Hence, the numerator of the fraction (8.19) is distributed as a $\frac{1}{2}\left(\chi^2 - 1\right)$ random variable. The denominator converges towards an integral of the square of a Brownian motion. A Brownian motion $(W_t)_{t\geq 0}$ is a continuous time stochastic process with independent, identically distributed normal increments: $W_s - W_t \sim N(0, t - s)$, for $s \leq t$. It can be shown that when $n \to \infty$,

$$\lim_{n \to \infty} \frac{1}{n^2\sigma_w^2}\sum_{t=1}^{n} x_{t-1}^2 = \int_0^1 W^2(s)\,ds$$

Therefore, the statistic $n(\widehat{\eta} - 1)$ converges in distribution toward:

$$n(\widehat{\eta} - 1) \xrightarrow{d} \frac{\frac{1}{2}\left(\chi^2 - 1\right)}{\int_0^1 W^2(s)ds}.$$

This statistic is known as the unit root test or Dickey–Fuller statistic. The distribution of this statistic does not admit a closed form expression but may be computed numerically. An alternative formulation of this test for an AR(1) process is obtained by subtracting x_{t-1} from $x_t = \eta x_{t-1} + w_t$. Using the difference operator, we infer that:

$$\nabla x_t = (\eta - 1) x_{t-1} + w_t.$$

If we denote $\gamma = \eta - 1$, the null hypothesis of the unit root test may be reformulated as $H_0 : \gamma = 0$. An estimator $\widehat{\gamma}$ of γ is in this case obtained by regressing ∇x_t on x_{t-1}. If $sd(\widehat{\gamma})$ is the standard deviation of the estimator $\widehat{\gamma}$, the hypothesis H_0 may then be tested with a Wald statistic $\frac{\widehat{\gamma}}{sd(\widehat{\gamma})}$. The limit distribution of this statistic is computed numerically. This test is extended to an AR(p) model:

$$x_t = \sum_{j=1}^{p} \eta_j x_{t-j} + w_t.$$

First, we subtract x_{t-1} from both sides of this last equation:

$$\nabla x_t = \gamma x_{t-1} + \sum_{j=1}^{p} \psi_j \nabla x_{t-j} + w_t,$$

where $\gamma = \sum_{j=1}^{p} \eta_j - 1$ and $\psi_j = -\sum_{j=i}^{p} \eta_i$ for $j = 2, \ldots, p$. To test the hypothesis that $\eta(z)$ admits a unit root (i.e. $\eta(z) = 0$ for $z = 1$) and therefore that x_t is not stationary, we test the following assumption:

$$\begin{cases} H_0 : & \gamma = 0 \\ H_1 : & |\eta| < 1 \end{cases}.$$

The estimator $\widehat{\gamma}$ of γ is obtained by regressing ∇x_t on $x_{t-1}, \nabla x_{t-1}, \ldots, \nabla x_{t-p}$ and the statistic of test is $\frac{\widehat{\gamma}}{sd(\widehat{\gamma})}$. Dickey and Fuller (1979) tabulated the distribution of this statistic. The lag length p has to be determined when applying the test. One possible approach is to test down from high orders and examine the t-values of coefficients ψ_j. The Augmented Dickey–Fuller (ADF) tests allow to check the presence of constant and trend terms in the AR(p) process:

$$\nabla x_t = \beta_0 + \beta_1 t + \gamma x_{t-1} + \sum_{j=1}^{p} \psi_j \nabla x_{t-j} + w_t.$$

Table 8.3 Results of the
Dickey–Fuller test applied to
the time series of detrended
minimal temperatures in
Brussels

Lag	DF statistics	p-Value
3	-16.49	<0.01
6	-13.54	<0.01
9	-12.45	<0.01
12	-11.05	<0.01

Under H_0 and if $\beta_1 = 0$, the process x_t is a random walk with a drift. If for a data set, the hypothesis of a unit root is not rejected, the time series is not considered to be stationary. Fitting a model directly to this time series is then useless and irrelevant since the statistical inference relies on the stationarity of observations. Fortunately, for most time series encountered in practice, first differences or logarithmic first differences, usually transforms these time series into stationary ones. By first differences, we mean to model the series $y_t = \nabla x_t$ instead of x_t. By first log-differences, we mean the time series $y_t = \ln x_t - \ln x_{t-1}$. Notice that an ARMA model of the differences of x_t are called autoregressive integrated moving average processes (ARIMA).

Table 8.3 presents the results of the Dickey–Fuller test applied to detrended temperatures measured in Brussels from the 1/8/2009 to the 8/8/2016 (see Sect. 8.1.1 for details about this time series). Whatever the time lag, the statistic is highly negative and leads to the rejection of the hypothesis of non-stationarity.

In our next example, we study the stationarity of the time series of S&P 500 daily values from the 13/11/2000 to 12/11/2018 (4528 observations). The S&P 500 is a market-capitalization-weighted index of the 500 largest U.S. publicly traded companies by market value, The index is widely regarded as the best single gauge of large capitalization U.S. equities.

The upper graph of Fig. 8.4 shows the evolution of this index over the considered period. The S&P 500 reached its lower levels in 2003 and 2008. In 2003, the S&P 500 fell due to uncertainties about the impact of the second Iraki war on the US economy. In 2008, the drop of the S&P 500 was triggered by the crisis of subprimes. The lower graph of the same figure presents the time series of log-differences. In finance, this log-difference may be interpreted as the daily return of investing in the S&P 500 equities.

Table 8.4 reports the statistics of the Dickey–Fuller test applied to the time series of daily S&P 500 values and to their log-differences. Whatever the chosen time lag, this test rejects the hypothesis that S&P 500 values are stationary. Trying to estimate directly of model explaining the S&P 500 evolution is therefore irrelevant. Fortunately, the time series of first order log-differences is well stationary with a very high confidence level. In the next sections, we will test several approaches for modelling the time-series of S&P 500 log-differences.

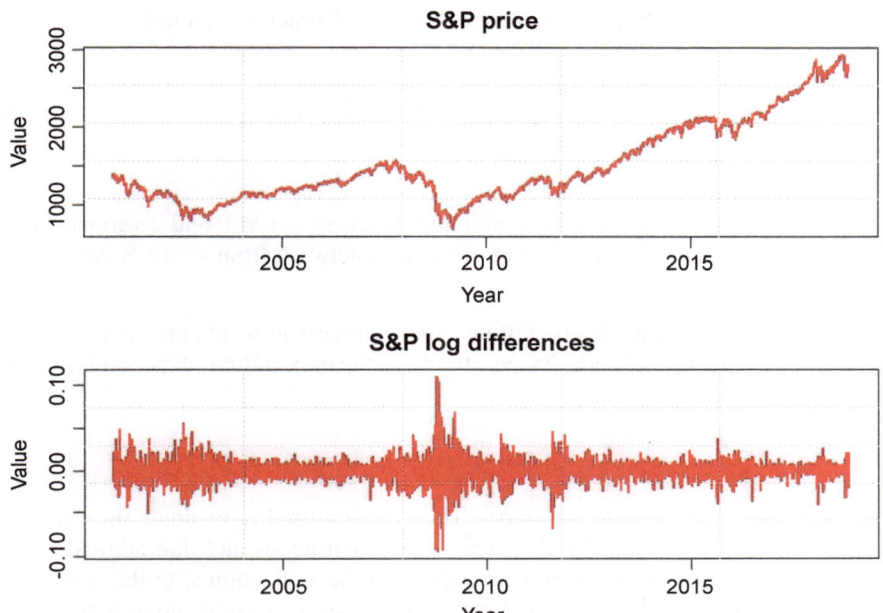

Fig. 8.4 First graph: values of the S&P500 from 2000 to 2018. Second graph: time series of log-differences

Table 8.4 Results of the Dickey–Fuller test applied to the time series of S&P 500 daily values and their log-differences

	Initial time series		Log-differences	
Lag	DF statistics	p-Value	DF statistics	p-Value
3	-1.87	0.63	-35.21	<0.01
6	-1.73	0.69	-27.65	<0.01
9	-1.68	0.72	-21.82	<0.01
12	-1.64	0.73	-18.49	<0.01

8.2 Autoregressive Neural Networks

In this section, we introduce a method for the analysis of time series that is based on neural systems. We start by studying autoregressive models in which there is a non-linear relationship between the current value of a process and its recent past realizations. As in previous developments, we do not differentiate anymore the stochastic process $(X_t)_{t \geq 0}$ from its time series $(x_t)_{t \geq 0}$ to lighten notations and we denote here by $\boldsymbol{x}_t = (x_t, x_{t-1}, \ldots, x_{t-p+1})$ the recent history of the process from time $t - p + 1$ up to t.

Definition 8.32 An autoregressive neural model of order p, denoted ARNN(p) is of the form

$$x_t = f_\Omega \left(x_{t-1}, \ldots, x_{t-p} \right) + w_t , \tag{8.20}$$
$$= f_\Omega \left(\boldsymbol{x}_{t-1} \right) + w_t ,$$

where w_t is a stochastic process with a null mean, $\mathbb{E}(w_t) = 0$ and a variance, σ_w^2. In Eq. (8.20), $f_\Omega(\boldsymbol{x}_{t-1})$ is the output of a neural network from $\mathbb{R}^p$ to $\mathbb{R}$ defined by a vector of parameters Ω.

The ARNN(p) is a generalization of the AR(p) process in which the linear trend is replaced by a neural network. By construction, the expectation of x_t conditionally to $\boldsymbol{x}_{t-1}$ is equal to

$$\widehat{x_t} = \mathbb{E}\left(x_t | \boldsymbol{x}_{t-1} \right) = f_\Omega \left(\boldsymbol{x}_{t-1} \right) ,$$

and its variance is $\mathbb{V}\left(x_t | \boldsymbol{x}_{t-1} \right) = \sigma_w^2$. The criterion used to estimate the network is a loss function denoted by $\mathcal{L} : \mathbb{R}^2 \to \mathbb{R}$, continuous and that admits a first order derivative. This loss function takes as input the realization x_t of the stochastic process and the prediction $\widehat{x_t}$, based on the information available up to time $t - 1$. The vector of weights Ω is next fitted in order to minimize the sum of losses:

$$\Omega = \arg\min_\Omega \frac{1}{T} \sum_{t=1}^{T} \mathcal{L}(x_t, \widehat{x_t}) , \tag{8.21}$$

$\mathcal{L}$ may be the opposite of the log-likelihood, a quadratic function or the deviance. Before detailing this point, we first develop the conditions that ensure the stationarity of a ARNN(p) process.

8.2.1 Stationarity of ARNN

A central question in linear time series theory is the stationarity of the model, i.e., whether the probabilistic structure of the series is stable over time or at least asymptotically stable (when not started in equilibrium). In this section, we review results of Leisch et al. (1999) about conditions that ensure the stationarity of ARNN.

Let us consider an ARNN(p) process. Equation (8.20) may be rewritten as a time series of vectors:

$$\boldsymbol{x}_t = g(\boldsymbol{x}_{t-1}) + \boldsymbol{\epsilon}_t \tag{8.22}$$

where $g(\boldsymbol{x}_{t-1}) = \left(f_\Omega(\boldsymbol{x}_{t-1}), x_{t-1}, \ldots, x_{t-p+1} \right)^\top$ and $\boldsymbol{\epsilon}_t = (w_t, 0, \ldots, 0)^\top$. Hence $(\boldsymbol{x}_t)_{t \in \mathbb{N}}$ is a p dimension Markov chain with the state space $(\mathbb{R}^p, \mathcal{B}, \lambda)$ where $\mathcal{B}$ is the Borel tribe of $\mathbb{R}^p$ and $\lambda(.)$ is the Lebesgue measure on $(\mathbb{R}^p, \mathcal{B})$. If we remember

Chap. 2, the chain x_t is stationary if $P(x_t \in A)$ for $A \in \mathcal{B}$ is independent from the time: $P(x_t \in A) = \pi(A)$ where $\pi(.)$ is called the stationary distribution of the series.

It is clear that a time series can only be stationary from the beginning if it is started with the stationary distribution such that $x_0 \sim \pi$. Otherwise, we call the series asymptotically stationary if it converges to its stationary distribution:

$$\lim_{t \to \infty} P(x_t \in A) = \pi(A) .$$

Let

$$P^n(x, A) = P(x_{t+n} \in A \mid x_t = x)$$

denotes the probability that the process x_t moves from x to the set $A \in \mathcal{B}$ in n steps. Here, $P(x, A)$ is the transition kernel of the Markov chain. This chain is φ-irreducible, if for some σ- finite measure φ on $(\mathbb{R}^p, \mathcal{B}, \lambda)$

$$\forall x \in \mathbb{R}^p : \sum_{n=1}^{\infty} P^n(x, A) > 0$$

with $\varphi(A) > 0$. As explained in Chap. 2, this means that all elements of the state space can be reached by the Markov chain, independently from the initial point. We now introduce the concept of geometrical ergodicity:

Definition 8.33 The chain $(x_t)_{t \in \mathbb{Z}}$ is called geometrically ergodic if there exists a probability measure $\pi(A)$ on $(\mathbb{R}^p, \mathcal{B}, \lambda)$ and a $\rho > 1$ such that

$$\forall x \in \mathbb{R}^p : \lim_{n \to \infty} \rho^n ||P^n(x, .) - \pi(.)|| = 0 \tag{8.23}$$

where $||.||$ is here the total variation norm.

By definition of geometrical ergodicity, Eq. (8.23) implies that $\pi(.)$ satisfies the invariance property:

$$\pi(A) = \int_{\mathbb{R}^p} P(x, A)\pi(dx) \quad \forall A \in \mathcal{B} .$$

If the Markov chain is geometrically ergodic, then its distribution will converge to $\pi(.)$ and the time series is asymptotically stationary. Furthermore, if the time series starts with $x_0 \sim \pi(.)$, the time series is strictly stationary.

We now consider the case where $f_\Omega(.)$ is one of the following standard network architectures:

1. Single hidden layer perceptron:

$$f_\Omega(x) = \gamma_0 + \sum_i \beta_i \phi \left(\alpha_i + \omega_i^\top x \right) , \tag{8.24}$$

where α_i, β_i and $\gamma_0 \in \mathbb{R}$, $\omega_i \in \mathbb{R}^p$. In Eq. (8.24), x is of dimension p and $\phi(.)$ is a bounded activation function (logistic, or tanh).

2. Single hidden layer perceptron with shortcut connections:

$$f_\Omega (x) = \gamma_0 + \eta^\top x + \sum_i \beta_i \phi \left(\alpha_i + \omega_i^\top x \right) , \qquad (8.25)$$

where $\eta \in \mathbb{R}^p$ is an additional weight vector for direct connections between inputs and output. The characteristic polynomial $\eta(z)$ of the linear part is defined as:

$$\eta (z) = 1 - \eta_1 z - \eta_2 z^2 - \ldots - \eta^p z^p , \quad z \in \mathbb{C} .$$

3. Radial basis function networks:

$$f_\Omega (x) = \gamma_0 + \sum_i \beta_i \phi \left(\omega_i^\top |x - \alpha_i| \right) , \qquad (8.26)$$

where $\alpha_i \in \mathbb{R}^p$ are center vectors and $\phi(.)$ is a radial basis function as $\phi(x) = \exp \left(-x^2 \right)$.

The next lemma says that the state space of the Markov chain cannot be reduced depending on the starting point.

Lemma 8.1 *Let the process $(x_t)_{t\in\mathbb{N}}$ be defined by Eq. (8.22). Let us assume that $\mathbb{E}(|w_t|) < \infty$ and that the probability density function of w_t is positive everywhere in $\mathbb{R}$. If $f_\Omega(.)$ is defined by any of equations (8.24), (8.25) or (8.26), the Markov chain $(x_t)_{t\in\mathbb{N}}$ is φ-irreducible and aperiodic.*

Proof It can be shown that x_t is φ-irreducible if the support of the probability density function (pdf) of w_t is positive everywhere in $\mathbb{R}$ (Chan and Tong 1985). In this case every non-null p-dimensional hypercube is reached in p steps with positive probability (and hence every non-null Borel set A).

A necessary and sufficient condition for x_t to be aperiodic is that there exists a set A and positive integer n such that $P^n(x, A) > 0$ and $P^{n+1}(x, A) > 0$ for all $x \in A$ (Tong 1990, p. 455). In our case this is true for all n due to the unbounded additive noise. $\qquad \square$

An example of reducible Markov chain is a series that is always positive if only $x_0 > 0$ (and negative otherwise). This cannot happen in the ARNN(p) case due to the unbounded additive noise term. Leisch et al. (1999) use the following result from nonlinear time series theory to analyze the stationarity of ARNN:

Theorem 8.4 (Chan and Tong 1985) *Let $(x_t)_{t\in\mathbb{N}}$ be defined by (8.20), (8.22) and let $g(.)$ be compact, i.e. preserve compact sets. If $g(.)$ can be decomposed as $g = g_h + g_d$ and $g_d(.)$ is of bounded range, $g_h(.)$ is continuous and homogeneous, i.e., $g_h(\alpha x) = \alpha g_h(x)$, the origin is a fixed point of $g_h(.)$ and $g_h(.)$ is uniform* *asymptotically stable, $\mathbb{E}|w_t| < \infty$ and the pdf of w_t is positive everywhere in $\mathbb{R}$, then $(x_t)_{t\in\mathbb{N}}$ is geometrically ergodic.*

We finally reproduce a result from Leisch et al. (1999) about the sufficient conditions for the stationarity of $(x_t)_{t\in\mathbb{N}}$:

Proposition 8.20 *Let the process $(x_t)_{t\in\mathbb{N}}$ be defined by Eq. (8.22). Let us assume that $\mathbb{E}(|w_t|) < \infty$ and that the probability density function of w_t is positive everywhere in $\mathbb{R}$. Then,*

- *If $f_\Omega(.)$ is a network without linear shortcuts as defined in (8.24) and (8.26), then $(x_t)_{t\in\mathbb{N}}$ is geometrically ergodic and $(x_t)_{t\in\mathbb{N}}$ is asymptotically stationary.*
- *If $f_\Omega(.)$ is a network with linear shortcuts as defined in (8.25) and additionally $\eta(z) \neq 0 \; \forall z \in \mathbb{C} : |z| < 1$, then $(x_t)_{t\in\mathbb{N}}$ is geometrically ergodic and $(x_t)_{t\in\mathbb{N}}$ is asymptotically stationary.*

Proof The proof relies on theorem 8.4. By construction, the noise process w_t fulfills the conditions and all networks are continuous compact functions. Standard neural networks without shortcut connections and radial base functions have a bounded range, hence $g_h(.) = 0$ and $g(.) = g_d(.)$. The series x_t is therefore asymptotically stationary.

If we allow for linear shortcut connections between the input and the outputs, we get

$$g_h(x) = \eta^\top x \,,$$

$$g_h(x) = \gamma_0 + \sum_i \beta_i \phi\left(\alpha_i + \omega_i^\top x\right) \,.$$

g_h is the linear part of the network, and g_d is a standard perceptron without shortcut connections. Clearly, g_h is continuous, homogeneous and has the origin as a fixed point. Hence, the series x_t is asymptotically stationary if g_h is asymptotically stable, i.e., when all characteristic roots of g_h have a magnitude less than unity. Obviously the same is true for radial bases function with shortcut connections. □

Notice that the time series remains stationary if we consider neural networks with more than one hidden layer or non-linear output units, as long as the overall mapping has bounded range. A neural network with shortcut connections combines a linear AR(p) process with a non-linear stationary neural network. Thus, the neural network is used for modelling non-linear fluctuations around a linear process. The stationarity of x_t depends only upon the linear shortcut connections. If there are no shortcuts, then the process is always stationary because of the boundedness of the activation function. As the stationarity of the process depends only on the linear part of the network, we can use the usual unit root test of Dickey–Fuller presented in Sect. 8.1.6 for checking the stationarity.

8.2.2 Gaussian ARNN

At this stage, we have not discussed yet the choice of the loss function $\mathcal{L}(x_t, \widehat{x}_t)$ involved in the calibration of ARNN(p). Remember that we consider a neural network defined by weights Ω with an output $f_\Omega(x_{t-1})$ and such that:

$$x_t = f_\Omega(x_{t-1}) + w_t.$$

Here x_t is stationary, w_t is a stochastic process with a null mean, $\mathbb{E}(w_t) = 0$ and a finite standard deviation, σ_w. The estimator of x_t with the information up to time $t-1$ is given by $\widehat{x}_t = \mathbb{E}(x_t|x_{t-1}) = f_\Omega(x_{t-1})$. If w_t is $N(0, \sigma_w^2)$, the probability density function of x_t, conditionally to x_{t-1} is

$$f_{x_t|x_{t-1}}(z) = \frac{1}{\sigma_w\sqrt{2\pi}} \exp\left(-\frac{1}{2}\left(\frac{z - \widehat{x}_t}{\sigma_w}\right)^2\right).$$

The log-likelihood of the observation x_t is therefore of the form

$$l(x_t|\Omega, x_{t-1}) = -\frac{1}{2\sigma_w^2}(x_t - \widehat{x}_t)^2 + c$$

where $\widehat{x}_t = f_\Omega(x_{t-1})$ and $c \in \mathbb{R}$ is a constant. Using the negative log-likelihood as loss function is then equivalent to minimize the mean squared error between estimated and observed values of the process. This is also equivalent to maximize the unscaled deviance. An estimate of the vector of weights $\widehat{\Omega}$ is in this case obtained as follows

$$\widehat{\Omega} = \arg\min_\Omega \frac{1}{T} \sum_{t=1}^{T} (x_t - \widehat{x}_t)^2 \tag{8.27}$$

$$= \arg\min_\Omega \frac{1}{T} \sum_{t=1}^{T} (x_t - f_\Omega(x_{t-1}))^2.$$

This loss function is the mean squared error of prediction and the most common in the literature. Once the network is fitted, the standard deviation, σ_w, of the random noise, is estimated by the standard deviation of residuals:

$$\widehat{\sigma}_w = \sqrt{\frac{1}{T - m} \sum (x_t - f_\Omega(x_{t-1}))^2},$$

where m is the number of parameters in the network. The neural networks can then be used for two purposes. The first one is the prediction one step ahead values of

x_t. The second one is the simulation of future sample paths of x_t. This point is illustrated in the following example.

To conclude this section, we compare the capacity of Gaussian ARMA(p,q) and ARNN(p) for explaining the dynamic of the S&P 500. This time series, introduced in Sect. 8.1.6, is an indicator of the financial wealth for the 500 largest U.S. publicly traded companies. As the Dickey–Fuller has revealed that the S&P 500 is not stationary, we focus on the modelling of the first order log-differences. If x_t is the S&P value at time t, the log-differences are equal to:

$$y_t = \ln x_t - \ln x_{t-1} \quad t = 1, \ldots, T \,.$$

The dataset contains daily observations from the 14/11/2000 to 12/11/2018 (4527 observations). To control the level of overfitting of ARNN(p), the time series is split into a training and a validation sample. Models are trained on observations from the 14/11/2000 to 10/10/2016 (4000 log-differences) whereas the validation is performed on the last 527 observations. We use a quadratic loss function for calibrating the neural network with the back-propagation algorithm (2000 iterations).

As explained in Sect. 1.5, activation functions of neurons like sigmoid, hyperbolic tangent and Gaussian cumulative functions quickly converge toward 0 (or -1) and 1 outside a relatively small interval centered around zero. Without scaling of initial data, the input signal may be far away from this interval. For this reason, we work with the rescaled time series of log-differences. We first calculate the average and standard deviation of log-differences:

$$\bar{y} = \frac{1}{T} \sum_{t=1}^{T} y_t \,,$$

$$s_y^2 = \frac{1}{T-1} \sum_{t=1}^{T} \left(y_t - \mu_y \right)^2 \,.$$

Next we center and normalize the data as follows:

$$y_t' = \frac{y_t - \bar{y}}{s_y} \quad t = 1, \ldots, T \,.$$

Table 8.5 reports the statistics of goodness of fit for ARMA(p,q) models adjusted to the training set of normalized log-differences. We have used the package "tseries" in R. The bold numbers indicate the models with the lowest AIC. According to this criterion, the best match is obtained with an ARMA(4,3) model. However, ARMA(2,3), ARMA(3,4) and ARMA(5,2) have very close AIC. We use these statistics for comparing with the performance of ARNN(p).

Table 8.6 reports the statistics of goodness of fit for ARNN(p) models with 2–6 neurons in their hidden layer. The activation function of hidden neurons is

Table 8.5 Statistics about the goodness of fit of ARMA(p, q) models to S&P 500 log-differences, training sample

	q	0	1	2	3	4
ARMA(1,.)	SSE	4308.93	4297.75	4294.28	4294.06	4288.04
	log-lik.	−5824.55	−5819.35	−5817.74	−5817.64	−5814.83
	AIC	11,655.09	11,646.71	11,645.48	11,647.27	**11,643.67**
	BIC	11,665.68	11,663.59	11,668.65	11,676.74	11,679.43
ARMA(2,.)	SSE	4294.85	4293.55	4292.4	4275.41	4285.08
	log-lik.	−5818	−5817.4	−5816.87	−5808.95	−5813.46
	AIC	11,644.01	11,644.8	11,645.73	**11,631.91**	11,642.91
	BIC	11,660.89	11,667.98	11,675.2	11,667.67	11,684.97
ARMA(3,.)	SSE	4293.56	4293.39	4293.55	4271.03	4268.49
	log-lik.	−5817.4	−5817.33	−5817.4	−5806.93	−5805.77
	AIC	11,644.81	11,646.65	11,648.8	11,629.85	**11,629.54**
	BIC	11,667.98	11,676.12	11,684.57	11,671.91	11,677.89
ARMA(4,.)	SSE	4292.48	4287.59	4284.16	4267.81	4267.69
	log-lik.	−5816.9	−5814.63	−5813.02	−5805.47	−5805.41
	AIC	11,645.8	11,643.26	11,642.05	**11,628.94**	11,630.82
	BIC	11,675.27	11,679.02	11,684.11	11,677.29	11,685.47
ARMA(5,.)	SSE	4282.63	4272.76	4270.03	4270.03	4267.22
	log-lik.	−5812.31	−5807.71	−5806.43	−5806.43	−5805.12
	AIC	11,638.63	11,631.43	**11,630.87**	11,632.86	11,632.24
	BIC	11,674.39	11,673.49	11,679.22	11,687.51	11,693.18

the hyperbolic tangent. We have implemented these networks in KERAS for R. The best model according to the AIC is the autoregressive network with 5 lags, ARNN(5), and 5 neurons, With this configuration the AIC, equal to 11,501.18, is clearly lower than 11,628.97, the AIC of the ARMA(4,3). Table 8.7 presents the sum of squared errors and log-likelihoods computed on the validation sample. Based on these statistics the ARNN(3) with 5 hidden neurons seems a good trade-off with good performances both on the training and validation samples. Even if the AIC on the training set for this configuration is higher than the one of the ARNN(5) with 5 neurons, it is still significantly less than the AIC of the ARMA(4,3). This confirms the superiority of autoregressive neural networks on ARMA models, at least for modeling financial time series. Autoregressive neural networks may be used for risk management purposes like one-step ahead previsions or Monte-Carlo simulations. Figure 8.5 compares the 1 day ahead forecast of the ARMA(4,3) and ARNN(3) with 5 neurons. This exercise is done over the period from the 23/12/2015 to 6/10/2016. The ARNN(3) yields more volatile forecasts than the ARMA(4,3). In many circumstances, both models predict the same trend for the next day (by trend we mean positive or negative log-differences) but the amplitude of forecasts may be very different.

Figure 8.6 shows simulated sample paths of the S&P 500 daily log-differences over a period of 50 days. The simulation is initialized with log-returns measured

Table 8.6 Statistics about the goodness of fit of ARNN(p) models to S&P 500 log-differences, training sample

		2	3	4	5	6
ARNN(1)	SSE	4292.73	4288.66	4285.47	4284.78	4292.53
	log-lik.	−5817.02	−5815.12	−5813.64	−5813.32	−5816.94
	AIC	**11,648.03**	11,650.24	11,653.27	11,658.64	11,671.88
	BIC	11,692.09	11,713.18	11,735.1	11,759.35	11,791.46
ARNN(2)	SSE	4296.51	4267.38	4280.45	4222.64	4207.42
	log-lik.	−5817.82	−5804.23	−5810.35	−5783.17	−5775.96
	AIC	11,653.65	11,634.45	11,654.69	11,608.34	**11,601.92**
	BIC	11,710.29	11,716.27	11,761.68	11,740.51	11,759.26
ARNN(3)	SSE	4287.74	4235.35	4200.3	4193.88	4190.75
	log-lik.	−5812.79	−5788.22	−5771.62	−5768.57	−5767.1
	AIC	11,647.57	11,608.44	**11,585.24**	11,589.15	11,596.2
	BIC	11,716.8	11,709.13	11,717.4	11,752.78	11,791.3
ARNN(4)	SSE	4284.47	4262.04	4229.32	4174.08	4176.19
	log-lik.	−5810.31	−5799.83	−5784.44	−5758.19	−5759.23
	AIC	11,646.62	11,637.67	11,618.89	**11,578.39**	11,592.45
	BIC	11,728.43	11,757.24	11,776.22	11,773.48	11,825.3
ARNN(5)	SSE	4278.38	4166.48	4178.86	4085.86	4092.75
	log-lik.	−5806.52	−5753.58	−5759.53	−5714.59	−5717.99
	AIC	11,643.03	11,551.17	11,577.06	**11,501.18**	11,521.98
	BIC	11,737.43	11,689.61	11,759.56	11,727.73	11,792.58

Bold values indicate the best fit

Table 8.7 Statistics about the goodness of fit of ARNN(p) models to S&P 500 log-differences, validation sample

		2	3	4	5	6
ARNN(1)	SSE	**185.9**	186.08	187.41	187.31	186.18
	log-lik.	**−472.84**	−473.12	−475.02	−474.93	−473.4
ARNN(2)	SSE	187.39	**184.76**	185.2	187.41	185.21
	log-lik.	−474.95	**−471.28**	−471.96	−475.16	−472.15
ARNN(3)	SSE	187.08	188.26	185.25	**182.77**	184.28
	log-lik.	−474.54	−476.26	−472.11	**−468.68**	−470.98
ARNN(4)	SSE	194.49	**182.9**	184.01	189.35	194.43
	log-lik.	−484.77	**−468.72**	−470.43	−478.13	−485.3
ARNN(5)	SSE	**187.46**	187.61	189.12	193.09	185.58
	log-lik.	**−475.13**	−475.46	−477.74	−483.45	−473.3

Bold values indicate the best fit

on the 3th, 4th and 5th of October 2016. The shaded area delimits the 0.5–99.5% confidence region of simulated log-differences. Table 8.8 reports some statistics computed on simulated sample paths and on the calibration data set. The average daily return is close to zero as observed on the training set. The standard deviations

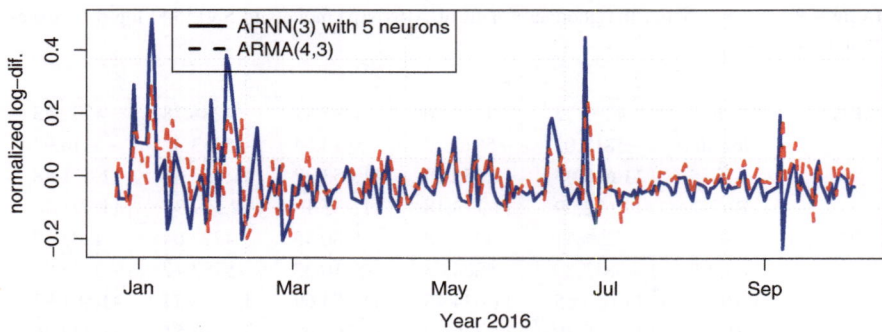

Fig. 8.5 Comparison of one-step ahead forecast computed by ARMA(4.3) and ARNN(3) with 5 neurons. Period: 23/12/2015 to 6/10/2016

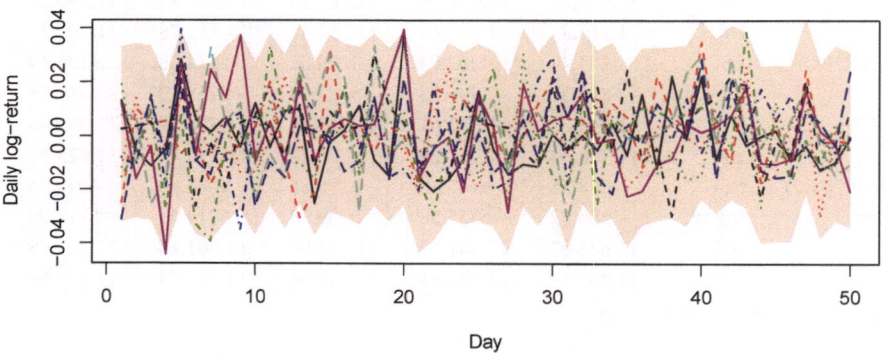

Fig. 8.6 Ten thousand simulations of S&P 500 log-differences over 50 days, with an ARNN(3). Starting day: 6/10/2016

Table 8.8 Statistics for simulated S&P 500 log-differences and real observations

	Simulations	Observations
Expectation	0.0001	−0.0001
St. deviation	0.0119	0.0127
$q_{0.05}$	−0.0185	−0.0201
$q_{0.95}$	0.0168	0.0211

of simulated and observed daily returns are close to 1.20%. We observe that 90% of simulated log-differences evolve in the interval of values $[-1.85\%, +1.68\%]$. The 5 and 95% percentiles of observed daily returns are slightly greater in absolute value than those of simulations. The reason is that the training set covers periods of deep crisis like the credit crunch of 2008 and the double dip recession of 2011.

8.3 Autoregressive Neural Models for Count Data

In actuarial sciences, the modeling of count data also interests us. This is a statistical type of data in which the observations can take only non-negative integer values, and where these integers arise from counting. In insurance, the time series of monthly claims per business lines is an example of count data that has received a lot of attention in the literature. To put the above ideas in the framework of count time series, we consider that $y_t \in \mathbb{N}$ is the observed response series for $t = 1, \ldots, T$. and set $\mathbf{y}_t = (y_t, \ldots, y_{t-p+1})$, for the past p values of the process.

In an autoregressive framework, the random component of the model is given by the conditional distribution of y_t given $\mathbf{y}_{t-1}$. Natural choices for the distribution of $y_t | \mathbf{y}_{t-1}$ are the Poisson or Binomial laws. If $y_t | \mathbf{y}_{t-1}$ has a Poisson distribution with intensity λ_t, we introduce a non-linear autoregressive feature between λ_t and the past realizations of the counting process by assuming that

$$\lambda_t = \exp\left(f_\Omega\left(\mathbf{y}_{t-1}\right)\right), \tag{8.28}$$

where $f_\Omega\left(\mathbf{y}_{t-1}\right)$ is a neural network from $\mathbb{R}^p \to \mathbb{R}$. If $y_t | \mathbf{y}_{t-1}$ is distributed as a binomial with parameters p_t, we link p_t to $\mathbf{y}_{t-1}$ with a logistic function:

$$p_t = \frac{1}{1 + \exp\left(-f_\Omega\left(\mathbf{y}_{t-1}\right)\right)}. \tag{8.29}$$

In both cases, the weights of the neural network are estimated by minimization of the unscaled Poisson or binomial deviance. The methodology does not differ from the one developed in Chap. 1: we use as explanatory variables the history of the process rather than characteristics of contracts. For this reason, we skip details related to the calibration of autoregressive neural models with classic Poisson and binomial distributions. Instead, we focus in the next section on a particular type of count data presenting what is called "overdispersion".

8.3.1 Modelling of Overdispersed Count Data

Modelling count data with a Poisson law is usually the preferred approach of actuaries, mainly due to its high analytical and numerical tractability. A Poisson random variable is defined by a single parameter which is equal to its expectation and variance. However, in many circumstances the variance of counted events is significantly higher than their average. For instance, we will see in the next subsection that the monthly number of claims caused by tornadoes in the U.S., presents this feature. Modelling these claims numbers with a Poisson law is therefore not appropriate and systematically leads to underestimate the variance of the counting process. This phenomenon is called overdispersion. One solution for managing this issue consists to work with exponential overdispersed random variable. To understand the origin of this type of distribution, we first remind the definition of exponential dispersed distributions, seen in Sect. 1.7 of Chap. 1. Y is an exponential dispersed random variable, if its density function admits the following representation:

$$f_Y(y; \theta_t, \phi) = \exp\left\{\frac{y\theta_t - a(\theta_t)}{\phi}\right\} c(y, \phi) \tag{8.30}$$

where θ_i is a parameter that depends on t, whereas the dispersion parameter ϕ is identical for all t. The $a(\theta_i)$ is called the cumulant function and is C^2 with an invertible second derivative. The function $c(.)$ is independent from θ_i. We now define the overdispersion as follows:

Definition 8.34 A regular exponential dispersed random variable Y, defined by parameters (θ, ϕ, a, c), is said to be overdispersed with coefficient $\delta \in (0, \infty)$ if the identity

$$\mathbb{V}(Y) = \delta \mathbb{E}(Y) \tag{8.31}$$

holds.

As stated in the next proposition, this constraint linking the variance to the expectation defines the cumulant function of exponential overdispersed law:

Proposition 8.21 *Assume that Y is an exponential overdispersed random variable defined by parameters (θ, ϕ, a, c) and such that $\mathbb{V}(Y) = \delta \mathbb{E}(Y)$. Then there exist $\beta \in (0, \infty)$ and $\gamma \in \mathbb{R}^+$ such that*

$$a(\theta) = \beta \frac{\phi}{\delta} \exp(\frac{\delta}{\phi}\theta) + \gamma \tag{8.32}$$

holds. Moreover the variance function satisfies

$$V(\mu) = \frac{\delta}{\phi}\mu$$

where $\mu = a'(\theta)$ is the expectation $\mathbb{E}(Y)$.

Proof As $\mathbb{E}(Y) = a'(\theta)$ and $\mathbb{V}(Y) = a''(\theta)$, we immediately infer from Eq. (8.31) that

$$\phi a''(\theta) = \delta a'(\theta),$$

and it is easy to check that $a(\theta)$ such as defined by Eq. (8.31) is solution of this ordinary differential equation.   □

If we remember Table 1.2 from Chap. 1, among the exponential dispersed distributions considered before, only the Poisson family forms an overdispersed Poisson family, and in this case we have $\delta = 1 = \phi$. The following result characterizes the overdispersed Poisson distributions.

Proposition 8.22 *Consider $\delta \in (0, \infty)$. Then, for a random variable Y, the following statements are equivalent:*

1. *The random variable $\frac{Y}{\delta}$ has a Poisson distribution of parameter $\frac{\lambda}{\delta}$.*
2. *The distribution of Y belongs to an overdispersed Poisson family with parameter δ.*

Proof It is straightforward to show that 1 implies 2. Indeed, if $\frac{Y}{\delta}$ has a Poisson distribution then

$$\mathbb{E}\left(\frac{Y}{\delta}\right) = \frac{\lambda}{\delta} = \frac{1}{\delta^2}\mathbb{V}(Y),$$

and $\mathbb{V}(Y) = \delta\lambda = \delta\mathbb{E}(Y)$. Notice also that for some $\epsilon \in (0, 1)$, the mgf of Y is

$$\mathbb{E}\left(e^{u\frac{Y}{\delta}}\right) = \exp\left(\frac{\lambda}{\delta}\left(e^u - 1\right)\right)$$

and therefore

$$\mathbb{E}\left(e^{vY}\right) = \exp\left(\frac{\lambda}{\delta}\left(e^{\delta v} - 1\right)\right). \tag{8.33}$$

for all $v \in (-\epsilon, \epsilon)$. Assume now that 2 holds. There exists some $\epsilon \in (0, \infty)$ such that $\theta + \phi t$ holds for all $t \in (-\epsilon, \epsilon)$. As Y is also a dispersed law, its moment generating function is given by Proposition 1.1:

$$\mathbb{E}(e^{tY}) = \exp\left(\frac{a(\theta + t\phi) - a(\theta)}{\phi}\right)$$

for all $t \in (-\epsilon, \epsilon)$. Inserting the identity $a(\theta) = \beta\frac{\phi}{\delta}\exp(\frac{\delta}{\phi}\theta) + \gamma$ from Proposition 8.21 and substituting $\lambda = a'(\theta) = \beta\exp\left(\frac{\delta}{\phi}\theta\right)$, we retrieve the expression (8.33) of the moment generating function of $\frac{Y}{\delta}$, a Poisson random variable of parameter $\frac{\lambda}{\delta}$. Therefore, the second statement implies the first one. $\square$

This last proposition emphasizes that counting data with an overdispersion coefficient δ can easily be modeled by a random variable Y, such that $\frac{Y}{\delta}$ has a Poisson distribution with parameter $\frac{\lambda}{\delta}$. In this case, the support of Y is the set $\{k\delta \mid k \in \mathbb{N}\}$. If $\delta \in \mathbb{N}$, Y is therefore a kind of counting process but by increments of size δ. We now explain how to estimate λ and δ. If y is a realization of Y, the log-likelihood of the Poisson variable $\frac{Y}{\delta}$ is:

$$l(\lambda) = -\frac{\lambda}{\delta} + \frac{y}{\delta} \ln\left(\frac{\lambda}{\delta}\right) - \ln\left(\left(\frac{y}{\delta}\right)!\right).$$

The scaled deviance is an alternative to the log-likelihood for assessing the goodness of fit of a model. It is defined as two times the difference between log-likelihoods of saturated and non-saturated models. In the saturated model, the frequency is replaced by the observation y. This model has no practical interest but since it perfectly fits data, its log-likelihood is the best one that we can obtain. If we denote by $\widehat{y}$ the estimator of $\lambda = \mathbb{E}(Y)$, the scaled deviance is here equal to

$$D^*(y, \widehat{y}) = \begin{cases} 2\left(-\frac{y}{\delta} + \frac{y}{\delta} \ln(y) + \frac{\widehat{y}}{\delta} - \frac{y}{\delta} \ln(\widehat{y})\right) & y > 0, \\ 2\frac{\widehat{y}}{\delta} & y = 0. \end{cases}$$

By multiplying this expression by δ, we get the unscaled deviance $D(y, \widehat{y}) = \delta D^*(y, \widehat{y})$ that is independent from δ. We also recognize the deviance of a Poisson random variable, presented in Table 1.3 of Chap. 1. We will use this Poisson deviance as loss function $\mathcal{L}(y, \widehat{y})$ for calibrating the neural network in the next subsection. On the other hand, the δ may be assimilated to the parameter ϕ of an exponential dispersed law. Hence, as explained in Sect. 1.7 of Chap. 1, an estimator $\widehat{\delta}$ of δ is:

$$\widehat{\delta} = \frac{1}{n - m} \sum_{t=1}^{T} \frac{(y_t - \widehat{y}_t)^2}{\widehat{y}_t},$$

where m is the number of parameters involved in the construction of the estimator $\widehat{y}$. The overdispersed Poisson distribution is combined in the next section with a neural network for explaining time series with overdispersion.

8.3.2 An Autoregressive Neural Model for Overdispersed Count Data

As in previous sections, we do not differentiate the stochastic counting process from the time series of y_t for $t = 1, \ldots, T$. Recall that $y_t \in \mathbb{N}$ and that we set $\boldsymbol{y}_t = (y_t, \ldots, y_{t-p+1})$, for the past p values of the process. In addition, we may imagine that we have additional explanatory information at time t contained in a vector of dimension d, denoted by $\boldsymbol{x}_t = (x_{t,1}, \ldots, x_{t,d})^\top$. We insist on the fact that $\boldsymbol{x}_t$ is not a time series in this section but instead a vector of covariates providing additional information about y_t.

The random component of the model is given by the conditional distribution of y_t given $\mathbf{y}_{t-1}$ and $\mathbf{x}_t$. As data can display overdispersion, we consider a coefficient of overdispersion such that:

$$\frac{y_t}{\delta} | \mathbf{y}_{t-1}, \mathbf{x}_t \sim Po\left(\frac{\lambda_t}{\delta}\right).$$

In order to introduce a non-linear autoregressive feature between the mean frequency and the past realizations of the counting process, we assume that λ_t is related to past realizations $\mathbf{y}_{t-1}$ and to information $\mathbf{x}_t$ by the next relation:

$$\lambda_t = \exp\left(f_\Omega\left(\mathbf{y}_{t-1}, \mathbf{x}_t\right)\right), \tag{8.34}$$

where $f_\Omega\left(\mathbf{y}_{t-1}, \mathbf{x}_t\right)$ is a neural network from $\mathbb{R}^{p+d} \to \mathbb{R}$. This neural network is characterized by a vector of neural weights Ω and takes as input a real $(p+d)$-vector. The exponential link between the output signal of the neural network and the frequency, ensures that the domain of λ_t is well in $\mathbb{R}^+$. According to results of the previous section, the unscaled deviance for the tth observation is equal to

$$D(y_t, \widehat{y}_t | \Omega, \mathbf{y}_{t-1}, \mathbf{x}_t) = \begin{cases} 2\left(y_t \ln y_t - y_t \ln \widehat{y}_t - y_t + \widehat{y}_t\right) & y_t > 0 \\ 2\widehat{y}_t & y_t = 0, \end{cases}$$

where $\widehat{y}_t = \exp\left(f_\Omega\left(\mathbf{y}_{t-1}, \mathbf{x}_t\right)\right)$. If we use the (unscaled) deviance as loss function, the weights of the neural network are estimated by minimizing the sum:

$$\widehat{\Omega} = \arg\min_\Omega \frac{1}{T} \sum_{t=1}^{T} D(y_t, \widehat{y}_t | \Omega, \mathbf{y}_{t-1}, \mathbf{x}_t).$$

If $m = \text{card}(\Omega)$ is the number of model parameters, the coefficient of overdispersion is next appraised as

$$\widehat{\delta} = \frac{1}{n-m} \sum_{t=1}^{T} \frac{(y_t - \widehat{y}_t)^2}{\widehat{y}_t}.$$

To illustrate this section, we show that this model is particularly powerful for the analysis and forecasting of recurrent meteorological events. Physical models for meteorological phenomena have a limited tractability for financial applications such as the pricing of weather derivatives, given their complexity. For this reason, the existing literature on the pricing of climatic products mainly relies on statistical models as in Hainaut (2012), Hainaut and Boucher (2014) but neural networks offer a reliable alternative. In particular, we study the time series of monthly numbers of tornadoes that have caused damages in the US from January 1975 to December

2008 (data retrieved on Sheldus[1]). The number of tornadoes per month is contained in the time series y_t where $t = 1, \ldots, 408$.

The frequency of phenomena like tornadoes exhibits seasonality combined with a huge volatility. As shown in Fig. 8.8, a peak of activity is observed from April to June. Nevertheless, tornadoes occur all year long. To manage this seasonality, we can think to detrend the time series by removing the average number of claims per month. Or we can try to model a trend with a mixture of sinus and cosinus functions as detailed in Sect. 8.1.1. However, the detrended time series with these methods has negative or non-integer realizations and therefore cannot be modeled by a Poisson law. For this reason, we prefer to work on raw data but we use as additional information, the month during which tornadoes are count.

This information is stored in a vector $\boldsymbol{x}_t$ of eleven binary variables that indicates the month corresponding to the step t. E.g. $x_{t,1} = 1$ and $x_{t,k} = 0$ for $k = 2, \ldots, 11$ if t is a time index for observations done in January. For observations of December, we set $x_{t,k} = 0$ for $k = 1, \ldots, 11$.

The number of tornadoes y_t at step t is assumed to be an overdispersed Poisson random variable with an intensity λ_t and dispersion parameter δ depending upon the p past occurrences $\boldsymbol{y}_{t-1} = \left(y_{t-1}, \ldots, y_{t-p}\right)$ and $\boldsymbol{x}_t$. The functional link between λ_t and $(\boldsymbol{\lambda}_{t-1}, \boldsymbol{x}_t)$ is the exponential of the output of a single layer neural network as in Eq. (8.34). The dataset is split into a validation set with observations from January 1975 to December 2006, and a training set with data of years 2007 and 2008. Tables 8.9 and 8.10 report the statistics of goodness of fit for autoregressive models of order 1–5 and shallow networks with 2–6 neurons in the hidden layer. Notice that the log-likelihood, AIC presented in these tables are computed with the Poisson model without overdispersion, $y_t \sim Po(\widehat{y_t})$. We also provide an estimate of the overdispersion parameter for each model.

On the calibration and validation sets, the lowest deviance and AIC is obtained with an ARNN(4) and 6 neurons. The rest of our analysis focuses therefore on this model. The estimate $\widehat{\delta}$ of the overdispersion parameter is equal to 33.29 and differs significantly from the overdispersion of a Poisson random variable. Ignoring that the variance of observations is significantly greater than their expectation would lead to underestimate the exposure of an insurer to damages caused by tornadoes.

Figure 8.7 shows the structure of the autoregressive network with 4 lags and 6 hidden neurons. The activation functions of the hidden and output layers are respectively hyperbolic tangent and linear functions. This network predicts the average number of tornadoes for the next month, based on observations over the last 4 months. Figure 8.8 compares the observed and estimated numbers of tornadoes for an autoregressive model with 4 lags and 6 neurons. We clearly see that the neural network captures well the seasonality of the tornadoes arrival process. We also observe the high volatility displayed by this process.

The neural network can also be used for simulating sample paths of the number of tornadoes. In this case, the network is initialized with recent observations

---

[1] https://cemhs.asu.edu/sheldus.

Table 8.9 Statistics about the goodness of fit of ARNN(p) models to the time series of monthly numbers of tornadoes, training sample

		2	3	4	5	6
ARNN(1)	Deviance	8526.16	8289.12	7823.7	7541.63	7744.87
	log-lik.	−5186.59	−5068.07	−4835.36	−4694.32	−4795.94
	AIC	10,431.17	10,222.14	9784.72	**9530.64**	9761.89
	$\widehat{\delta}$	34.21	30.39	28.86	31.29	27.55
ARNN(2)	Deviance	11,930.35	7983.07	7743.33	7631.63	7426.18
	log-lik.	−6885.8	−4912.17	−4792.29	−4736.44	−4633.72
	AIC	13,833.61	9916.33	9706.59	9624.89	**9449.44**
	$\widehat{\delta}$	32.28	29.45	28.03	28.96	34.02
ARNN(3)	Deviance	9328.42	8106.88	8684.22	7688.14	7616.09
	log-lik.	−5581.65	−4970.88	−5259.55	−4761.51	−4725.48
	AIC	11,229.29	10,039.76	10,649.09	9685.02	**9644.97**
	$\widehat{\delta}$	34.24	25.99	34.99	31.01	26.89
ARNN(4)	Deviance	8689.13	9828.44	7824.64	7668.09	6942.87
	log-lik.	−5258.41	−5828.06	−4826.16	−4747.89	−4385.28
	AIC	10,586.82	11,760.13	9790.32	9667.78	**8976.56**
	$\widehat{\delta}$	29.65	27.45	34.23	37.25	33.29
ARNN(5)	Deviance	13,649.75	8205.11	8647.38	8467.1	8154.22
	log-lik.	−7735.32	−5013	−5234.14	−5143.99	−4987.55
	AIC	15,544.63	**10,135.99**	10,614.27	10,469.98	10,193.11
	$\widehat{\delta}$	27.96	31.83	33.77	36.38	35.33

Log-likelihood and AIC are computed with a standard Poisson model. Bold values indicate the best fit

Table 8.10 Statistics about the goodness of fit of ARNN(p) models to the time series of monthly numbers of tornadoes, validation sample

		2	3	4	5	6
ARNN(1)	Deviance	881.49	854.12	877.28	**761.14**	856.13
	log-lik.	−510.03	−496.35	−507.93	−449.86	−497.35
ARNN(2)	Deviance	1388.47	912.42	**798.8**	963.37	819.21
	log-lik.	−763.52	−525.5	−468.69	−550.97	−478.89
ARNN(3)	Deviance	939.82	**713.83**	892.02	740.61	828.51
	log-lik.	−539.2	−426.2	−515.3	−439.59	−483.54
ARNN(4)	Deviance	853.46	997.08	928.28	912.77	**753.81**
	log-lik.	−496.02	−567.83	−533.43	−525.67	−446.19
ARNN(5)	Deviance	1543.86	955.33	890.76	**814.36**	836.9
	log-lik.	−841.22	−546.95	−514.66	−476.47	−487.74

Bold values indicate the best fit

$(y_{t-1}, \ldots, y_{t-4})$ and estimates $\widehat{y}_t$, the expected number of events for the next month. Next, we draw a random number, noted $\tilde{s}_t$, from a Poisson distribution with parameter $\frac{\widehat{y}_t}{\delta}$. The simulated number of tornadoes for the next month, $\tilde{y}_t$, is obtained

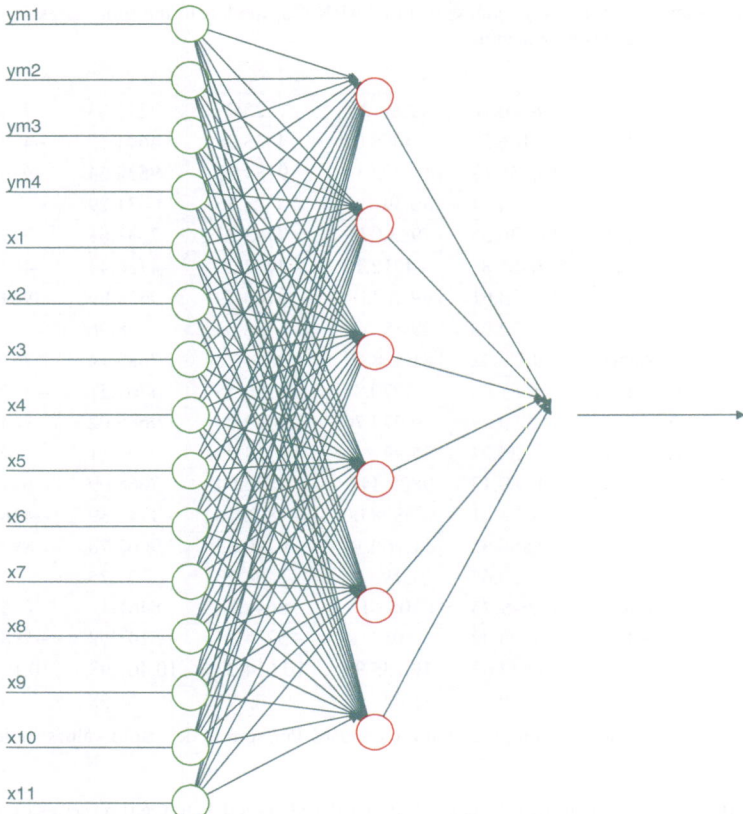

Fig. 8.7 Structure of the autoregressive neural network with 4 lags and 6 neurons in the hidden layer

by multiplying this random number by the parameter of overdispersion: $\tilde{y}_t = \widehat{\delta}\tilde{s}_t$. A simulated value at time $t + 1$, $\tilde{y}_{t+1}$, is next computed by repeating these steps with an updated initial input: $(\tilde{y}_t, y_{t-1}, \ldots, y_{t-3})$. As previously mentioned, $\tilde{y}_t$ takes its values in the set $\{0, \widehat{\delta}, 2\widehat{\delta}, 3\widehat{\delta}, \ldots\}$.

Figure 8.9 shows some simulated sample paths of monthly numbers of tornadoes over a period of 10 years. The shaded area delimits the 0.5–99.5% confidence region for 10,000 simulations. This graph reveals that the overdispersed Poisson law replicates well the high volatility displayed by observations. This point is confirmed by Table 8.11 that compares the first two moments of simulated and observed monthly number of tornadoes. Averages, standard deviations, minimums and maximums computed over simulated and observed data sets have the same range.

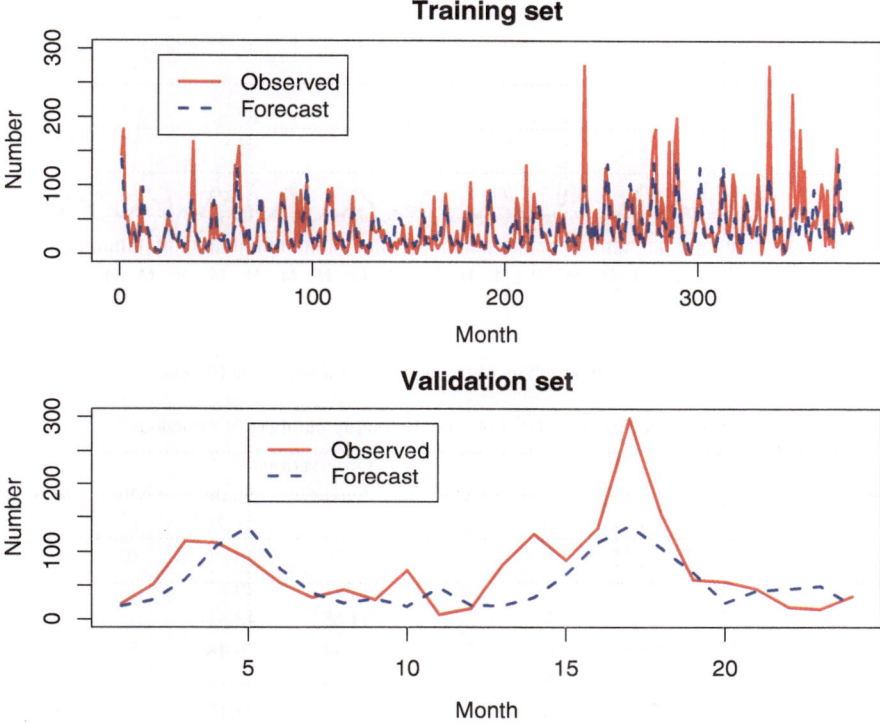

Fig. 8.8 Upper graph: comparison of y_t and $\widehat{y}_t$ on the training dataset (January 1975 to December 2006), for an ARNN(2) with 4 neurons. Lower graph: same comparison but on the validation sample (years 2007 and 2008)

8.4   Multivariate Normal Autoregressive Neural Networks

In this section we assume that the realizations of the stochastic process is a vector of dimension l: $\boldsymbol{x}_t \in \mathbb{R}^l$. We denote by

$$X_t = (\boldsymbol{x}_t, \boldsymbol{x}_{t-1}, \ldots, \boldsymbol{x}_{t-p+1})$$

the history of the multivariate process from time $t - p + 1$ up to t. Notice that $\boldsymbol{x}_t$ has not the same interpretation as in the previous section where it was a vector of explanatory variables. Here, $\boldsymbol{x}_t$ is the value at time t of X_t. To ease the reading, we also do not differentiate anymore the stochastic processes $(\boldsymbol{x}_t)_{t \in \mathbb{N}}$ from its time series $(x_t)_{t \in \mathbb{N}}$.

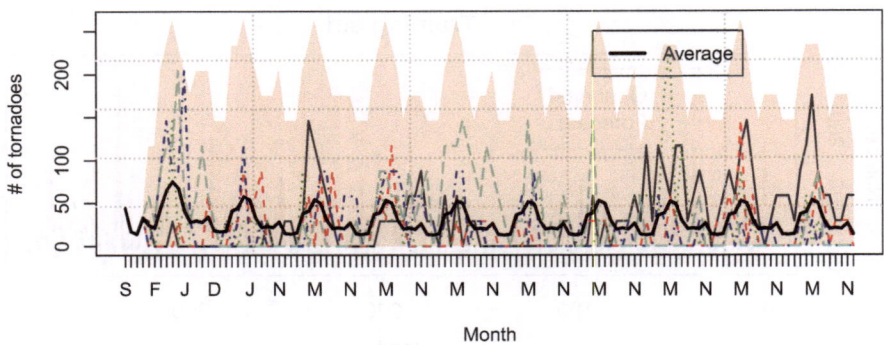

Fig. 8.9 Simulations of monthly numbers of tornadoes over a period of 10 years

Table 8.11 Statistics about simulated and observed monthly numbers of tornadoes

Month	Simulations				Observations			
	Average	St. dev.	Min	Max	Average	St. dev.	Min	Max
1	17.84	27.38	0	110.37	18.65	31.78	0	164
2	17.6	26.25	0	110.37	21.29	26.91	0	124
3	34.67	40.37	0	165.55	47.71	29.87	4	114
4	51.44	50.6	0	220.73	74.65	48.83	7	211
5	59.37	62.84	0	275.91	97.44	76.48	13	297
6	55.67	49.99	0	220.73	73.74	46.14	13	182
7	34.81	39.8	0	193.14	38.74	21.17	2	87
8	23.34	33.55	0	165.55	27.62	19.89	2	102
9	23.02	35.53	0	193.14	28.71	34.57	2	182
10	23.59	34.72	0	165.55	26.38	26.95	0	116
11	27.6	38.5	0	193.14	35.41	37.96	0	129
12	16.46	28.15	0	137.96	14.97	18.2	0	81

Definition 8.35 A multivariate normal autoregressive neural model of order p, denoted MARNN(p) is of the form

$$x_t = f_\Omega \left(x_{t-1}, \ldots, x_{t-p} \right) + w_t , \tag{8.35}$$
$$= f_\Omega \left(X_{t-1} \right) + w_t$$

where w_t is here a multivariate Normal random variable of dimension l with null mean and a covariance matrix Σ of dimension $l \times l$. $f_\Omega(X_{t-1})$ is the output of a neural network from $\mathbb{R}^p$ to $\mathbb{R}^l$ defined by a vector of parameters Ω.

As in the 1 dimension case, the expectation of x_t conditionally to X_{t-1} is equal to

$$\widehat{x}_t = \mathbb{E} \left(x_t | X_{t-1} \right) = f_\Omega \left(x_{t-1}, \ldots, x_{t-p} \right) .$$

The multivariate pdf of x_t conditionally to X_{t-1} is given by

$$f(x_t|X_{t-1}) = \frac{1}{(2\pi)^{l/2}|\Sigma|^{1/2}} \, e^{-\frac{1}{2}(x_t-\widehat{x}_t)^\top \Sigma^{-1}(x_t-\widehat{x}_t)} .$$

where $|\Sigma|$ is the determinant of Σ. The criterion used to estimate the network is a loss function denoted by $\mathcal{L} : \mathbb{R}^{l\times 2} \to \mathbb{R}$, continuous and that admits a first order derivative. This loss function takes as input the realization x_t of the stochastic process and the prediction $\widehat{x}_t$, based on the information available up to time $t-1$.

The log-likelihood is therefore proportional to

$$l(x_t|\Omega, \Sigma, X_{t-1}) \propto -\frac{1}{2}(x_t - \widehat{x}_t)^\top \Sigma^{-1}(x_t - \widehat{x}_t) ,$$

where $\widehat{x}_t = f_\Omega(X_{t-1})$. If the matrix of covariance is known then we can choose the negative log-likelihood as loss function for calibrating the weights of the neural network.

$$\widehat{\Omega} = \arg\min_\Omega \frac{1}{T}\sum_{t=1}^{T} l(x_t|\Omega, \Sigma, X_{t-1}).$$

In practice, this matrix is nevertheless undetermined. For this reason, the calibration of the neural networks in done iteratively. The ith iteration takes as input a covariance matrix $\Sigma^{(i)}$. We run a few iteration of the back-propagation algorithm to estimate weights of the neural network. We use as loss criterion in this algorithm, the sum of negative log-likelihoods computed with $\Sigma^{(i)}$. Next we update $\Sigma^{(i+1)}$ with the empirical covariance matrix of residuals. The procedure is summarized in Algorithm 8.1. The algorithm can eventually be stopped when the variation of the loss criterion falls below a predetermined threshold.

To illustrate this section, we fit a bivariate autoregressive Gaussian neural network to the financial time series of S&P 500 and CAC 40. The CAC 40, as the S&P 500, is an index equal to the weighted sum of stock prices for the 40st biggest French listed companies. The dataset covers common days of trading for the French and US markets from the 13/11/2000 to 12/11/2018 (4479 days). The Dickey–Fuller test reveals that the CAC 40 and the S&P 500 are not stationary. For this reason, we study their first order log-differences, that may be interpreted as the daily log-return of these indexes. The data set is split into a training and validation sample. The training set contains information about 4000 days of trading from the 13/11/2000 to 9/12/2016. The validation is performed with data from the 9/12/2016 to 12/11/2018.

If z_t^1 and z_t^2 respectively denote the S&P and CAC values at time t, the time series of log-differences are equal to $y_t^k = \ln z_t^k - \ln z_{t-1}^k$, for $t = 1$ to T and $k = 1, 2$. We next rescale the time series as follows:

$$x_t^k = \frac{y_t^k - \bar{y}^k}{s_y^k} \quad t = 1, \ldots, T \; k = 1, 2$$

Algorithm 8.1 Procedure for calibrating a MARNN

Initialization:
        Randomly attribute weights to each neurons: Ω_0.
        Set $\boldsymbol{\Sigma}^{(0)}$ as the empirical covariance matrix of $(\boldsymbol{x}_t)_{t=0,...,T}$.

Main procedure:
        For $j = 1$ to maximum epoch, M

            1. Find

$$\Omega_j = \arg\min_{\Omega} \frac{1}{T} \sum_{t=1}^{T} l\left(\boldsymbol{x}_t | \Omega, \boldsymbol{\Sigma}^{(j-1)} X_{t-1}\right).$$

            2. Forecast $\widehat{\boldsymbol{x}}_t^{(j)} = f_{\Omega_j}(X_{t-1})$ and calculate the time series
            of residuals

$$(\epsilon_t)_{t=0,...,T} = (\boldsymbol{x}_t - \widehat{\boldsymbol{x}}_t)_{t=0,...,T}$$

            3. update $\boldsymbol{\Sigma}^{(j)}$ with the covariance matrix of residuals:

$$\boldsymbol{\Sigma}^{(j)} = \left(\mathbb{C}\left(\boldsymbol{\epsilon}_{.,u}, \boldsymbol{\epsilon}_{.,v}\right)\right)_{u,v=1,...,m}$$

        End loop on epochs

---

where $\bar{y}^k$ and s_y^k are the averages and standard deviations of the series $\left(y_t^k\right)_{t=1,...,T}$ for $k = 1, 2$. The bivariate vector of centered and normalized log-returns is denoted by $\boldsymbol{x}_t = \left(x_t^1, x_t^2\right)^\top$. At each epoch of Algorithm 8.1, neural weights are obtained by minimizing the objective:

$$\frac{1}{T} \sum_{t=1}^{T} (\boldsymbol{x}_t - \widehat{\boldsymbol{x}}_t)^\top \boldsymbol{\Sigma}^{-1} (\boldsymbol{x}_t - \widehat{\boldsymbol{x}}_t) , \tag{8.36}$$

where $\boldsymbol{\Sigma}$ is the covariance matrix of $\boldsymbol{x}_t$. For practical reasons, the implementation of this objective function is easier if we change of basis for representing vectors $\boldsymbol{x}_t$. Let us detail this point. We denote by $\boldsymbol{\Psi}$ the Choleski decomposition of the covariance matrix: $\boldsymbol{\Sigma} = \boldsymbol{\Psi}\boldsymbol{\Psi}^\top$. Since, $\boldsymbol{\Sigma}^{-1} = \boldsymbol{\Psi}^{\top-1}\boldsymbol{\Psi}^{-1}$, defining the rotated vector $\boldsymbol{x}_t'$ by

$$\boldsymbol{x}_t' = \boldsymbol{\Psi}^{-1}\boldsymbol{x}_t ,$$

allows us to rewrite the objective (8.36) as the sum of two mean squared errors (MSE):

$$\sum_{k=1}^{2} \left(\frac{1}{T} \sum_{t=1}^{T} \left(x_t^{k'} - \widehat{x}_t^{k'}\right)^2\right). \tag{8.37}$$

Table 8.12 Statistics about the goodness of fit of bivariate MARNN(p) models to S&P 500 and CAC 40 log-differences, training sample

		2	3	4	5	6
MARNN(1)	SSE	8133.53	8146.67	8073.61	8078.09	8068.5
	log-lik.	−10,324.15	−10,319.1	−10,311.34	−10,299.71	−10,295.13
	AIC	20,672.31	20,672.2	20,666.68	**20,653.41**	20,654.26
	BIC	20,747.84	20,779.2	20,805.15	20,823.35	20,855.67
MARNN(2)	SSE	8122.32	8089.42	8080.94	7979.17	7975.46
	log-lik.	−10,257.48	−10,239.06	−10,239.26	−10,185.98	−10,194.03
	AIC	20,546.96	20,524.12	20,538.53	**20,445.96**	20,476.07
	BIC	20,647.66	20,668.87	20,727.34	20,678.84	20,752.99
MARNN(3)	SSE	8096.28	8089.52	8031.15	7854.95	7901.61
	log-lik.	−10,240.6	−10,224.42	−10,188.81	−10,139.94	−10,183.93
	AIC	20,521.2	20,506.83	20,453.62	**20,373.88**	20,479.86
	BIC	20,647.07	20,689.35	20,692.77	20,669.68	20,832.29
MARNN(4)	SSE	8096.13	8031.44	7982.55	7891.47	7738.6
	log-lik.	−10,237.28	−10,220.87	−10,180.13	−10,180.8	−10,129.67
	AIC	20,522.56	20,511.73	20,452.26	20,475.61	**20,395.34**
	BIC	20,673.6	20,732	20,741.75	20,834.32	20,823.28
MARNN(5)	SSE	8071.19	8034.24	7989.43	7608.18	7646.18
	log-lik.	−10,219.53	−10,182.03	−10,155.89	−10,078	−10,067.72
	AIC	20,495.06	20,446.07	20,419.78	**20,290.01**	20,295.44
	BIC	20,671.26	20,704.08	20,759.61	20,711.64	20,798.89

The Keras library in R accepts objective functions of the form (8.37) (just the standard MSE applied to a network with a bivariate output) whereas directly implementing the function (8.36) requires to program a new type of neural layer in Python. Table 8.12 presents the statistics about the goodness of fit of bivariate autoregressive models with 2–6 neurons in the hidden layer. The activation function of these neurons is the tangent hyperbolic. We run 40 iterations of Algorithm 8.1 and at each iterations, we adjust neural weights with 100 steps of the back-propagation procedure. Figure 8.10 presents the evolution of the sum of squared errors (SSE) after each iteration of Algorithm 8.1, for autoregressive models of orders 3 to 5, with 5 neurons. These graphs confirm that the algorithm converges toward a minimum SSE, both on the training and validation datasets. Tables 8.12 and 8.13 present the statistics of calibration on the training and validation datasets. The bold numbers indicate the lowest AIC for a given order of autoregression. According the AIC, the best model is the model of order 5, with 5 neurons. Its performance on the validation test measured by the log-likelihood is however less good than all other models. The second best model is the MARNN(3) with 5 neurons. Its SSE on the validation set being acceptable, the rest of our analysis focuses on this configuration.

Figure 8.11 shows the one-step ahead expected normalized log-differences $\left(x_t^{k'} \right)_{k=1,2}$ computed by the MARNN(3) with 5 neurons, for the period from the

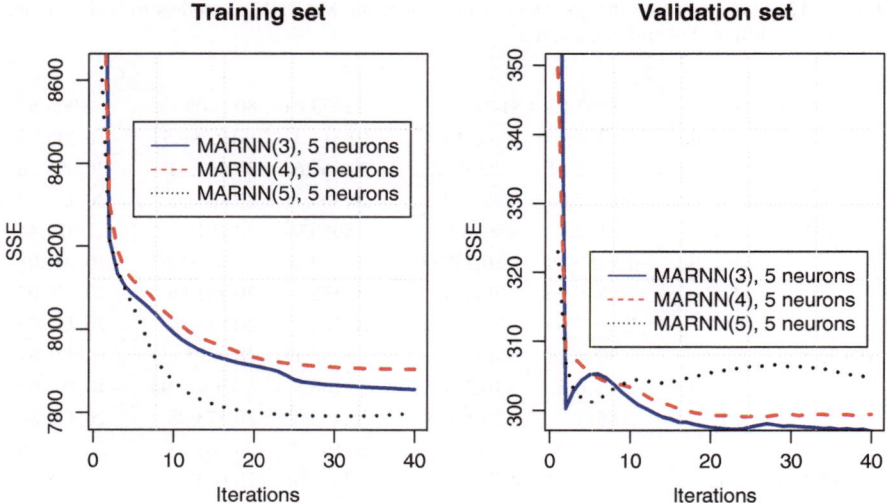

Fig. 8.10 Evolution of the total loss, Eq. (8.37), on the training and validation datasets

Table 8.13 Statistics about the goodness of fit of bivariate MARNN(p) models to S&P 500 and CAC 40 log-differences, validation sample

		2	3	4	5	6
MARNN(1)	SSE	292.49	292.41	295.96	293.40	292.70
	log-lik.	−719.68	−718.87	−726.99	−719.53	**−718.76**
MARNN(2)	SSE	294.61	293.48	292.36	296.62	295.8
	log-lik.	−722.30	−721.05	**−718.38**	−725.30	−723.88
MARNN(3)	SSE	295.3	294.8	297.38	295.37	296.05
	log-lik.	−723.84	**−720.32**	−729.49	−731.61	−727.37
MARNN(4)	SSE	295.22	298.27	301.28	305.14	293.17
	log-lik.	−723.77	−731.30	−733.10	−738.00	**−723.59**
MARNN(5)	SSE	301.64	302.13	303.46	311.79	302.17
	log-lik.	−738.83	−742.29	−743.72	−754.49	**−740.3**

25/2/2016 to 6/12/2016. The predicted CAC 40 normalized log-returns are more volatile than the S&P 500: 0.245 for the S&P versus 0.390 for the CAC. This trend is also observed for the initial time series and is explainable by the wider effect of diversification between the 500 equities of the S&P than between the 40 equities of the CAC. The correlation between predicted normalized returns is equal to 16.91%. The estimate of the matrix Σ of covariance for normalized log-differences is equal to

$$\Sigma = \begin{pmatrix} 1.001 & 0.619 \\ 0.619 & 0.915 \end{pmatrix}.$$

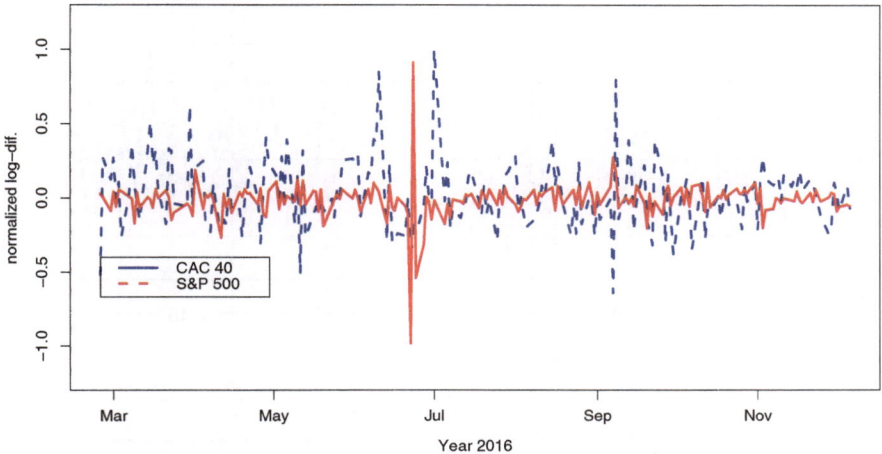

Fig. 8.11 Comparison of one-step ahead forecast computed with the MARNN(3) and 5 neurons. Period: 25/2/2016 to 6/12/2016

Figure 8.12 shows simulated sample paths for the S&P 500 and CAC 40 with the MARNN(3). We use as input the quotes of indexes from the 7 to the 9 of December 2016. The shaded area delimits the 0.5–99.5% confidence region of simulated log-differences. Table 8.14 reports some statistics about simulated sample paths and log-returns in the training data set. The observed and simulated average daily returns are similar and close to zero. The standard deviations are comparable and around 1.20% for simulated returns. The 90% confidence interval for simulated S&P 500 log-returns is slightly wider than the one of observed returns. For the CAC40, the 90% confidence interval is slightly wider for observations than for simulated returns. The correlation between simulated log-differences is equal to 54.91%, which is the correlation of observed log-returns. These statistics confirm that a bivariate MARNN is an efficient alternative to classical econometric models for simulating joint movements of several financial markets.

8.5 Recurrent Neural Networks

Previous sections focus on autoregressive neural networks, ARNN, which are feed-forward structures taking as input recent realizations of a process. We have seen that ARNNs may have a modeling performance comparable to ARMA. However, ARNNs do not include any moving average mechanism which in practice is needed for explaining memory effects observed in certain time series. Recurrent neural networks solve this problem and may be seen as non-linear ARMA models. A recurrent neural network, also called Elman network (Elman 1990) named after its inventor, makes use in a similar fashion to MA of lagged as well as current values of

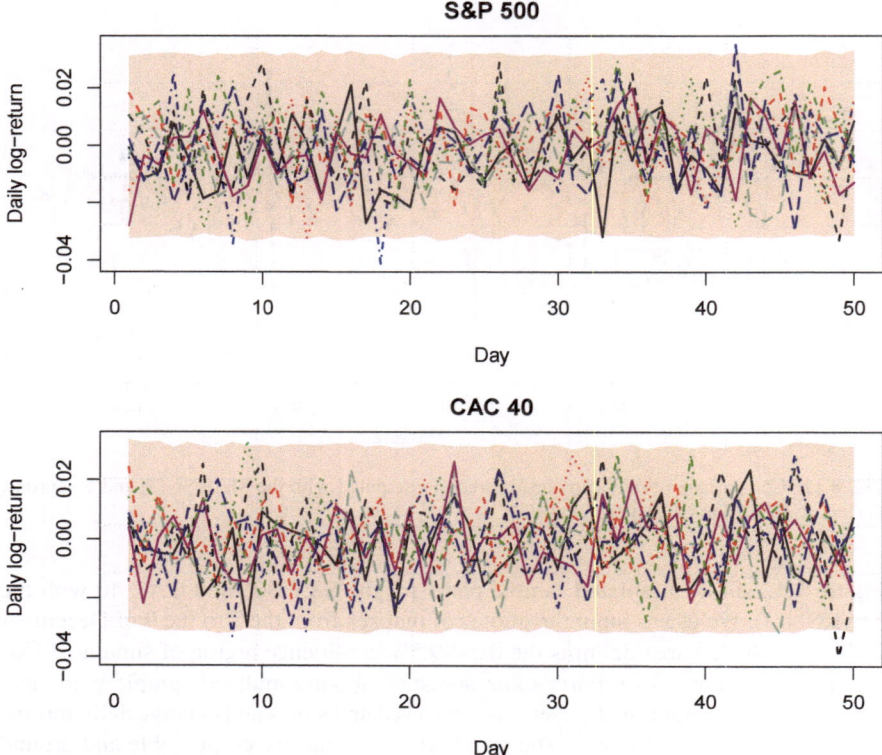

Fig. 8.12 Ten thousand simulations of S&P 500 log-differences over 50 days, with a bivariate MARNN(3) and 5 neurons. Starting day: 9/12/2016

Table 8.14 Statistics about simulated (MARNN(3) with 5 neurons) and real S&P 500/CAC 40 log-differences

	S&P 500		CAC 40	
Log-returns	Observed	Simulated	Observed	Simulated
Mean	0.02%	0.01%	0.00%	0.01%
St. deviation	1.20%	1.25%	1.45%	1.19%
5% percentile	−1.87%	−1.99%	−2.31%	−1.99%
95% percentile	1.69%	2.10%	2.16%	1.93%

outputs from neurons located in the hidden layer. In the remainder of this section, we focus on Elman networks with a single hidden layer (shallow network) but nothing prevents us to include several recurrent layers in a more complex architecture.

As in previous sections, we consider a stochastic process for which the flow of information is carried by a $\mathbb{R}^p$ vector, $\boldsymbol{x}_t = (x_t, x_{t-1}, \dots, x_{t-p+1})$, that contains the recent history of the process from time $t - p + 1$ up to t. Figure 8.13 shows an

Fig. 8.13 Illustration of a
recurrent neuron

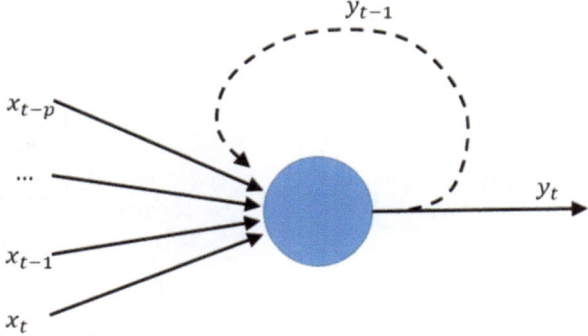

example of a recurrent neuron that is the building block of an Elman network. This
neuron is fed by the vector of input x_t and by the output signal y_{t-1}, result of
the previous activation. An Elman network is a combination of multiple recurrent
neurons and is mathematically characterized by the following definition.

Definition 8.36 A shallow recurrent autoregressive neural model of order p with q
hidden neurons, denoted RARNN(p,q) is of the form

$$x_{t+1} = f_\Omega\left(x_t, y_{t-1}\right) + w_t. \tag{8.38}$$

$\left(y_t\right)_{t\in\mathbb{Z}} \in \mathbb{R}^q$ is the output of the intermediate layer of neurons:

$$y_t = \left(\phi\left(\alpha_k^\top y_{t-1} + \beta_k^\top \begin{pmatrix} 1 \\ x_t \end{pmatrix}\right)\right)_{i=1,\dots,q}$$

where $\phi(.)$ is a bounded activation function and $\alpha_k \in \mathbb{R}^q$, $\beta_k \in \mathbb{R}^{p+1}$ for $k = 1, \dots, q$. Whereas the output of the neural network is computed by:

$$f_\Omega\left(x_t, y_{t-1}\right) = \delta^\top \begin{pmatrix} 1 \\ y_t \end{pmatrix},$$

where $\delta \in \mathbb{R}^{q+1}$. Furthermore, w_t is a stochastic process with a null mean,
$\mathbb{E}(w_t) = 0$ and a standard deviation noted σ_w^2.

The function $f_\Omega(x_t, y_{t-1})$ is the output of a neural network from $\mathbb{R}^{p+q}$ to $\mathbb{R}$
defined by a set of parameters $\Omega = \{(\alpha_k)_{k=1\dots q}, (\beta_k)_{k=1\dots q}, \delta\}$. In practice, we
consider a Gaussian distribution for the noise $w_t \sim N\left(0, \sigma_w^2\right)$. The RARNN(p,q)
is a generalization of the ARMA process. By construction, the expectation of x_{t+1}
conditionally to x_t and y_{t-1} is equal to

$$\widehat{x}_{t+1} = \mathbb{E}\left(x_{t+1}|x_t, y_{t-1}\right) = f_\Omega\left(x_t, y_{t-1}\right).$$

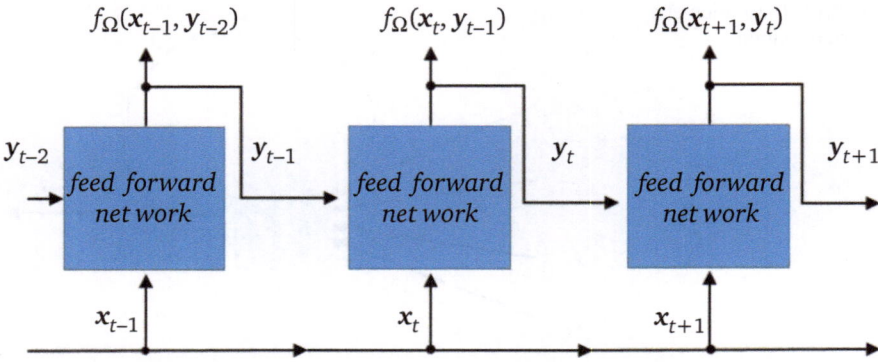

Fig. 8.14 Illustration of a recurrent network

In an Elman network, the lagged signal from the intermediate neural layer passes through the activation function. Therefore, it has an indirect feedback effect on the output of the network. This is an important difference with the MA model in which the feedback is direct. Indeed, in MA models, lagged random disturbances modify linearly the time series. In Elman networks, lagged signals are instead 'squashed' by the activation function. Figure 8.14 presents an alternative representation of an Elman network in which we unroll the time axis. This approach is useful for understanding Long Short-Term Memory layers that are introduced in the next section.

During the training procedure of an Elman network, similar to the case of a multilayer perceptron training, the network's output is compared with the target output and a total loss function $\mathcal{R}(\Omega)$ is used to update the network's weights according to, for instance, the backpropagation algorithm. As the noise w_t is Gaussian, a natural choice for this loss function is the average of deviances:

$$\mathcal{R}(\Omega) = \frac{1}{T} \sum_{t=1}^{T} \left(x_{t+1} - f_\Omega\left(\boldsymbol{x}_t, \boldsymbol{y}_{t-1}\right)\right)^2 .$$

Estimates of weights, $\widehat{\Omega}$, are found by minimizing this loss function : $\widehat{\Omega} = \arg\min_\Omega \mathcal{R}(\Omega)$. However, a multistep estimation procedure is needed since the output of intermediate neurons fully depends on neural weights. We start by initializing the vector of lagged neural output $\boldsymbol{y}_t^{(0)}$ with lagged proxies from a simple feed-forward network. Then, we estimate neural weights and recalculate the vector of lagged intermediate signals $\boldsymbol{y}_t^{(1)}$. Parameters are next estimated again in a recursive fashion as detailed in Algorithm 8.2 with a back-propagation of errors. The process stops when the total loss converges. The Elman network with a lag of one time step is too simplistic but it is an excellent introduction to Long Short-Term

---

Algorithm 8.2 Back-propagation procedure for calibrating a recurrent neural network

---

Initialization:

Randomly attribute weights to each neurons: Ω_0.

Calculate intermediate outputs $\left(y_t^{(0)}\right)_{t=1,\dots,T}$

Select an initial step size, ρ_0.

Main procedure:

For $e = 0$ to maximum epoch, E

1. Calculate the gradient $\nabla \mathcal{R}(\Omega_e)$ where

$$\mathcal{R}(\Omega_e) = = \frac{1}{T} \sum_{t=1}^{T} \left(x_{t+1} - f_{\Omega_e}\left(x_t, y_{t-1}^{(e)}\right)\right)^2$$

2. Update the step size

$$\rho_{e+1} = \rho_0 e^{-\alpha e}$$

3. Modify the vector of weights:

$$\Omega_{e+1} = \Omega_e - \rho_{e+1} \nabla \mathcal{R}(\Omega_e). \tag{8.39}$$

4. Recalculate intermediate outputs $y_t^{(e+1)}$

End loop on epochs

Memory (LSTM) networks. This type of architecture is presented in the next section and captures long term time dependencies.

8.6 Long Short-Term Memory (LSTM)

Although it is theoretically possible to build a recurrent network that retains information about output signals of hidden neurons, seen many time steps before, such long term dependencies are impossible to learn in practice. This phenomenon is studied by Hochreiter et al. (2001) and is due to the 'vanishing gradient problem'. This effect is also observed for deep neural networks with too many layers: as you keep adding layers to a network, the network becomes untrainable by back-propagation. The reason is that in the back-propagation algorithm, the neural network weight are updated proportionally to the partial derivative of the error function with respect to current weights, at each iteration of training. The problem is that for deep networks or recurrent networks with many time lags, the gradient will be vanishingly small, effectively preventing the weights from changing their value. Long Short-Term Memory (LSTM) networks solve this problem.

A LSTM memory layer is a variant of a single recurrent layer in which an information vector $(c_t)_{t \in \mathbb{Z}}$ with $c_t \in \mathbb{R}^q$ carries the information across many time

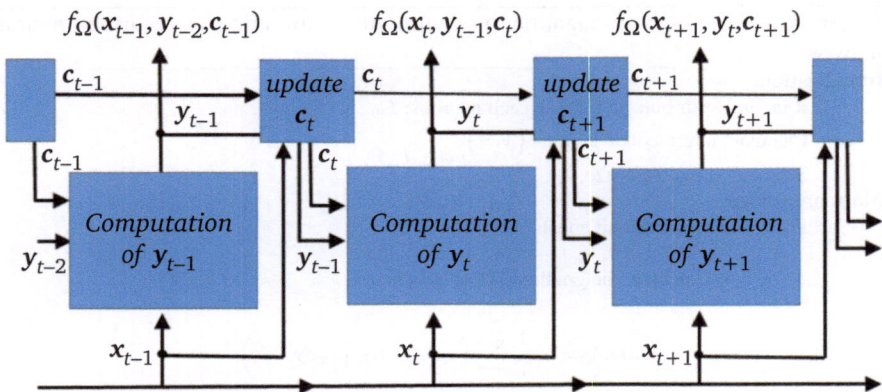

Fig. 8.15 Illustration of a LSTM network

steps. As illustrated in Fig. 8.15, The information c_t flows in parallel to the sequence that is processed. The output signal y_t from hidden neurons at time t is saved in the information flow c_{t+1} and keeps intact for the next activation. The next definition provides a mathematical definition of the autoregressive LSTM network.

Definition 8.37 A LSTM autoregressive neural model of order p with q hidden neurons, denoted LSTM(p,q) is of the form

$$x_{t+1} = f_\Omega \left(x_t, y_{t-1}, c_t \right) + w_t , \qquad (8.40)$$

where w_t is a stochastic process with a null mean, $\mathbb{E}(w_t) = 0$ and a standard deviation, σ_w^2. We define the forget rate $(f_t)_{t \in \mathbb{Z}} \in \mathbb{R}^q$, the input rate $(i_t)_{t \in \mathbb{Z}} \in \mathbb{R}^q$ and the marginal information at time t, $(\tilde{c}_t)_{t \in \mathbb{Z}}$. These quantities are respectively computed in what is called the forget, input and information gates as follows:

$$f_t = \phi_{sig} \left(\alpha_{f,k}^\top y_{t-1} + \beta_{f,k}^\top \begin{pmatrix} 1 \\ x_t \end{pmatrix} \right)_{k=1,\dots,q} ,$$

$$i_t = \phi_{sig} \left(\alpha_{i,k}^\top y_{t-1} + \beta_{i,k}^\top \begin{pmatrix} 1 \\ x_t \end{pmatrix} \right)_{k=1,\dots,q} ,$$

$$\tilde{c}_t = \tanh \left(\alpha_{c,k}^\top y_{t-1} + \beta_{c,k}^\top \begin{pmatrix} 1 \\ x_t \end{pmatrix} \right)_{k=1,\dots,q} ,$$

where $\phi_{sig}(.)$ is the sigmoid activation function, $\alpha_{f,k}, \alpha_{i,k}, \alpha_{c,k} \in \mathbb{R}^q$ and $\beta_{f,k}$, $\beta_{i,k}, \beta_{c,k} \in \mathbb{R}^{p+1}$ for $k = 1, \dots, q$. If $\otimes$ is the Hadamard product (element-wise product), the information flow $(c_t)_{t \in \mathbb{Z}} \in \mathbb{R}^q$ is the sum:

$$c_t = f_t \otimes c_{t-1} + i_t \otimes \tilde{c}_t . \qquad (8.41)$$

The output of the intermediate layer of q neurons is denoted by $(y_t)_{t\in\mathbb{Z}} \in \mathbb{R}^q$ and is equal to the product of an output gate $(o_t)_{t\in\mathbb{Z}} \in \mathbb{R}^q$ computed as

$$o_t = \phi_{sig}\left(\alpha_{o,k}^\top y_{t-1} + \beta_{o,k}^\top \begin{pmatrix} 1 \\ x_t \end{pmatrix}\right)_{k=1,\ldots,q},$$

and of the information squashed by the hyperbolic tangent function:

$$y_t = o_t \otimes \tanh(c_t) . \tag{8.42}$$

$\alpha_{o,k}$ and $\beta_{0,k} \in \mathbb{R}^{p+1}$ for $k = 1, \ldots, q$. Finally, the function $f_\Omega(.)$ in the definition (8.40) of the LSTM model is the scalar product:

$$f_\Omega\left(x_t, y_{t-1}, c_t\right) = \delta^\top \begin{pmatrix} 1 \\ y_t \end{pmatrix}, \tag{8.43}$$

where $\delta \in \mathbb{R}^{q+1}$.

Figure 8.16 details the computation process of the output from the LSTM network, presented in Fig. 8.15. As in a simple recurrent network, we calculate the output of q neurons taking as input the previous output y_{t-1} and recent values of the time series, contained in x_t. But this output is weighted by a coefficient $\tanh(c_t) \in [-1, 1]$, function of the information flow. The function $f_\Omega\left(x_t, y_{t-1}, c_t\right)$ is finally a linear function of output signals. By construction $f_\Omega\left(x_t, y_{t-1}, c_t\right)$ is a non linear function from $\mathbb{R}^{p+2\times q} \to \mathbb{R}$ which takes as input the history of the process $(x_t)_{t\in\mathbb{Z}}$ from $t-p$

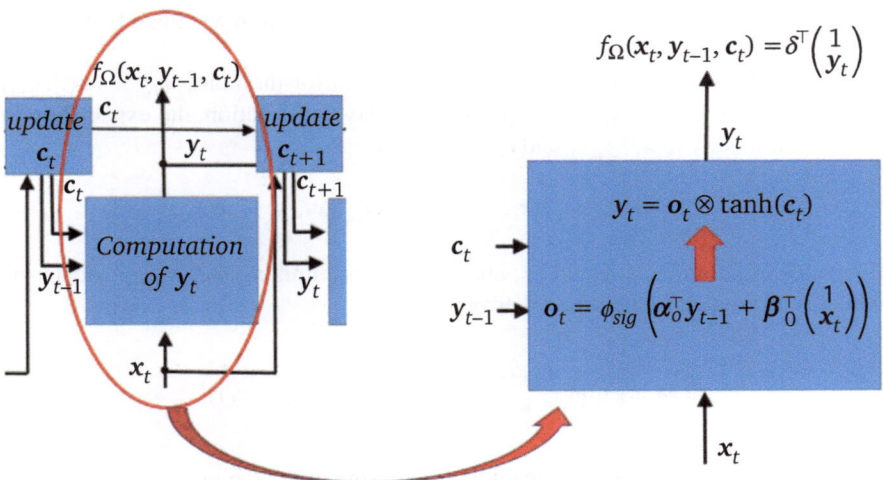

Fig. 8.16 Zoom on the process for computing the output of the neural network

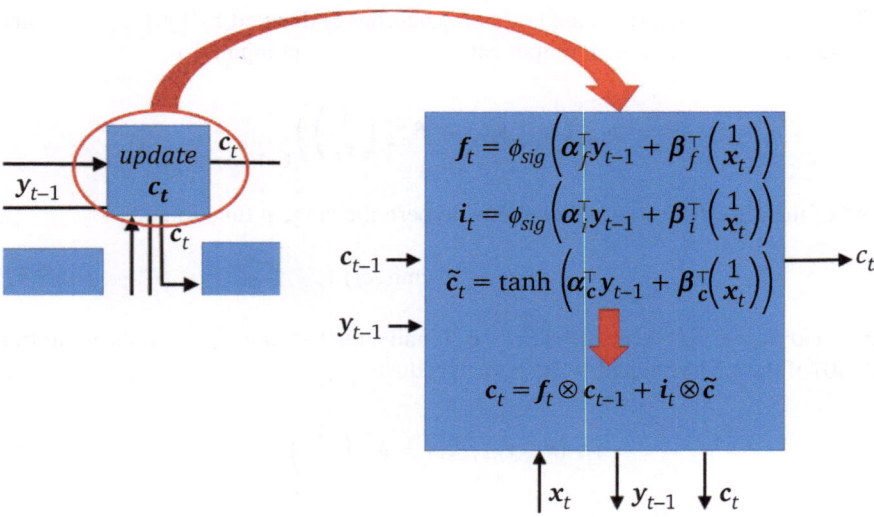

Fig. 8.17 Zoom on the process for updating the flow of information

to t, the recent output of q hidden neurons y_{t-1} and the information flow, c_t. The set of all parameters is denoted by Ω and contains vectors: $\alpha_{f,k}$, $\alpha_{i,k}$, $\alpha_{c,k}$, $\alpha_{o,k}$, $\beta_{f,k}$, $\beta_{i,k}$, $\beta_{c,k}$, $\beta_{c,k}$ for $k = 1, \ldots, q$ and δ.

Figure 8.17 shows the mechanism for updating the flow of information. Since $\phi_{sig}(.)$ is a sigmoid function, the forget and input rates, f_t and i_t, take their value in $[0, 1]$ whereas the new marginal information $\tilde{c}_t$ is in $[-1, 1]$. According to Eq. (8.41), the information flow, c_t, is the sum of the previous information weighted by the forget rate, $f_t \otimes c_{t-1}$, and of the marginal information weighted by the input rate, $i_t \otimes \tilde{c}_t$.

In practice, we consider a Gaussian distribution for the noise $w_t \sim N\left(0, \sigma_w^2\right)$ that is added to $f_\Omega\left(x_t, y_{t-1}, c_t\right)$ in Eq. (8.40). By construction, the expectation of x_{t+1} conditionally to x_t, y_{t-1} and c_t is equal to

$$\widehat{x}_{t+1} = \mathbb{E}\left(x_{t+1} | x_t, y_{t-1} c_t\right) = f_\Omega\left(x_t, y_{t-1}, c_t\right) .$$

Estimates of parameters, noted $\widehat{\Omega}$, are found by minimizing the average deviance, which is here equal to the squared error:

$$\widehat{\Omega} = \arg\min_{\Omega} \frac{1}{T} \sum_{t=1}^{T} \left(x_{t+1} - f_\Omega\left(x_t, y_{t-1}, c_t\right)\right)^2 .$$

This minimization problem is solved with the same multistep procedure as the one used for calibrating recurrent networks and described in Algorithm 8.2.

Table 8.15 Statistics about the goodness of fit of ARNN(p) models to normalized exchange rates, training sample

		2	3	4	5	6
ARNN(1)	SSE	13.32	9.36	10.86	10.39	9.47
	log-lik.	5805.89	6516.88	6218.24	6306.56	6494.89
	AIC	−11,597.79	**−13,013.77**	−12,410.49	−12,581.11	−12,951.78
	BIC	−11,553.66	−12,950.74	−12,328.54	−12,480.26	−12,832.02
ARNN(2)	SSE	12.1	10.41	9.68	9.9	10.84
	log-lik.	5996.65	6300.78	6448.1	6402.67	6219.74
	AIC	−11,975.29	−12,575.55	**−12,862.21**	−12,763.35	−12,389.48
	BIC	−11,918.57	−12,493.62	−12,755.05	−12,630.99	−12,231.91
ARNN(3)	SSE	9.85	11.48	9.96	10.55	9.54
	log-lik.	6409.95	6101.98	6387.14	6272.44	6474.26
	AIC	−12,797.9	−12,171.97	−12,732.27	−12,492.87	**−12,886.52**
	BIC	−12,728.57	−12,071.12	−12,599.91	−12,329	−12,691.13
ARNN(4)	SSE	21.28	11.85	11.62	10.6	11.09
	log-lik.	4854.64	6036	6074.95	6260.74	6169.39
	AIC	−9683.29	−12,034.01	−12,099.9	**−12,459.48**	−12,264.78
	BIC	−9601.35	−11,914.26	−11,942.33	−12,264.1	−12,031.59
ARNN(5)	SSE	12.89	17.18	11.57	11.21	10.1
	log-lik.	5864.37	5284.68	6081.14	6144.97	6356.24
	AIC	−11,698.75	−10,525.35	−12,104.28	−12,217.94	**−12,626.48**
	BIC	−11,604.21	−10,386.7	−11,921.51	−11,991.06	−12,355.48

Bold values indicate the best fit

To illustrate this section, we fit LSTM networks to the time series of daily foreign exchange (fx) rate Euro against Dollar. The training set contains the daily €/\$ rates from the 17/5/2000 to the 11/2/2016 (4029 observations). For the validation, we use data from the 12/2/2016 to the 3/12/2018 (670 fx rates). The fx rates are centered around zero by subtracting the average fx rate and normalized by dividing by the standard deviation of fx rate.

All tests are performed with the Keras package in R. We consider a "lookback" period of 50 days for calculating the information flow c_t. This means that we run the LSTM over 50 observations before considering the network output as relevant. Tested models contain one single layer of LSTM cells and one linear output neuron. We use as benchmark autoregressive neural networks (ARNN) with a single hidden layer of neurons. Tables 8.15, 8.16, 8.17, and 8.18 report the statistics about the goodness of fit of LSTM and ARNN models with 2 to 6 neurons and for regression orders from 1 to 5. On the training sample, the lowest AIC (−13013) and SSE (9.36) are obtained with an autoregressive model with 3 neurons and one time lag. The LSTM with 2 cells and 1 time lag has an AIC of −13139 whereas the lowest SSE (8.71) is reached by a LSTM(2) model with 3 neurons (and its AIC is very close to the optimal one). On the validation set, the autoregressive model that achieves the lowest SSE (0.83) is the ARNN(3) with 6 neurons. The LSTM network that

Table 8.16 Statistics about the goodness of fit of LSTM(p) models to normalized exchange rates, training sample

		2	3	4	5	6
LSTM(1)	SSE	8.9	9.53	9.15	9.77	8.98
	log-lik.	6604.73	6467.07	6547.72	6415.7	6582.96
	AIC	**−13,139.47**	−12,806.13	−12,893.45	−12,539.41	−12,767.91
	BIC	−12,918.91	−12,402.84	−12,257	−11,619.38	−11,513.91
LSTM(2)	SSE	9.32	8.71	8.78	9.23	10.08
	log-lik.	6508.16	6644.96	6627.58	6527.25	6348.37
	AIC	−12,930.33	**−13,137.92**	−13,021.15	−12,722.5	−12,250.74
	BIC	−12,659.37	−12,659.03	−12,283.9	−11,676.49	−10,845.55
LSTM(3)	SSE	10.75	10.62	9.97	9.31	9.75
	log-lik.	6219.19	6244.55	6370.4	6506.25	6411.58
	AIC	−12,336.38	−12,313.1	−12,474.8	**−12,640.5**	−12,329.17
	BIC	−12,015.02	−11,758.61	−11,636.76	−11,468.51	−10,772.81
LSTM(4)	SSE	9.67	13.54	9.73	9.4	10.29
	log-lik.	6430.92	5752.98	6416.57	6484.89	6300.43
	AIC	**−12,743.84**	−11,305.96	−12,535.15	−12,557.77	−12,058.86
	BIC	−12,372.1	−10,675.88	−11,596.33	−11,259.81	−10,351.35
LSTM(5)	SSE	12.63	11.81	11.1	10.5	10.36
	log-lik.	5892.62	6026.96	6150.76	6261.65	6287.22
	AIC	−11,651.24	−11,829.92	−11,971.53	**−12,071.31**	−11,984.45
	BIC	−11,229.08	−11,124.24	−10,931.9	−10,647.33	−10,125.72

Bold values indicate the best fit

Table 8.17 Statistics about the goodness of fit of ARNN(p) models to normalized exchange rates, validation sample

		2	3	4	5	6
ARNN(1)	SSE	1.33	0.95	1.22	1.13	**0.86**
	log-lik.	1226.96	1347.49	1257.99	1283.16	**1382.6**
ARNN(2)	SSE	1.27	1.14	0.9	**0.89**	1.56
	log-lik.	1244.01	1280.3	1365.44	**1367.58**	1169.18
ARNN(3)	SSE	0.89	1.52	0.93	1.32	**0.83**
	log-lik.	1369.73	1178.9	1354.2	1228.15	**1395.49**
ARNN(4)	SSE	2.04	1.31	1.62	**0.89**	1.37
	log-lik.	1074.34	1232.38	1155.42	**1369.5**	1214.81
ARNN(5)	SSE	1.78	2.06	1.22	1.71	**0.94**
	log-lik.	1122.22	1070.35	1255.44	1137.48	**1348.78**

Bold values indicate the best fit

minimizes the SSE on the same sample is the model with two time lags and three memory cells (SSE of 0.71). These statistics confirm that long short term memory models outperform pure autoregressive neural networks.

Table 8.18 Statistics about the goodness of fit of LSTM(p) models to normalized exchange rates, validation sample

		2	3	4	5	6
LSTM(1)	SSE	0.76	0.71	**0.7**	0.73	0.72
	log-lik.	1322.67	1341.31	**1342.24**	1324.66	1323.04
LSTM(2)	SSE	0.73	**0.71**	0.72	0.82	0.76
	log-lik.	1334.11	**1343.73**	1333.72	1281.39	1295.89
LSTM(3)	SSE	0.94	0.95	0.79	**0.73**	0.76
	log-lik.	1249.55	1243.54	1298.88	**1318.49**	1291.66
LSTM(4)	SSE	0.76	1.19	0.79	**0.75**	0.86
	log-lik.	1317.94	1167.32	1300.13	**1303.79**	1241.74
LSTM(5)	SSE	0.93	0.9	0.88	0.79	**0.77**
	log-lik.	1250.55	1257.49	1255.99	1279.9	**1267.89**

Bold values indicate the best fit

Figure 8.18 shows observed (centered and normed) €/$ fx rates and the 1 day ahead prevision computed by the LSTM(2) model with 3 memory cells. Both on the validation an training sets, it is clearly visible that forecast rates are very close to observed ones.

The LSTM models are not only useful for forecasting fx rates over the next days, they may also be used for simulating future exchange rates over long periods. For this purpose, we first estimate the standard deviation, σ_w, of the random noise in Eq. (8.40), by the standard deviation of residuals between forecast and observed fx rates:

$$\widehat{\sigma}_w = \sqrt{\frac{1}{T-m} \sum_{t=1}^{T} \left(x_{t+1} - f_\Omega \left(x_t, y_{t-1}, c_t\right)\right)^2},$$

where m is the number of network parameters. The algorithm of simulation consists to forecast the €/$ rate for the next date and to sum up a random noise. This simulated exchange rate is next taken as input by LSTM model for predicting the next €/$ rate. By iterating this loop, we simulate samples path for €/$ rates as illustrated in Fig. 8.19. This figure shows 10 simulated sample paths over a period of three trading years ($255 \times 3 = 765$ days of trading). To initialize the simulation and for calculating the information flow, we have considered the information from the 24/9/2018 to 3/12/2018 (lookback period of 50 trading days). The shaded area delimits the 0.5 and 99.5% confidence interval of simulated fx rates. Table 8.19 reports statistics about simulated rates. On average, the LSTM model predicts a €/$ rate of 1.2052. The fx rates evolve in the interval [0.8391 , 1.4771] which is a plausible range with regard to the history of the €/$ exchange rate.

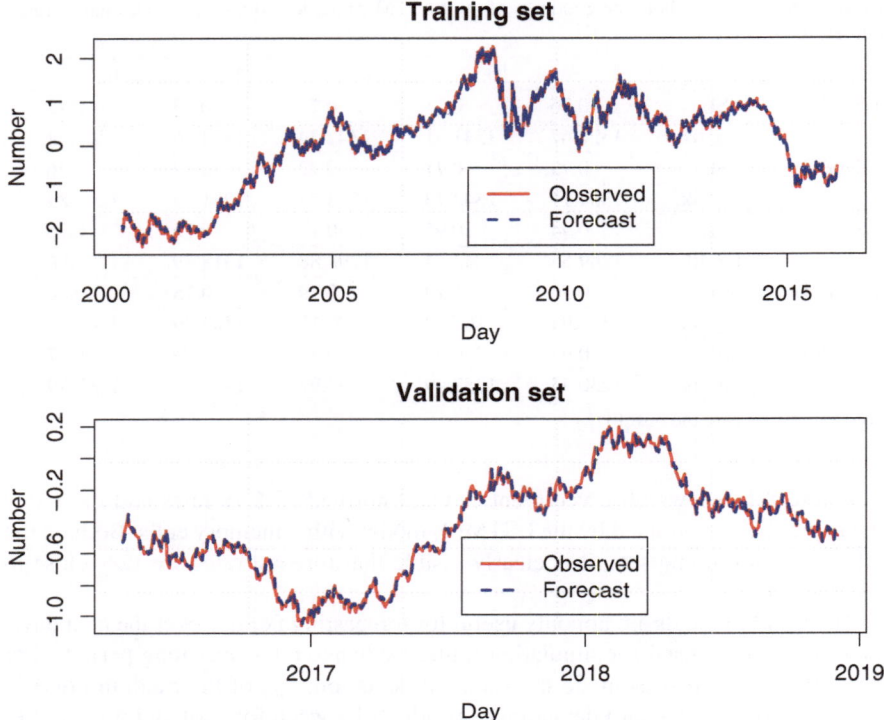

Fig. 8.18 Comparison of observed €/$ fx rates (centered and normed) and 1 day ahead forecast computed by the LSTM model with 3 memory cells and 2 time lags

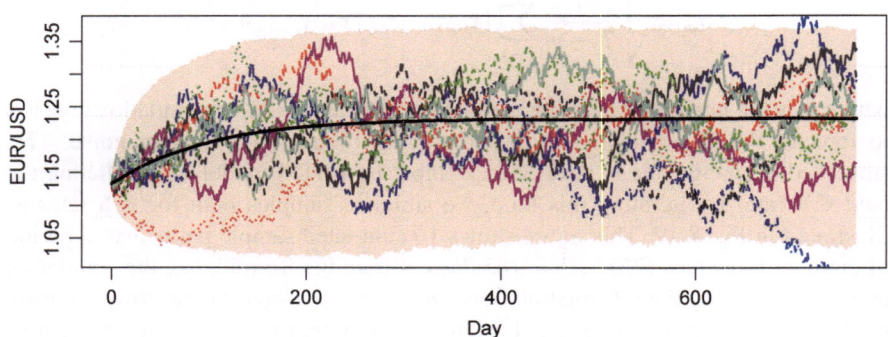

Fig. 8.19 Zoom on the process for updating the flow of information

Table 8.19 Statistics about simulated $/€ exchange rates

Simulated €/$ FX rates			
Mean	1.2052	St. Deviation	0.05838
$q_{5\%}$	1.1153	$q_{95\%}$	1.2952
Min	0.8391	Max	1.4771

8.7   Further Readings

The literature on applications of neural networks to economic time series modelling is abundant and emphasizes their ability to produce reliable forecasts. White (1988) was among the first researchers to propose a feed-forward network for economic predictions. He focuses on the case of IBM common stock daily returns. In White (1989), he presents a statistical test of the hypothesis that a multilayer feed-forward network exactly represents some unknown mapping against the alternative that the network neglects some nonlinear structure. Kuan and White (1994) studies neural networks from an econometric point of view. Kaastra and Boyd (1996) provide a practical introductory guide in the design of a neural network for forecasting economic time series data. Swanson and White (1997) question whether artificial neural networks are useful for predicting future values of nine macroeconomic variables like the gross national income, the corporate profits after taxes or the industrial production index. Anders et al. (1998) build neural network models which explain the prices of call options written on the German stock index DAX. Medeiros et al. (2006) study the modeling of time series by single hidden layer feed-forward neural network models. Dietz (2011) studies in his PhD dissertation the properties of univariate and multivariate time autoregressive networks and apply them to the German automobile industry. Gers et al. (1999) introduces an adaptive "forget gate" that enables an LSTM cell to release internal information. They also show that standard LSTM outperforms other recurrent network algorithms. The book of Mandic and Chambers (2001) shows how recurrent neural networks can be implemented to expand the range of traditional signal processing techniques. Aymen et al. (2011) propose a model combining self-organizing maps and recurrent neural networks for local predictions.

References

Anders U, Korn O, Schmitt C (1998) Improving the pricing of options: a neural network approach. J Forecast 17:369–388

Aymen C, Cardot H, Boné R (2011) SOM time series clustering and prediction with recurrent neural networks. Neurocomputing 74(11):1936–1944

Chan KS, Tong H (1985) On the use of the deterministic Lyapunov function for the ergodicity of stochastic difference equations. Adv Appl Probab 17:666–678

Dickey DA, Fuller WA (1979) Distribution of the estimators for autoregressive time series with a unit root. J Am Stat Assoc 74(366):427–431

Dietz S (2011) Autoregressive neural network processes: univariate, multivariate and cointegrated models with application to the German automobile industry. Ph.D. Dissertation, University of Passau

Durbin J (1960) The fitting of time-series models. Rev Int Stat Inst 28:233–224

Elman JL (1990) Finding structure in time. Cogn Sci 14(2):179–211

Gers FA, Schmidhuber J, Cummins F (1999) Learning to forget: continual prediction with LSTM. In: 9th international conference on artificial neural networks: ICANN'99, pp 850–855

Hainaut D (2012) Pricing of an insurance bond, with stochastic seasonal effects. Bulletin Français d'actuariat 12(23):129–150

Hainaut D, Boucher JP (2014) Pricing of catastrophe bonds with Multifractal processes: an application to US tornadoes. Environ Model Assess 19(3):207–220

Hochreiter S, Bengio Y, Frasconi P, Schmidhuber J (2001) Gradient flow in recurrent nets: the difficulty of learning long-term dependencies. In: Kremer SC, Kolen JF (eds) A field guide to dynamical recurrent neural networks. IEEE Press, Piscataway

Kaastra I, Boyd M (1996) Designing a neural network for forecasting financial and economic time series. Neurocomputing 10:215–236

Kuan CM, White H (1994) Artificial neural networks: an econometric perspective. Econ Rev 13:1–91

Leisch F, Trapletti A, Hornik K (1999) Stationarity and stability of autoregressive neural network processes. Adv Neural Inf Proces Syst 11:267–273

Mandic DP, Chambers J (2001) Recurrent neural networks for prediction: learning algorithms, architectures and stability. Wiley, New York

Medeiros MC, Teräsvirta T, Rech G (2006) Building neural network models for some series: a statistical approach. J Forecast 25:49–75

Schumway RH, Stoffer DS (2011) Time series analysis and its applications, with R examples, 3rd edn. Springer, Berlin

Swanson NR, White H (1997) A model selection approach to real-time macroeconomic forecasting using linear models and artificial neural networks. Rev Econ Stat 79:540–550

Tong H (1990) Non-linear time series: a dynamical system approach. Oxford University Press, New York

White H (1988) Economic prediction using neural networks: the case of IBM daily stock returns. IEEE Int Conf Neural Netw 2:451–458

White H (1989) An additional hidden unit test for neglected nonlinearity in multilayer feedforward networks. In: Proceedings of the international joint conference on neural networks, Washington, vol 2, pp 90–131

Conclusions

Recent developments of computing technologies offer new horizons to actuaries and neural networks are probably one of the best tool to explore them. The aim of this book was not to provide an exhaustive description of all neural models but rather to provide a self-contained and rigorous presentation of recent neural models for actuarial sciences. An important part of this work is devoted to the study of feed-forward neural networks. According to our results, this category of model outperforms generalized linear models for predicting claims frequency or size. On the other hand, we show in the second chapter that MCMC algorithms offer an alternative to traditional optimisation methods for calibrating neural networks. Furthermore, MCMC algorithms deliver information about the uncertainty around parameter estimates. Finally, combining MCMC and back-propagation algorithm seems an excellent solution to avoid local minimums during the calibration.

Over the last half century, the processing power has grown exponentially. In parallel, the development of the internet and digital communications opened multiple new channels for collecting data. The combination of these two revolutions is at the origin of the booming of deep learning. Chapter 3 gives an overview of possibilities of deep neural networks for non-life insurance pricing. Deep networks usually performs better than shallow models but a particular attention must be granted to overfitting. In that respect, Lasso or ridge regularization techniques allow keeping this phenomenon under control.

The use of feed-forward neural networks is not limited to non-life insurance. The fourth chapter demonstrates that neural nets offer a promising alternative to traditional mortality models. In particular, bottleneck networks are powerful tools to reduce the dimension of mortality tables. They summarize the surface of log-forces of mortality in a limited number of latent factors that are next extrapolated. The term structures of future mortality rates are obtained by an inverse transform. Given the important number of parameters, a genetic algorithm combined to a gradient descent, is used to calibrate the network. Numerical tests performed on the French, UK and US log-forces of mortality, emphasizes that this new approach

© Springer Nature Switzerland AG 2019
M. Denuit et al., *Effective Statistical Learning Methods for Actuaries III*,
Springer Actuarial, https://doi.org/10.1007/978-3-030-25827-6

outperforms LC model and its multi-factor extensions, fitted by SVD or log-likelihood maximization.

The fifth chapter focuses on self-organizing maps. They are artificial neural networks that do not ask any a priori information on the relevancy of variables. For this reason, they belong to the family of unsupervised algorithms. This method developed by Kohonen aims to analyze the original information by simplifying the amount of rough data, in order to produce a low dimensional representation of the input space. Initially, this algorithm was exclusively designed for the analysis of quantitative data. We extend this approach in two direction. First, we introduce a metric in order to adapt SOM to categorical variables. Secondly, we modify the calibration framework to regress claims frequency on characteristics of policyholders. We finally emphasize the similarities of self-organizing maps with the k-means clustering method.

In Chap. 6, we study pricing methods based on an ensemble of neural networks. The most frequent strategy to data-driven modeling consists to estimate only a single strong predictive model. A different approach consists to build a bucket of models for some particular learning task. We can build a set of weak models, like small shallow neural networks, which can be combined altogether to produce a reliable prediction. The most prominent examples of such machine-learning ensemble techniques are random forests. These algorithms use as base learner, regression trees instead of neural networks.

Whereas ensemble techniques rely on the averaging of series of models, gradient boosting machines combine sequentially a family of base learners. To summarize, at each iteration, a new weak base-learner is trained with respect to the error of the whole ensemble built so far. Chapter 7 explores the performance of this approach when the base learner is a neural network. We also underline the potential bias induced by the linear regression of gradients on explanatory variables in non Gaussian cases.

The last chapter explores the possibilities of neural networks for time series analysis and simulation. Autoregressive neural networks offer an interesting alternative to classical econometric models and are able to explain non-linear behaviours. Extending these models to study multivariate time-series is possible by adapting the calibration procedure. In actuarial sciences, we are also interested by the modeling of count series which is a statistical type of data in which the observations can take only non-negative integer values. One section is dedicated to the modelling of this category of time-series. We also explain how to manage the overdispersion of observations. Finally, we present recurrent and long short term memory (LSTM) networks. The LTSM cells capture long term dependencies and outperforms basic autoregressive models for the modelling of financial time series. The multivariate extension of this approach is not developed in this chapter but does not present any particular difficulty.